Birkhäuser

Progress in Mathematics
Volume 289

Fabrizio Colombo
Irene Sabadini
Daniele C. Struppa

Noncommutative Functional Calculus

Theory and Applications of Slice Hyperholomorphic Functions

Birkhäuser

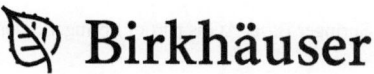

Fabrizio Colombo
Dipartimento di Matematica
Politecnico di Milano
Via Bonardi 9
20133 Milano
Italy
fabrizio.colombo@polimi.it

Irene Sabadini
Dipartimento di Matematica
Politecnico di Milano
Via Bonardi 9
20133 Milano
Italy
irene.sabadini@polimi.it

Daniele C. Struppa
Schmid College of Science
Chapman University
Orange, CA 92866
USA
struppa@chapman.edu

2010 Mathematical Subject Classification: 30G35, 47A10, 47A60

ISBN 978-3-0348-0324-3 ISBN 978-3-0348-0110-2 (eBook)
DOI 10.1007/978-3-0348-0110-2

Softcover reprint of the hardcover 1st edition 2011

Cover design: deblik, Berlin

Printed on acid-free paper

Springer Basel AG is part of Springer Science+Business Media

www.birkhauser-science.com

Contents

Chapter 1

Introduction

1.1 Overview

In this book we propose a novel approach to two important problems in the theory of functional calculus: the construction of a general functional calculus for not necessarily commuting n-tuples of operators, and the construction of a functional calculus for quaternionic operators. The approach we suggest is made possible by a series of recent advances in Clifford analysis, and in the theory of quaternion-valued functions (see, e.g., [26] and [49]).

After the success, and recognized importance, of the classical Riesz–Dunford functional calculus, it became apparent that there was a need for a functional calculus for several operators. The necessity of such a calculus was pointed out by Weyl already in the 1930s, see [103], and this issue was first addressed by Anderson in [4] using the Fourier transform and n-tuples of self-adjoint operators satisfying suitable Paley–Wiener estimates.

In his early and seminal work [99], Taylor introduces a new approach which works successfully for n-tuples of commuting operators, while in [100] he considers the Weyl calculus for noncommuting, self-adjoint operators. These works have set the stage for different possible outgrowth of this research.

A promising and successful idea was to address the noncommutativity by exploiting the setting of Clifford algebra-valued functions. This idea has been fruitfully followed in the works of Jefferies, McIntosh and their coworkers, see, e.g., [60], [61], [65], [77], and the book [62] with the references therein for a complete overview of this setting. Note that, despite the noncommutative setting which is useful in the case of several operators, one may still have restriction on the n-tuples of operators and on their spectrum.

Of course, for the sake of generality, one would like to abandon these restrictions. To this purpose we have come to understand that one could attempt the development of a functional calculus based on the use of slice monogenic functions.

These functions were first introduced by the authors in [26], but their theory is by now very well developed, as made evident by the rich literature which is available (see, e.g., [15], [18], [24], [26], [27], [28], [29], [30] and [53], [55]).

As it is well known, in order to construct a functional calculus associated to a class of functions, one of the crucial results is the existence of a suitable integral formula which, for the case of slice monogenic functions, we state and prove in Chapter 2. Such a formula was originally proved by Colombo and Sabadini in [15] (for more details see [18]). It is worth noticing that this integral formula is computed over a path which lies in a complex plane. Moreover, despite what happens with the classical monogenic functions, [7], in the slice monogenic case the analog of the Cauchy kernel is a function which is left or right slice monogenic in a given variable. For this reason, we will need two different kernels when dealing with left or right slice monogenic functions. The Cauchy formula we obtain in the case of slice monogenic functions turns out to be perfectly suited to the definition of a functional calculus for bounded or unbounded n-tuples of not necessarily commuting operators, see Chapter 3.

In the first part of this book therefore, we will develop the main results of the theory of slice monogenic functions and the associated functional calculus for n-tuples of not necessarily commuting operators. This calculus has been introduced in the paper [25] for a particular class of functions and then extended to the general case in [18].

In the second part of the book we deal with a related, and yet independent, problem which has been of interest for many years and which, so far, has proved to be rather difficult to tackle. Specifically, we are interested in attempting to define a function of a single quaternionic linear operator. It is clear that, at least in some sense, there are similarities with the problems discussed above: the setting is noncommutative, and the space of quaternions is a Clifford algebra. Nevertheless, the actual problem is different from the case analyzed before.

When dealing with the functional calculus for n-tuples of operators, our approach is to embed the n-tuple of linear operators (over the real field) into the Clifford algebra setting; in this second case, however, we are given an operator which is quaternionic linear. Since the setting is noncommutative, the operator is either left or right linear, and we shall see that our approach differentiates these two cases. The study of this type of operators is needed to deal, for example, with quaternionic quantum mechanics, see [1].

The first natural issue, of course, is to define the space of functions for which we can construct such a functional calculus. Traditionally, the best understood space of functions defined on quaternions is the space of regular functions as defined by Fueter in his fundamental works [43], [44]. Those functions are differentiable on the space of quaternions and they satisfy a system of first-order linear partial differential equations known as the Cauchy–Fueter system. Note that the Cauchy–Fueter system deals with functions defined in $\mathbb{R}^4$ and hence in $\mathbb{R}^3$ as well. Historically, this last case was introduced before the former one, see [79], by G. Moisil and N. Theodorescu. One may therefore attempt to define a functional calculus in which the functions are regular in the sense of Fueter (and the authors have outlined how this would work in [11]). It turns out, however, that such a functional calculus does not perform as well as one would hope, for a variety of reasons that are described in [11] but that can be easily surmised by noticing, for example, that even the simple function $f(q) = q^2$ is not regular in the sense of Fueter.

However, in a recent series of papers, see, e.g., [9], [12], [48], [49] the authors and some of their collaborators have introduced a completely different notion of regularity, the so-called slice regularity, which was in fact the inspiration for the notion of slice monogenicity. This notion is different from the original one of Fueter, and therefore the second part of this book will show how a functional calculus for quaternionic linear operators over the quaternions can be obtained through the use of slice regular functions. The quaternionic functional calculus, at least for functions admitting a power series expansion, was first introduced in [10], [13] and [14], however the exposition in Chapter 4 is inspired by the more recent papers [16] and [17] which are based on a new Cauchy formula, which becomes the natural tool to define the quaternionic functional calculus for quaternionic bounded or unbounded operators (with components that do not necessarily commute). As an application of the quaternionic functional calculus we define and we study the properties of the quaternionic evolution operator, limiting ourselves to the case of bounded linear operators. The evolution operator is studied in [21] where it is proved that the Hille–Phillips–Yosida theory can be extended to the quaternionic setting. This, it seems to us, is the first step in demonstrating the importance, in physics, of this new functional calculus.

It is worth pointing out that while the definitions and some of the properties of slice monogenic and of slice regular functions appear to be quite similar, there are in fact several important differences, that force an independent treatment for the two cases. Those differences are mainly due to the different algebraic nature of quaternions and of Clifford numbers in higher dimensions, when the number of imaginary units which generate the Clifford algebra is greater than two.

1.2 Plan of the book

Almost all the material presented in this book comes from the recent research of the authors. The only exceptions are the basic notions on Clifford algebras, the Appendix, in which we provide some basic facts on the classical Riesz–Dunford functional calculus, and a few results appearing in some of the notes. To illustrate the central results of this book we provide a quick description.

Slice monogenic functions. Consider the universal Clifford algebra $\mathbb{R}_n$ generated by n imaginary units $\{e_1, \dots, e_n\}$ satisfying $e_ie_j + e_je_i = -2\delta_{ij}$ and a function f defined on the Euclidean space $\mathbb{R}^{n+1}$, identified with the set of paravectors in $\mathbb{R}_n$, with values in $\mathbb{R}_n$. The notion of slice monogenic function is based on the requirement that all the restrictions of the function f to suitable complex planes be holomorphic functions. To describe the complex planes we will consider the sphere of the unit 1-vectors, i.e.,

$$\mathbb{S} = \{\underline{x} = e_1x_1 + \ldots + e_nx_n \in \mathbb{R}^{n+1} \mid x_1^2 + \ldots + x_n^2 = 1\}.$$

From a geometric point of view, $\mathbb{S}$ is an $(n-1)$-sphere in $\mathbb{R}^{n+1}$. Note that an element $I \in \mathbb{S}$ is again an imaginary unit since $I^2 = -1$. If we take any element $I \in \mathbb{S}$ we can construct the plane $\mathbb{R} + I\mathbb{R}$ passing through 1 and I: it is a two-dimensional real subspace of $\mathbb{R}^{n+1}$ isomorphic to the complex plane and for this reason, we will denote it by $\mathbb{C}_I$. This isomorphism is an algebra isomorphism, thus we will refer to a plane $\mathbb{C}_I$ as a "complex plane" and an element in $\mathbb{C}_I$ will be often denoted by $\mathbf{x} = u + Iv$.

Any element in $\mathbb{R}^{n+1}$ belongs to a complex plane so, in other words, the Euclidean space $\mathbb{R}^{n+1}$ is the union of all the complex planes $\mathbb{C}_I$ as above when I varies in $\mathbb{S}$. Let $U \subseteq \mathbb{R}^{n+1}$ be an open set and let $f : U \to \mathbb{R}^{n+1}$ be a function differentiable in the real sense. Let $I \in \mathbb{S}$ and let f_I be the restriction of f to the complex plane $\mathbb{C}_I$. We say that f is a left slice monogenic function if, for every $I \in \mathbb{S}$, we have

$$\frac{1}{2}\left(\frac{\partial}{\partial u} + I\frac{\partial}{\partial v}\right) f_I(u + Iv) = 0.$$

Because of the noncommutativity we also have the right version of this notion and we say that f is a right slice monogenic function if, for every $I \in \mathbb{S}$, we have

$$\frac{1}{2}\left(\frac{\partial}{\partial u} f_I(u + Iv) + \frac{\partial}{\partial v} f_I(u + Iv)I\right) = 0.$$

From the definition, it immediately appears that a slice monogenic function is not necessarily harmonic (but its restrictions to any complex plane $\mathbb{C}_I$ are harmonic) and this is a major difference between this theory and the theory of classical monogenic functions, see [7]. However, with this definition of monogenicity we gain the good property that all convergent power series $\sum_{n\geq 0} \mathbf{x}^n a_n$ are left slice monogenic in their domain of convergence and this property will be crucial to construct a functional calculus.

To better understand the nature of slice monogenic functions, it is necessary to consider them on axially symmetric slice domains which turn out to be their natural domains of definition. We say that a domain U in $\mathbb{R}^{n+1}$ is a slice domain (s-domain for short) if $U \cap \mathbb{R}$ is nonempty and if $U \cap \mathbb{C}_I$ is a domain in $\mathbb{C}_I$ for all $I \in \mathbb{S}$. We say that $U \subseteq \mathbb{R}^{n+1}$ is an axially symmetric domain if, for all $u + Iv \in U$, the whole $(n-1)$-sphere $u + v\mathbb{S}$ is contained in U.

The class of slice monogenic functions over axially symmetric s-domains is characterized by the following Representation Formula proved in [15] (which in some papers is referred to as the Structure Formula):

Representation Formula. *Let $U \subseteq \mathbb{R}^{n+1}$ be an axially symmetric s-domain and let f be a left slice monogenic function on U. For any vector $\mathbf{x} = x_0 + I_{\mathbf{x}}|\underline{x}| \in U$ and for all $I \in \mathbb{S}$, we have*

$$f(\mathbf{x}) = \frac{1}{2}\Big[f(x_0 + I\,|\underline{x}|) + f(x_0 - I|\underline{x}|) + I_{\mathbf{x}}I[f(x_0 - I|\underline{x}|) - f(x_0 + I|\underline{x}|)]\Big]. \quad (1.1)$$

The Representation Formula states that if we know the value of a slice monogenic function on the intersection of an axially symmetric s-domain U with a plane $\mathbb{C}_I$, then we can reconstruct the function on all of U.

An analogous formula, with suitable modifications, holds for right slice monogenic functions. The first step, in constructing a functional calculus, is to prove a Cauchy integral formula with a slice monogenic kernel. Note that it is possible to prove an integral representation formula, see [26], using the standard Cauchy kernel $(\mathbf{x} - \mathbf{x}_0)^{-1}$. This approach, however, is limited by the fact that the kernel is not slice monogenic. Thus, let us consider the Cauchy kernel series for left slice monogenic functions: take $\mathbf{x}$, $\mathbf{s} \in \mathbb{R}^{n+1}$ (which, in general, do not commute). We say that

$$S_L^{-1}(\mathbf{s},\mathbf{x}) := \sum_{n\geq 0} \mathbf{x}^n \mathbf{s}^{-1-n}$$

is the left noncommutative Cauchy kernel series; note that this series is convergent for $|\mathbf{x}| < |\mathbf{s}|$ and that it is slice monogenic in $\mathbf{x}$. It is actually possible to compute the sum of the Cauchy kernel series, and it turns out that

$$\sum_{n\geq 0} \mathbf{x}^n \mathbf{s}^{-1-n} = -(\mathbf{x}^2 - 2\,\mathrm{Re}[\mathbf{s}]\,\mathbf{x} + |\mathbf{s}|^2)^{-1}(\mathbf{x} - \bar{\mathbf{s}}), \quad \text{for} \quad |\mathbf{x}| < |\mathbf{s}|,$$

where $\mathrm{Re}[\mathbf{s}]$ is the real part of the paravector $\mathbf{s}$ and $|\mathbf{s}|$ denotes its Euclidean norm. The function $-(\mathbf{x}^2 - 2\,\mathrm{Re}[\mathbf{s}]\,\mathbf{x} + |\mathbf{s}|^2)^{-1}(\mathbf{x} - \bar{\mathbf{s}})$, which we still denote by $S_L^{-1}(\mathbf{s},\mathbf{x})$ is therefore a good candidate to be the Cauchy kernel for a Cauchy formula for left slice monogenic functions because when we restrict it to the plane $\mathbb{C}_I$ where the variables $\mathbf{x}$ and $\mathbf{s}$ now commute, we get the usual Cauchy kernel of complex analysis. Note that the function $S^{-1}(\mathbf{s},\mathbf{x})$ is left slice monogenic in the variable $\mathbf{x}$ and right slice monogenic in the variable $\mathbf{s}$ in its domain of definition. Analogous considerations can be repeated for right slice monogenic functions. In this case, we call

$$S_R^{-1}(\mathbf{s},\mathbf{x}) := \sum_{n\geq 0} \mathbf{s}^{-n-1}\mathbf{x}^n$$

a right noncommutative Cauchy kernel series; it is convergent for $|\mathbf{x}| < |\mathbf{s}|$. The sum of the series this time is given by the function

$$\sum_{n\geq 0} \mathbf{s}^{-n-1}\mathbf{x}^n = -(\mathbf{x} - \bar{\mathbf{s}})(\mathbf{x}^2 - 2\mathrm{Re}[\mathbf{s}]\mathbf{x} + |\mathbf{s}|^2)^{-1}, \quad \text{for} \quad |\mathbf{x}| < |\mathbf{s}|.$$

Moreover, $S_R^{-1}(\mathbf{s},\mathbf{x})$ is right (resp. left) slice monogenic in the variable $\mathbf{x}$ (resp. $\mathbf{s}$). We will call $-(\mathbf{x} - \bar{\mathbf{s}})(\mathbf{x}^2 - 2\mathrm{Re}[\mathbf{s}]\mathbf{x} + |\mathbf{s}|^2)^{-1}$ the Cauchy kernel for right slice monogenic functions and we will use the same symbol $S_R^{-1}(\mathbf{s},\mathbf{x})$ to denote such kernel. Even though $S_L^{-1}(\mathbf{s},\mathbf{x})$ and $S_R^{-1}(\mathbf{s},\mathbf{x})$ are different they satisfy a remarkable relation:

$$S_L^{-1}(\mathbf{x},\mathbf{s}) = -S_R^{-1}(\mathbf{s},\mathbf{x}), \qquad \text{for} \quad \mathbf{x}^2 - 2\mathrm{Re}[\mathbf{s}]\mathbf{x} + |\mathbf{s}|^2 \neq 0.$$

Using these kernels it is possible to prove the following result:

The Cauchy formulas with slice monogenic kernel. *Let $U \subset \mathbb{R}^{n+1}$ be an axially symmetric s-domain. Suppose that $\partial(U \cap \mathbb{C}_I)$ is a finite union of continuously differentiable Jordan curves for every $I \in \mathbb{S}$. Set $d\mathbf{s}_I = -d\mathbf{s}I$ for $I \in \mathbb{S}$. If f is a (left) slice monogenic function on a set that contains $\overline{U}$, then*

$$f(\mathbf{x}) = \frac{1}{2\pi} \int_{\partial(U\cap\mathbb{C}_I)} S_L^{-1}(\mathbf{s},\mathbf{x}) d\mathbf{s}_I f(\mathbf{s}), \qquad \mathbf{x} \in U. \tag{1.2}$$

Similarly, if f is a right slice monogenic function on a set that contains $\overline{U}$, then

$$f(\mathbf{x}) = \frac{1}{2\pi} \int_{\partial(U\cap\mathbb{C}_I)} f(\mathbf{s}) d\mathbf{s}_I S_R^{-1}(\mathbf{s},\mathbf{x}), \qquad \mathbf{x} \in U,$$

and the integrals above do not depend on the choice of the imaginary unit $I \in \mathbb{S}$ nor on U.

The fact that the integrals are independent of the choice of the plane $\mathbb{C}_I$ seems surprising, but if one keeps in mind the Representation Formula and the fact that the two quantities appearing in it,

$$\frac{1}{2}[f(x_0 + I|\underline{x}|) + f(x_0 - I|\underline{x}|)] \quad \text{and} \quad I\frac{1}{2}[f(x_0 - I|\underline{x}|) - f(x_0 + I|\underline{x}|)],$$

do not depend on $I \in \mathbb{S}$, the independence from the plane $\mathbb{C}_I$ becomes clear.

These results are the basic tools to introduce the functional calculus for n-tuples of operators.

The functional calculus for n-tuples of (not necessarily commuting) operators. The operators we will consider act on a Banach space V over $\mathbb{R}$ with norm $\|\cdot\|$. In general, it is possible to endow V with an operation of multiplication by elements of $\mathbb{R}_n$ which gives a two-sided module over $\mathbb{R}_n$. By V_n we indicate the two-sided Banach module over $\mathbb{R}_n$ corresponding to $V \otimes \mathbb{R}_n$. Since we want to construct a functional calculus for n-tuples of not necessarily commuting operators, we will consider the auxiliary operator

$$T = T_0 + \sum_{j=1}^{n} e_j T_j,$$

where $T_\mu \in \mathcal{B}(V)$ for $\mu = 0, 1, \dots, n$, and where $\mathcal{B}(V)$ is the space of all bounded $\mathbb{R}$-linear operators acting on V. By considering the operator T as above we have a theory which will be slightly more general than we need; in fact, to study n-tuples of operators it is sufficient to consider operators T of the form $T = \sum_{j=1}^{n} e_j T_j$. Since our theory allows us to treat also this more general case, suitable for example if one wishes to consider linear operators T acting on modules over $\mathbb{R}_n$, which are not necessarily constructed from linear operators acting on V, we will study this case. When dealing with n-tuples of operators we always mean that $T_0 = 0$. The set of bounded operators of the form $T_0 + \sum_{j=1}^{n} e_j T_j$ will be denoted by $\mathcal{B}_n^{0,1}(V_n)$. Let $T_A \in \mathcal{B}(V)$ and define the operator

$$T = \sum_A e_A T_A$$

and its action on

$$v = \sum v_B e_B \in V_n$$

as

$$T(v) = \sum_{A,B} T_A(v_B) e_A e_B.$$

The operator $\sum_A e_A T_A$ is a right-module homomorphism which is a bounded linear map on V_n: the set of all such bounded operators is denoted by $\mathcal{B}_n(V_n)$ and is endowed with the norm

$$\|T\|_{\mathcal{B}_n(V_n)} = \sum_A \|T_A\|_{\mathcal{B}(V)}.$$

We obviously have the inclusion $\mathcal{B}_n^{0,1}(V_n) \subset \mathcal{B}_n(V_n)$. To construct a functional calculus for n-tuples of noncommuting operators using the theory of left slice monogenic functions, we define the left S-resolvent operator series for $T \in \mathcal{B}_n^{0,1}(V_n)$ as

$$S^{-1}(\mathbf{s}, T) := \sum_{n \geq 0} T^n \mathbf{s}^{-1-n}, \quad \text{for} \quad \|T\| < |\mathbf{s}|.$$

In the Cauchy formula for slice monogenic functions it is always possible to replace, at least formally, the variable $\mathbf{x}$ by an operator $T = T_0 + T_1 e_1 + \ldots + T_n e_n$. This substitution is not always possible in other function theories. In our case, we have proved that the sum of the left S-resolvent operator series

$$\sum_{n \geq 0} T^n \mathbf{s}^{-1-n} = -(T^2 - 2\,\mathrm{Re}[\mathbf{s}]\,T + |\mathbf{s}|^2 \mathcal{I})^{-1}(T - \overline{\mathbf{s}}\mathcal{I}),$$

for $\|T\| < |\mathbf{s}|$ is exactly equal to the left Cauchy kernel in which we have replaced the paravector $\mathbf{x}$ by the operator $T = T_0 + e_1 T_1 + \ldots + e_n T_n$. This replacement can be done even when the components of T do not commute. This observation is the main reason why our functional calculus can be developed in a natural way starting from the Cauchy formula (1.2). The sum of the series in which we have replaced $\mathbf{x}$ by operator T suggests the notions of S-*spectrum set*, of S-*resolvent set* and of *S-resolvent operator.* Taking $T \in \mathcal{B}_n^{0,1}(V_n)$ we define: the S-spectrum $\sigma_S(T)$ of T as

$$\sigma_S(T) = \{\mathbf{s} \in \mathbb{R}^{n+1} \quad : \quad T^2 - 2\,\mathrm{Re}[\mathbf{s}]\,T + |\mathbf{s}|^2\mathcal{I} \quad \text{is not invertible}\},$$

the S-resolvent set as

$$\rho_S(T) = \mathbb{R}^{n+1} \setminus \sigma_S(T),$$

and the S-resolvent operator as

$$S^{-1}(\mathbf{s},T) := -(T^2 - 2\mathrm{Re}[\mathbf{s}]T + |\mathbf{s}|^2\mathcal{I})^{-1}(T - \overline{\mathbf{s}}\mathcal{I}).$$

Observe that if $T\mathbf{s} = \mathbf{s}T$, then we have $S^{-1}(\mathbf{s},T) = (\mathbf{s}\mathcal{I} - T)^{-1}$, i.e., we obtain the classical resolvent operator. To define the functional calculus, we have to introduce the set of admissible functions that are defined on the S-spectrum, as in the classical case of the Riesz–Dunford functional calculus.

Let $U \subset \mathbb{R}^{n+1}$ be an axially symmetric s-domain that contains the S-spectrum $\sigma_S(T)$ of T and such that $\partial(U \cap \mathbb{C}_I)$ is a finite union of continuously differentiable Jordan curves for every $I \in \mathbb{S}$. Suppose that $\overline{U}$ is contained in a domain of slice monogenicity of a function f. Then such a function f is said to be locally slice monogenic on $\sigma_S(T)$.

For those functions, setting $d\mathbf{s}_I = -d\mathbf{s}I$ for $I \in \mathbb{S}$, we define

$$f(T) = \frac{1}{2\pi}\int_{\partial(U\cap\mathbb{C}_I)} S^{-1}(\mathbf{s},T)\; d\mathbf{s}_I\; f(\mathbf{s}).$$

The functional calculus is well defined because we can prove that the integral does not depend on the open set U and on the choice of the imaginary unit $I \in \mathbb{S}$.

With all these new definitions, one may wonder which classical properties on the spectrum can be proved also in this case. An important result is that the S-spectrum of bounded operators is a compact nonempty set contained in $\{\mathbf{s} \in \mathbb{R}^{n+1} : |\mathbf{s}| \le \|T\|\ \}$ just as in the classical Riesz–Dunford case. The S-spectrum has a particular structure: if $T \in \mathcal{B}_n^{0,1}(V_n)$ and $\mathbf{p} = \mathrm{Re}[\mathbf{p}] + \underline{p} \in \sigma_S(T)$, then all the elements of the sphere $\mathbf{s} = \mathrm{Re}[\mathbf{s}] + \underline{s}$ with $\mathrm{Re}[\mathbf{s}] = \mathrm{Re}[\mathbf{p}]$ and $|\underline{s}| = |\underline{p}|$ belong to the S-spectrum of T. In other words, the S-spectrum is made of real points or entire $(n-1)$-spheres. The structure of the spectrum allows us to explain, from an intuitive point of view, why the integral $\int_{\partial(U\cap\mathbb{C}_I)} S^{-1}(\mathbf{s},T)\; d\mathbf{s}_I\; f(\mathbf{s})$ is independent of I: observe that the structure of the S-spectrum of T has a symmetry such that on each plane $\mathbb{C}_I$, for every $I \in \mathbb{S}$, we see the "same set" of points $\sigma_S(T) \bigcap \mathbb{C}_I$, and that the functions f satisfy the Representation Formula.

With our definition of functional calculus we can prove several results, among which the algebraic rules on the sum, product, composition of functions (when defined). Moreover, it is possible to prove that the spectral radius theorem, the spectral mapping theorem, the theorem of bounded perturbations of the S-resolvent operators hold. Thus the theory we obtain is quite rich.

The functional calculus can also be extended to linear closed densely defined operators $T : \mathcal{D}(T) \to V_n$ with $\rho_S(T) \cap \mathbb{R} \neq \emptyset$ and for slice monogenic functions f defined on the extended S-spectrum $\overline{\sigma}_S(T) := \sigma_S(T) \cup \{\infty\}$. The function of operator $f(T)$ can be defined as follows: take $k \in \mathbb{R}$ and define the homeomorphism $\Phi : \overline{\mathbb{R}}^{n+1} \to \overline{\mathbb{R}}^{n+1}$ as

$$\mathbf{p} := \Phi(\mathbf{s}) = (\mathbf{s} - k)^{-1}, \qquad \Phi(\infty) = 0, \qquad \Phi(k) = \infty.$$

Let $T : \mathcal{D}(T) \to V_n$ be a linear closed densely defined operator and suppose that f is slice monogenic on an open set with "suitable properties" over $\overline{\sigma}_S(T)$. Let us set $\phi(\mathbf{p}) := f(\Phi^{-1}(\mathbf{p}))$ and $A := (T - k\mathcal{I})^{-1}$, for some $k \in \rho_S(T) \cap \mathbb{R} \neq 0$. Note that A is now a bounded operator for which we have a functional calculus. The operator $f(T)$ is then defined as follows:

$$f(T) = \phi(A).$$

We have proved that the operator $f(T)$ is independent of $k \in \rho_S(T) \cap \mathbb{R}$ and we have the representation

$$f(T)v = f(\infty)\mathcal{I}v + \frac{1}{2\pi}\int_{\partial(U\cap\mathbb{C}_I)} S_L^{-1}(\mathbf{s}, T)\,d\mathbf{s}_I\, f(\mathbf{s})v, \quad v \in V_n,$$

where U need not be connected and contains $\overline{\sigma}_S(T)$.

Slice regular functions. In this book we do not dwell on the theory of slice regular functions over the algebra $\mathbb{H}$ of quaternions whose results are similar to those obtained for slice monogenic functions. We introduce its main results only in order to develop the quaternionic functional calculus.

Let $U \subseteq \mathbb{H}$ be an open set and let $f : U \to \mathbb{H}$ be a real differentiable function. Denote by $\mathbb{S}$ the sphere of purely imaginary quaternions, i.e.,

$$\mathbb{S} = \{x_1 i + x_2 j + x_3 k \ : \ x_1^2 + x_2^2 + x_3^2 = 1\}.$$

Let $I \in \mathbb{S}$ and let f_I be the restriction of f to the complex plane $\mathbb{C}_I := \mathbb{R} + I\mathbb{R}$ passing through 1 and I and denote by $x + Iy$ an element on $\mathbb{C}_I$. We say that f is a left slice regular function if, for every $I \in \mathbb{S}$, we have

$$\frac{1}{2}\left(\frac{\partial}{\partial x} + I\frac{\partial}{\partial y}\right) f_I(x + Iy) = 0.$$

An analogous definition can be given for right slice regular functions. Functions left (resp. right) slice regular on U form a set denoted by $\mathcal{R}^L(U)$ (resp. $\mathcal{R}^R(U)$). The advantage of dealing with quaternions, instead of general Clifford algebras, is that $\mathbb{H}$ has a richer algebraic structure. For example, $\mathbb{H}$ is helpful when we want to determine the sum of the Cauchy kernel series, which is defined, for q and s quaternions, by

$$S_L^{-1}(s, q) := \sum_{n\geq 0} q^n s^{-1-n}.$$

One can prove that the inverse $S_L(s,q)$ of $S_L^{-1}(s,q)$ is the nontrivial solution to the equation

$$S_L^2 + S_L q - sS_L = 0.$$

In particular, an application of Niven's algorithm [82] gives

$$S_L(s,q) = -(q-\overline{s})^{-1}(q^2 - 2qRe[s] + |s|^2).$$

Note that this approach would not be possible in the Clifford algebra setting, where no analog of the Niven's algorithm is known (in fact, in the Clifford algebras setting, one cannot even guarantee the existence of solutions to a polynomial equation). Also note that in the case of quaternions the term $q^n s^{-1-n}$ in the Cauchy kernel series is a quaternion for every $n \in \mathbb{N}$ while in the Clifford algebra setting one starts with paravectors $\mathbf{x}$ and $\mathbf{s}$ but the terms $\mathbf{x}^n \mathbf{s}^{-1-n}$ do not contain only paravectors but also terms of the form $e_i e_j$.

These preliminaries allow us to find the left and right Cauchy formulas in the quaternionic setting as follows. Let $\overline{U} \subset W$ be an axially symmetric s-domain (the definition is as in the Clifford algebras case), and let $\partial(U \cap \mathbb{C}_I)$ be a finite union of continuously differentiable Jordan curves for every $I \in \mathbb{S}$. Set $ds_I = -dsI$. Let f be a left slice regular function on $W \subset \mathbb{H}$. Then, if $q \in U$, we have

$$f(q) = \frac{1}{2\pi} \int_{\partial(U\cap\mathbb{C}_I)} S_L^{-1}(s,q)ds_I f(s).$$

Let f be a right slice regular function on $W \subset \mathbb{H}$. Then, if $q \in U$, we have

$$f(q) = \frac{1}{2\pi} \int_{\partial(U\cap\mathbb{C}_I)} f(s)ds_I S_R^{-1}(s,q)$$

and the integrals do not depend on the choice of the imaginary unit $I \in \mathbb{S}$ nor on U. The left and the right slice regular kernels are defined by

$$S_L^{-1}(s,q) := -(q^2 - 2\,\mathrm{Re}[s]\,q + |s|^2)^{-1}(q-\overline{s}),$$

and

$$S_R^{-1}(s,q) := -(q-\overline{\mathbf{s}})(q^2 - 2\,\mathrm{Re}[s]\,q + |s|^2)^{-1}.$$

These Cauchy formulas will be the basis to define a quaternionic functional calculus.

The quaternionic functional calculus. The work of Adler [1] suggests the importance of the development of functional calculus for quaternionic operators. The fundamental question, pointed out in [1], is what function theory should be used to develop such a functional calculus if we are to obtain a calculus which shares the basic properties of the Riesz–Dunford functional calculus. In order to be able to do so, one needs a function theory simple enough to include polynomials and yet developed enough to allow a Cauchy like formula. The theory of slice regular functions that we develop in Chapter 4 satisfies both requirements.

When dealing with quaternionic operators there is a major difference with respect to the case treated in Chapter 3 for n-tuples of $\mathbb{R}$-linear operators. In fact, there are four cases of interest for a functional calculus: left and right linear quaternionic operators and left and right slice regular functions. Even though the majority of our results will be stated and proved in the case of right linear operators and for left slice monogenic functions, it is worth describing the differences among the various cases, since they have to be taken into account, especially when dealing with the case of unbounded operators.

Let V be a right vector space on $\mathbb{H}$. An operator $T : V \to V$ is said to be a right linear operator if $T(u+v) = T(u) + T(v)$, $T(us) = T(u)s$, for all $s \in \mathbb{H}$ and for all $u, v \in V$. In the sequel, we will consider only two-sided vector spaces V, otherwise the set of right linear operators is not a (left or right) vector space. With this assumption, the set of right linear operators $\text{End}^R(V)$ on V is both a left and a right vector space on $\mathbb{H}$ with respect to the operations $(sT)(v) := sT(v)$, $(Ts)(v) := T(sv)$, for all $s \in \mathbb{H}$, and for all $v \in V$. Similarly, a map $T : V \to V$ is said to be a left linear operator if $T(u+v) = T(u) + T(v)$, $T(su) = sT(u)$, for all $s \in \mathbb{H}$ and for all $u, v \in V$. The set $\text{End}^L(V)$ of left linear operators on V is both a left and a right vector space on $\mathbb{H}$ with respect to the operations $(Ts)(v) := T(v)s$, $(sT)(v) := T(vs)$, for all $s \in \mathbb{H}$ and for all $v \in V$.

A crucial fact is that the composition of left and right linear operators acts in an opposite way with respect to the composition of maps. In fact the two rings $\text{End}^R(V)$ and $\text{End}^L(V)$ with respect to the addition and composition of operators are opposite rings of each other. This fact has important consequences in the definition of the S-resolvent operators for unbounded operators. Similarly, we will have the two-sided vector space $\mathcal{B}^R(V)$ of all right linear bounded operators on V and the two-sided vector space $\mathcal{B}^L(V)$ of all left linear bounded operators on V. When it is not necessary to specify if a bounded operator is left or right linear on V, we use the symbol $\mathcal{B}(V)$ and we call an element in $\mathcal{B}(V)$ a "linear operator". As before, we introduce, for $T \in \mathcal{B}(V)$ the left Cauchy kernel operator series, or S-resolvent operator series, as

$$S_L^{-1}(s,T) = \sum_{n\geq 0} T^n s^{-1-n},$$

and the right Cauchy kernel operator series as

$$S_R^{-1}(s,T) = \sum_{n\geq 0} s^{-1-n} T^n,$$

for $\|T\| < |s|$. The fundamental point of this theory and its importance for physical applications is the fact that we can replace the variable q, whose components are commuting real numbers, with a linear quaternionic operator T whose components are, in general, noncommuting operators. It is also important to note that the action of the S-resolvent operators series $S_L^{-1}(s,T)$ and $S_R^{-1}(s,T)$ in the case of left linear operators T is on the right, i.e., for every $v \in V$ we have $v \mapsto vS_L^{-1}(s,T)$

and $v \mapsto vS_R^{-1}(s,T)$. For example for the left Cauchy kernel operator series we have $v \mapsto \sum_{n\geq 0} T^n(v)s^{-n-1} = \sum_{n\geq 0} vT^n s^{-n-1}$. Thus, even though $S_L^{-1}(s,T)$ is formally the same operator used for right linear operators, $S_L^{-1}(s,T)$ acts in a different way. Note that the following important results hold for both left and right linear quaternionic operators.

Let $T \in \mathcal{B}(V)$. Then, for $\|T\| < |s|$, we have

$$\sum_{n\geq 0} T^n s^{-1-n} = -(T^2 - 2\operatorname{Re}[s]\, T + |s|^2\mathcal{I})^{-1}(T - \overline{s}\mathcal{I}),$$

and

$$\sum_{n\geq 0} s^{-1-n} T^n = -(T - \overline{s}\mathcal{I})(T^2 - 2\operatorname{Re}[s]\, T + |s|^2\mathcal{I})^{-1}.$$

Observe that the quaternionic operators treated in Section 4 act on a quaternionic Banach space while the n-tuples of noncommuting operators act on Banach modules over a Clifford algebra.

We point out that, when T is a quaternionic operator, the Cauchy operator series $\sum_{n\geq 0} T^n s^{-1-n}$ is a quaternionic operator because T^n are quaternionic operators. In the Clifford setting when we consider $T = T_0 + e_1 T_1 + \ldots + e_n T_n$, $n \geq 3$, then T^n contains not only the terms with the units $e_1, \ldots, e_n$ but also those with $e_i e_j, \ldots, e_1 e_2 e_3, \ldots$ and so on. Thus the powers T^n are not anymore operators in the form $A_0 + e_1 A_1 + \ldots + e_n A$.

The S-spectrum and the S-resolvent sets can be defined as for the case of n-tuples of noncommuting operators. Let $T \in \mathcal{B}(V)$, then the S-spectrum $\sigma_S(T)$ of $T \in \mathcal{B}(V)$ is

$$\sigma_S(T) = \{s \in \mathbb{H} \ : \ T^2 - 2\ \operatorname{Re}[s]T + |s|^2\mathcal{I} \quad \text{is not invertible}\}.$$

The S-resolvent set $\rho_S(T)$ is defined by

$$\rho_S(T) = \mathbb{H} \setminus \sigma_S(T).$$

For $s \in \rho_S(T)$ we define the left S-resolvent operator as

$$S_L^{-1}(s,T) := -(T^2 - 2\operatorname{Re}[s]T + |s|^2\mathcal{I})^{-1}(T - \overline{s}\mathcal{I}),$$

and the right S-resolvent operator as

$$S_R^{-1}(s,T) := -(T - \overline{s}\mathcal{I})(T^2 - 2\operatorname{Re}[s]T + |s|^2\mathcal{I})^{-1}.$$

They satisfy the equations:

$$S_L^{-1}(s,T)s - TS_L^{-1}(s,T) = \mathcal{I},$$

and

$$sS_R^{-1}(s,T) - S_R^{-1}(s,T)T = \mathcal{I}.$$

To define a functional calculus we need to introduce the admissible domains on which it can be formulated.

Let $T \in \mathcal{B}(V)$ and let $U \subset \mathbb{H}$ be an axially symmetric s-domain that contains the S-spectrum $\sigma_S(T)$ and such that $\partial(U \cap \mathbb{C}_I)$ is a finite union of continuously differentiable Jordan curves for every $I \in \mathbb{S}$. Let W be an open set in $\mathbb{H}$. A function $f \in \mathcal{R}^L(W)$ is said to be locally left regular on $\sigma_S(T)$ if there exists a domain $U \subset \mathbb{H}$, as above and such that $\overline{U} \subset W$, on which f is left regular. A function $f \in \mathcal{R}^R(W)$ is said to be locally right regular on $\sigma_S(T)$ if there exists a domain $U \subset \mathbb{H}$, as above and such that $\overline{U} \subset W$, on which f is right regular.

The quaternionic functional calculus can now be defined as follows. Let $U \subset \mathbb{H}$ be a domain as above and set $ds_I = -dsI$. We define

$$f(T) = \frac{1}{2\pi} \int_{\partial(U \cap \mathbb{C}_I)} S_L^{-1}(s,T)\, ds_I\, f(s), \quad \text{for} \quad f \in \mathcal{R}^L_{\sigma_S(T)},$$

and

$$f(T) = \frac{1}{2\pi} \int_{\partial(U \cap \mathbb{C}_I)} f(s)\, ds_I\, S_R^{-1}(s,T), \quad \textit{for} \quad f \in \mathcal{R}^R_{\sigma_S(T)}.$$

The definitions are well posed because the integrals do not depend on the open set U and on the imaginary unit $I \in \mathbb{S}$. Note that when $T \in \mathcal{B}^L(V)$ we have $f(T)(v) = vf(T)$ while if $T \in \mathcal{B}^R(V)$ we have $f(T)(v) = f(T)v$.

One can also define a quaternionic functional calculus for closed densely defined linear quaternionic operators. Here we must pay attention to the differences between the cases of left and right linear operators. Denote by $\mathcal{K}^R(V)$ ($\mathcal{K}^L(V)$ resp.) the set of right (left resp.) linear closed operators $T : \mathcal{D}(T) \subset V \to V$, such that: $\mathcal{D}(T)$ is dense in V, $\mathcal{D}(T^2) \subset \mathcal{D}(T)$ is dense in V, $T - \overline{s}\mathcal{I}$ is densely defined in V. We will use the symbol $\mathcal{K}(V)$ when we do not distinguish between $\mathcal{K}^L(V)$ and $\mathcal{K}^R(V)$. Since T is a closed operator, then $T^2 - 2\,\mathrm{Re}[s]\,T + |s|^2\mathcal{I} : \mathcal{D}(T^2) \subset V \to V$ is a closed operator. In analogy with the case of bounded operators, we denote by $\rho_S(T)$ the S-resolvent set of T, i.e., the set

$$\rho_S(T) = \{s \in \mathbb{H} \ : \ (T^2 - 2\,\mathrm{Re}[s]\,T + |s|^2\mathcal{I})^{-1} \in \mathcal{B}(V)\},$$

and, as a consequence, we define the S-spectrum $\sigma_S(T)$ of T as

$$\sigma_S(T) = \mathbb{H} \setminus \rho_S(T).$$

For any $T \in \mathcal{K}(V)$ and $s \in \rho_S(T)$, we denote by $Q_s(T)$ the operator

$$Q_s(T) := (T^2 - 2\,\mathrm{Re}[s]\,T + |s|^2\mathcal{I})^{-1} : \quad V \to \mathcal{D}(T^2). \tag{1.3}$$

The definition of the S-resolvent operators S_L^{-1}, S_R^{-1} relies on a deep difference between the case of left and right linear operators. To start with, consider the left S-resolvent operator used in the bounded case, that is

$$S_L^{-1}(s,T) = -Q_s(T)(T - \overline{s}\mathcal{I}). \tag{1.4}$$

Note that in the case of right linear unbounded operators, $S_L^{-1}(s,T)$ turns out to be defined only on $\mathcal{D}(T)$, while in the case of left linear unbounded operators it is defined on V. This is a striking difference between the two cases due to the presence of the term $Q_s(T)T$. However, for $T \in \mathcal{K}^R(V)$, observe that the operator $Q_s(T)T$ is the restriction to the dense subspace $\mathcal{D}(T)$ of V of a bounded linear operator defined on V. This fact follows from the commutation relation $Q_s(T)Tv = TQ_s(T)v$ which holds for all $v \in \mathcal{D}(T)$ since the polynomial operator $T^2 - 2\operatorname{Re}[s]\, T + |s|^2\mathcal{I} : \mathcal{D}(T^2) \to V$ has real coefficients. Since $TQ_s(T) : V \to \mathcal{D}(T)$ and it is continuous for $s \in \rho_S(T)$, the left S-resolvent operators for unbounded right linear operators is defined as

$$S_L^{-1}(s,T)v := -Q_s(T)(T - \overline{s}\mathcal{I})v, \quad \text{for all } \; v \in \mathcal{D}(T),$$

and we will call

$$\hat{S}_L^{-1}(s,T)v = Q_s(T)\overline{s}v - TQ_s(T)v, \quad \text{for all } \; v \in V,$$

the extended left S-resolvent operator. The right S-resolvent operator is

$$S_R^{-1}(s,T)v := -(T - \mathcal{I}\overline{s})Q_s(T)v,$$

and it is already defined for all $v \in V$. Observe also that for the right S-resolvent operator $S_R^{-1}(s,T)$ we have that for $s \in \rho_S(T)$ the operator $Q_s(T) : V \to \mathcal{D}(T^2)$ is bounded so also $(T - \mathcal{I}\overline{s})Q_s(T) : V \to \mathcal{D}(T)$ is bounded.

The discussion of this case shows that, in the case of unbounded linear operators, the S-resolvent operators (left and right) have to be defined in a different way for left and right linear operators and motivates the following definition.

Let A be an operator containing the term $Q_s(T)T$ (resp. $TQ_s(T)$). We define $\hat{A}$ to be the operator obtained from A by substituting each occurrence of $Q_s(T)T$ (resp. $TQ_s(T)$) by $TQ_s(T)$ (resp. $Q_s(T)T$).

In the case of a left linear operator, i.e., $T \in \mathcal{K}^L(V)$ and $s \in \rho_S(T)$, we define the left S-resolvent operator as (compare with the case $T \in \mathcal{K}^R(V)$)

$$vS_L^{-1}(s,T) := -vQ_s(T)(T - \overline{s}\mathcal{I}), \quad \text{for all } \; v \in V,$$

and the right S-resolvent operator as

$$vS_R^{-1}(s,T) := -v(T - \mathcal{I}\overline{s})Q_s(T), \quad \text{for all } \; v \in \mathcal{D}(T).$$

To have an operator defined on the whole V we introduce

$$v\hat{S}_R^{-1}(s,T) = vQ_s(T)\overline{s} - vQ_s(T)T, \quad \text{for all } \; v \in V,$$

which is called the extended right S-resolvent operator.

A second difference between the functional calculus for left linear operators and for right linear operators is given by the S-resolvent equations which, to hold on V, need different extensions of the operators involved. Specifically, we have that: if $T \in \mathcal{K}^R(V)$ and $s \in \rho_S(T)$, then the left S-resolvent operator satisfies the equation

$$\hat{S}_L^{-1}(s,T)sv - T\hat{S}_L^{-1}(s,T)v = \mathcal{I}v, \quad \text{for all } \ v \in V,$$

while the right S-resolvent operator satisfies the equation

$$sS_R^{-1}(s,T)v - (\widehat{S_R^{-1}(s,T)T})v = \mathcal{I}v, \quad \text{for all } \ v \in V.$$

If $T \in \mathcal{K}^L(V)$ and $s \in \rho_S(T)$, then the left S-resolvent operator satisfies the equation

$$v\hat{S}_L^{-1}(s,T)s - v\widehat{TS_L^{-1}}(s,T) = v\mathcal{I}, \quad \text{for all } \ v \in V.$$

Finally, the right S-resolvent operator satisfies the equation

$$vs\hat{S}_R^{-1}(s,T) - v(\hat{S}_R^{-1}(s,T)T) = v\mathcal{I}, \quad \text{for all } \ v \in V.$$

Another issue which requires the use of extended operators is the treatment of unbounded operators. In the classical case of a complex unbounded linear operator $B : \mathcal{D}(B) \subset X \to X$, where X is a complex Banach space, the resolvent operator

$$R(\lambda, B) := (\lambda\mathcal{I} - B)^{-1}, \quad \text{for } \ \lambda \in \rho(B),$$

satisfies the relations

$$(\lambda\mathcal{I} - B)R(\lambda, B)x = x, \quad \text{for all } \ x \in X,$$

$$R(\lambda, B)(\lambda\mathcal{I} - B)x = x, \quad \text{for all } \ x \in \mathcal{D}(B).$$

It is then natural to ask what happens in the quaternionic case for unbounded operators. Again, one has to use suitable extensions and the results, in the case $T \in \mathcal{K}^R(V)$, are:

$$\hat{S}_L(s,T)\hat{S}_L^{-1}(s,T)v = \mathcal{I}v, \quad \text{for all } \ v \in V,$$

$$\hat{S}_L^{-1}(s,T)\hat{S}_L(s,T)v = \mathcal{I}v, \quad \text{for all } \ v \in \mathcal{D}(T),$$

and

$$S_R(s,T)S_R^{-1}(s,T)v = \mathcal{I}v, \quad \text{for all } \ v \in V,$$

$$S_R^{-1}(s,T)S_R(s,T)v = \mathcal{I}v, \quad \text{for all } \ v \in \mathcal{D}(T).$$

The corresponding results, with suitable modifications, are proved also for $T \in \mathcal{K}^L(V)$.

We are now ready to present the functional calculus in the four cases of unbounded operators. Let $T \in \mathcal{K}^R(V)$ and let W be an open set as defined above such that $\overline{\sigma}_S(T) \subset W$ and let f be a regular function on $W \cup \partial W$. Let $I \in \mathbb{S}$ and $W \cap \mathbb{C}_I$ be such that its boundary $\partial(W \cap \mathbb{C}_I)$ is positively oriented and consists of a finite number of rectifiable Jordan curves. If $T \in \mathcal{K}^R(V)$ with $\rho_S(T) \cap \mathbb{R} \neq \emptyset$, then the operator $f(T)$, defined in an analogous way as we did for the case of an n-tuple of noncommuting operators, is independent of the real number $k \in \rho_S(T)$, and, for $f \in \mathcal{R}^L_{\overline{\sigma}_S(T)}$ and $v \in V$, we have

$$f(T)v = f(\infty)\mathcal{I}v + \frac{1}{2\pi} \int_{\partial(W \cap \mathbb{C}_I)} \hat{S}_L^{-1}(s,T)\; ds_I\; f(s)v,$$

and for $f \in \mathcal{R}^R_{\overline{\sigma}_S(T)}$ and $v \in V$, we have

$$f(T)v = f(\infty)\mathcal{I}v + \frac{1}{2\pi} \int_{\partial(W \cap \mathbb{C}_I)} f(s)\; ds_I\; S_R^{-1}(s,T)v.$$

If $T \in \mathcal{K}^L(V)$ we can define two analogous functional calculi, according to the use of left or right regular functions.

We conclude the overview of the book with an important application of the quaternionic functional calculus to the theory of quaternionic semigroups. A surprising result is the remarkable relation of the semigroup e^{tT} with the S-resolvent operator: let $T \in \mathcal{B}(V)$ and let $s_0 > \|T\|$. Then the right S-resolvent operator $S_R^{-1}(s,T)$ is given by

$$S_R^{-1}(s,T) = \int_0^{+\infty} e^{-ts}\, e^{tT}\, dt.$$

Let $T \in \mathcal{B}(V)$ and let $s_0 > \|T\|$. Then the left S-resolvent operator $S_L^{-1}(s,T)$ is given by

$$S_L^{-1}(s,T) = \int_0^{+\infty} e^{tT}\, e^{-ts}\, dt.$$

Note that, as in the classical case, we have a characterization result: if $\mathcal{U}(t)$ is a quaternionic semigroup on a quaternionic Banach space V, then $\mathcal{U}(t)$ has a bounded infinitesimal quaternionic generator if and only if it is uniformly continuous.

Chapter 2

Slice monogenic functions

2.1 Clifford algebras

Clifford algebras will be the setting in which we will work throughout this book. They were introduced under the name of geometric algebras by Clifford in 1878. Since then, several people have extensively studied them and nowadays there are, in the literature, several possible ways to introduce Clifford algebras: for example one can use exterior algebras, or present them as a quotient of a tensor algebra or by means of a universal property (see [23], [31], [34], or [75] for a survey on the various possible definitions). In this book, we will adopt an equivalent but more direct approach, using generators and relations.

Definition 2.1.1. *Given n elements $e_1, \dots, e_n$, $n = p + q$, $p, q \geq 0$, which will be called* imaginary units, *together with the* defining relations

$$\begin{aligned} e_i^2 &= +1, \quad \textit{for} \quad i = 1, \dots, p, \\ e_i^2 &= -1, \quad \textit{for} \quad i = p+1, \dots, n, \\ e_i e_j + e_j e_i &= 0, \quad i \neq j. \end{aligned}$$

Assume that

$$e_1 e_2 \dots e_n \neq \pm 1 \quad \text{if } p - q \equiv 1 (\operatorname{mod} 4). \tag{2.1}$$

We will call (universal) Clifford algebra *the algebra over $\mathbb{R}$ generated by $e_1, \dots, e_n$ and we will denote it by $\mathbb{R}_{p,q}$.*

Remark 2.1.2. It is immediate that $\mathbb{R}_{p,q}$, as a real vector space and has dimension 2^n, $n = p + q$.

An element in $\mathbb{R}_{p,q}$, called a *Clifford number*, can be written as

$$a = a_0 + a_1 e_1 + \dots + a_n e_n + a_{12} e_1 e_2 + \dots + a_{123} e_1 e_2 e_3 + \dots + a_{12\dots n} e_1 e_2 \dots e_n.$$

Denote by A an element in the power set $\wp(1,\ldots,n)$. If $A = i_1 \ldots i_r$, then the element $e_{i_1} \ldots e_{i_r}$ can be written as $e_{i_1\ldots i_r}$ or, in short, e_A. Thus, in a more compact form, we can write a Clifford number as

$$a = \sum_A a_A e_A.$$

Possibly using the defining relations, we will order the indices in A as $i_1 < \ldots < i_r$. When $A = \emptyset$ we set $e_\emptyset = 1$.

We now give some examples of real Clifford algebras $\mathbb{R}_n$ of low dimension.

Example 2.1.3. First of all, we point out that the index $n = 0$ is allowed in the definition, and in this case we obtain the real numbers. For $n = 1$ we have that $\mathbb{R}_{0,1}$ is the algebra generated by e_1 over $\mathbb{R}$ with the relation $e_1^2 = -1$. Hence there is an $\mathbb{R}$-algebra isomorphism $\mathbb{R}_{0,1} \cong \mathbb{C}$ where $\mathbb{C}$ denotes, as customary, the algebra of complex numbers.

Example 2.1.4. For $n = 2$, the Clifford algebra $\mathbb{R}_{0,2}$ is generated by e_1 and e_2. This real algebra is the so-called algebra of quaternions and it is usually denoted by the symbol $\mathbb{H}$. A quaternion q is traditionally written as $q = x_0 + ix_1 + jx_2 + kx_3$ where the imaginary units i, j, k anti-commute among them and satisfy $i^2 = j^2 = k^2 = -1$. With the identification

$$e_1 \to i, \quad e_2 \to j,$$

(and the consequent $e_1e_2 \to k$), it is immediate to identify $\mathbb{R}_{0,2}$ with $\mathbb{H}$.

Example 2.1.5. We now compare the two Clifford algebras $\mathbb{R}_{1,1}$ generated by the elements e_1 and ϵ_1 such that $e_1^2 = -1$ and $e_2^2 = +1$, and $\mathbb{R}_{2,0}$ generated by the elements ε_1 and ε_2 both having square $+1$. These two Clifford algebras are isomorphic. In fact, let us consider the matrices

$$\eta_0 = \begin{bmatrix} 1 & 0 \\ 0 & 1 \end{bmatrix} \qquad \eta_1 = \begin{bmatrix} 0 & 1 \\ 1 & 0 \end{bmatrix}$$

$$\eta_2 = \begin{bmatrix} 0 & -1 \\ 1 & 0 \end{bmatrix} \qquad \eta_3 = \begin{bmatrix} 1 & 0 \\ 0 & -1 \end{bmatrix}.$$

They form a basis for the vector space $M(2,\mathbb{R})$ of 2×2 real matrices. The map

$$\varphi : \ \mathbb{R}_{1,1} \mapsto M(2,\mathbb{R})$$

defined by $\varphi(e_1) = \eta_2$, $\varphi(e_2) = \eta_1$ can be extended to an isomorphism for which $\varphi(1) = \eta_0$, and $\varphi(e_2e_1) = \eta_3$. The map

$$\psi : \ \mathbb{R}_{2,0} \mapsto M(2,\mathbb{R})$$

defined by $\psi(\varepsilon_1) = \eta_1$, $\psi(\varepsilon_2) = \eta_3$ can be extended to an isomorphism for which $\psi(1) = \eta_0$, $\psi(\varepsilon_1\varepsilon_2) = \eta_2$. Thus the Clifford algebras $\mathbb{R}_{1,1}$ and $\mathbb{R}_{2,0}$ are isomorphic but, as the reader can verify, they are not isomorphic to $\mathbb{R}_{0,2}$.

The case of $\mathbb{R}_{0,n}$ will be the only case we will use in this book. For this reason, we will write $\mathbb{R}_n$ instead of $\mathbb{R}_{0,n}$.

Definition 2.1.6. *Let $k \in \mathbb{N}$ and $0 \leq k \leq n$. The linear subspace of $\mathbb{R}_n$ generated by the $\binom{n}{k}$ elements of the form $e_A = e_{i_1} \dots e_{i_k}$, $i_\ell \in \{1, \dots, n\}$, $i_1 < \dots < i_k$, will be denoted by $\mathbb{R}_n^k$. The elements in $\mathbb{R}_n^k$ are called k-vectors.*

For $k = 0$, the subspace $\mathbb{R}_n^0$ is identified with the space of scalars $\mathbb{R}$; for $k = 1$ we have the subspace $\mathbb{R}_n^1$ of 1-vectors, also called vectors for short and denoted by $\underline{x}$, with basis $\{e_1, \dots, e_n\}$; an element $(x_1, x_2, \dots, x_n) \in \mathbb{R}^n$ can be identified with a vector $\underline{x} \in \mathbb{R}_n^1$ in the Clifford algebra using the map:

$$(x_1, x_2, \dots, x_n) \mapsto \underline{x} = x_1 e_1 + \dots + x_n e_n.$$

The subspace $\mathbb{R}_n^2$ consists of 2-vectors or bivectors, and has basis $\{e_{ij} = e_i e_j,\ i < j\}$. In general, for any subset $A = \{i_1, \dots, i_k\}$ of $N = \{1, \dots, n\}$ of cardinality $|A| = k$, the elements $e_A = e_{i_1} \dots e_{i_k}$, $i_1 < \dots < i_k$, form a basis for the $\binom{n}{k}$-dimensional vector space $\mathbb{R}_n^k$ of the k-vectors. Every element belonging to $\mathbb{R}_n^0 \oplus \mathbb{R}_n^1$ is a sum of a scalar and a vector. It is called paravector. An element $(x_0, x_1, \dots, x_n) \in \mathbb{R}^{n+1}$ can be identified with a paravector $\mathbf{x} \in \mathbb{R}_n^0 \oplus \mathbb{R}_n^1$ by the map:

$$(x_0, x_1, \dots, x_n) \mapsto \mathbf{x} = x_0 + x_1 e_1 + \dots + x_n e_n.$$

Note also that every element $a \in \mathbb{R}_n$ may also be uniquely written as

$$a = [a]_0 + [a]_1 + \dots + [a]_k + \dots + [a]_n$$

where $[\cdot]_k : \mathbb{R}_n \to \mathbb{R}_n^k$ denotes the projection of $\mathbb{R}_n$ onto the space of k-vectors. Finally, a can be written in the form

$$a = a_+ + a_-$$

where $[a]_+ = [a]_0 + [a]_2 + \dots$, and $[a]_- = [a]_1 + [a]_3 + \dots$. We hence have a direct sum decomposition

$$\mathbb{R}_n = \mathbb{R}_{n,+} \oplus \mathbb{R}_{n,-}$$

where $\mathbb{R}_{n,+}$ is the even subalgebra generated by the bivectors e_{ij}, while $\mathbb{R}_{n,-}$ contains all the elements a that may be written in the form $a = -e_1(e_1 a)$, $e_1 a \in \mathbb{R}_{n,+}$. Note that $\mathbb{R}_{n,-}$ is not an algebra while $\mathbb{R}_{n,+}$ is an algebra isomorphic to $\mathbb{R}_{n-1}$.

Among the elements in the Clifford algebra $\mathbb{R}_n$, we can consider the product of all the imaginary units e_i:

Definition 2.1.7. *The product $e_N := e_1 \dots e_n$ is called pseudoscalar.*

Remark 2.1.8. If n is odd the pseudoscalar commutes with any element of the Clifford algebra $\mathbb{R}_n$ since it can be verified that

$$e_j e_N = e_N e_j,$$

while when n is even e_N anticommutes with any imaginary unit in the Clifford algebra:

$$e_j e_N = -e_N e_j.$$

As a consequence of the remark, we immediately have the following result:

Proposition 2.1.9. *The center of a Clifford algebra $\mathbb{R}_n$ is $\mathbb{R}$ for n even, while it is $\mathbb{R} \oplus e_N\mathbb{R} = \{x + e_N y \mid x, y \in \mathbb{R}\}$ for n odd.*

Proposition 2.1.10. *The Clifford algebra $\mathbb{R}_n$, $n \geq 3$, contains zero divisors.*

Proof. Since $n \geq 3$, $\mathbb{R}_n$ contains the element e_{123}. We have

$$(1 - e_{123})(1 + e_{123}) = 1 - e_{123} + e_{123} - e_{123}e_{123} = 1 - e_{123}^2 = 0. \qquad \square$$

In a Clifford algebra it is possible to introduce several involutions, but for our purposes we will simply consider the so-called conjugation:

Definition 2.1.11. *Let $a, b \in \mathbb{R}_n$. The conjugation is defined by*

$$\bar{e}_j = -e_j, \quad j = 1, \dots, n, \qquad \overline{ab} = \bar{b}\bar{a}.$$

As a consequence of the definition, for any $a \in \mathbb{R}_n$, $a = \sum a_A e_A$, we have

$$\bar{a} = \sum a_A \bar{e}_A = [a]_0 - [a]_1 - [a]_2 + [a]_3 + [a]_4 - \dots$$

i.e., for any $a \in \mathbb{R}_n^k$ we have the 4-periodicity

$$\bar{a} = a \quad \text{for} \quad k \equiv 0, 3 \bmod 4,$$
$$\bar{a} = -a \quad \text{for} \quad k \equiv 1, 2 \bmod 4.$$

The following properties of the conjugation can be easily verified by direct computation:

Proposition 2.1.12. *The conjugation of Clifford numbers satisfies:*

(1) $\bar{\bar{a}} = a$ *for all* $a \in \mathbb{R}_n$;

(2) $\overline{a + b} = \bar{a} + \bar{b}$ *for all* $a, b \in \mathbb{R}_n$;

(3) $a + \bar{a} = 2[a]_0$ *for all paravectors* a.

The conjugation allows us to introduce an inner product defined on the real linear space of Clifford numbers:

Proposition 2.1.13. *Let $a, b \in \mathbb{R}_n$. Then*

$$\langle a, b \rangle := [\bar{a}b]_0 = [b\bar{a}]_0 = [\bar{b}a]_0,$$

is a positive definite inner product on $\mathbb{R}_n$.

Proof. Let $a = \sum_A a_A e_A$, $b = \sum_B b_B e_B$. We have

$$\bar{a}b = (\overline{\sum_A a_A e_A})(\sum_B b_B e_B) = \sum_{A,B} a_A b_B \bar{e}_A e_B$$

and since $\bar{e}_A e_A = (-1)^{|A|(|A|+1)/2} e_A e_A = 1$ we obtain

$$\bar{a}b = \sum_A a_A b_A + \sum_{A \neq B} a_A b_B \bar{e}_A e_B$$

and so $[\bar{a}b]_0 = \sum_A a_A b_A$. Thus $[\bar{a}b]_0$ coincides with the scalar product of the vectors in $\mathbb{R}^{2^n}$ corresponding to the real components of a and b and it defines a scalar product. The fact that it coincides with $[b\bar{a}]_0$ and $[\bar{b}a]_0$ can be proved by similar computations. □

We note that the inner product defined by Proposition 2.1.13 behaves like a scalar product on the space of vectors and, if $\underline{x}$ and $\underline{y}$ are two vectors we have

$$\langle \underline{x}, \underline{y} \rangle = \frac{1}{2}(\underline{x}\underline{y} + \underline{y}\underline{x}).$$

The wedge product of two vectors $\underline{x}$ and $\underline{y}$ is defined by

$$\underline{x} \wedge \underline{y} = \frac{1}{2}(\underline{x}\underline{y} - \underline{y}\underline{x}).$$

Note that the wedge product represents the directed and oriented surface measure of the parallelogram individuated by $\underline{x}$ and $\underline{y}$. It is also immediate that the product of two vectors can be written as

$$\underline{x}\,\underline{y} = \frac{1}{2}(\underline{x}\underline{y} + \underline{y}\underline{x}) + \frac{1}{2}(\underline{x}\underline{y} - \underline{y}\underline{x}) = \langle \underline{x}, \underline{y} \rangle + \underline{x} \wedge \underline{y}. \tag{2.2}$$

Note also that, in the case of vectors, the scalar product can be written as

$$\langle \underline{x}, \underline{y} \rangle = \sum_{j=1}^{n} x_i y_i,$$

and, if by $|\underline{x}|$ we denote the Euclidean norm of a vector $\underline{x}$, we have

$$|\underline{x}| = \sqrt{\langle \underline{x}, \underline{x} \rangle} \tag{2.3}$$

which is the length of the vector $\underline{x}$.

We will say that two nonzero vectors $\underline{x}$, $\underline{y}$ are orthogonal if $\langle \underline{x}, \underline{y} \rangle = 0$. As customary, a basis $\{u_1, \dots u_s\}$ of a subspace U of the Euclidean space $\mathbb{R}^n$ is said to be orthonormal if $|u_i| = 1$ and $\langle u_i, u_j \rangle = 0$ for every u_i, u_j, such that $u_i \neq u_j$.

In general, given an element $a = \sum_A a_A e_A \in \mathbb{R}_n$ we can define its modulus as

$$|a| = (\sum_A a_A^2)^{\frac{1}{2}}.$$

The proof of Proposition 2.1.13 shows that

$$|a|^2 = [a\bar{a}]_0 = \langle a, a\rangle,$$

thus generalizing formula (2.3) to the case of a general Clifford number. We have the following properties:

Proposition 2.1.14. *The modulus of Clifford numbers satisfies:*

(1) $|\lambda a| = |\lambda|\,|a|$ *for all* $\lambda \in \mathbb{R}$, $a \in \mathbb{R}_n$*;*

(2) $||x| - |y|| \leq |x - y| \leq |x| + |y|$*;*

However, the modulus is not multiplicative, as shown in the next result.

Proposition 2.1.15. *For any two elements* $a, b \in \mathbb{R}_n$ *we have*

$$|ab| \leq C_n |a|\,|b|$$

where C_n *is a constant depending only on the dimension of the Clifford algebra* $\mathbb{R}_n$*. Moreover, we have* $C_n \leq 2^{n/2}$*.*

Remark 2.1.16. The modulus is multiplicative in the case of complex numbers and quaternions. To have a multiplicative modulus when enlarging the field of real numbers one has to abandon the notion of order to get $\mathbb{C}$ and then the notion of commutativity to get $\mathbb{H}$. There is another possibility to enlarge further the dimension: by abandoning associativity one obtains the (division) algebra of octonions. In fact, Hurwitz' theorem shows that the only algebras over the real field with multiplicative modulus are the field of real numbers, the field of complex numbers, the quaternion skew field and the alternative algebra of octonions.

Inside a Clifford algebra there is the possibility, in some special cases, to have that the modulus is multiplicative. These cases are described in the following result:

Proposition 2.1.17. *Let* $b \in \mathbb{R}_n$ *be such that* $b\bar{b} = |b|^2$*. Then*

$$|ab| = |a||b|.$$

Proof. Consider $|ab|$. We have:

$$|ab|^2 = [ab\overline{ab}]_0 = [ab\bar{b}\bar{a}]_0 = [a|b|^2\bar{a}]_0 = [a\bar{a}]_0|b|^2 = |a|^2|b|^2. \qquad \square$$

Note that the result holds, in particular, when a is paravector $\mathbf{x}$. Moreover any nonzero paravector $\mathbf{x}$ admits an inverse, the so-called Kelvin inverse, defined by

$$\mathbf{x}^{-1} = \frac{\bar{\mathbf{x}}}{|\mathbf{x}|^2}.$$

2.2 Slice monogenic functions: definition and properties

As mentioned in Section 2.1, an element $(x_1, x_2, \dots, x_n) \in \mathbb{R}^n$ can be identified with a vector $\underline{x} = x_1e_1 + \dots + x_ne_n \in \mathbb{R}_n^1$ while an element $(x_0, x_1, \dots, x_n) \in \mathbb{R}^{n+1}$ can be identified with the paravector

$$\mathbf{x} = x_0 + x_1e_1 + \dots + x_ne_n = x_0 + \underline{x} \in \mathbb{R}_n^0 \oplus \mathbb{R}_n^1.$$

In the sequel, with an abuse of notation, we will write $\underline{x} \in \mathbb{R}^n$ and $\mathbf{x} \in \mathbb{R}^{n+1}$. Thus, if $U \subseteq \mathbb{R}^{n+1}$ is an open set, a function $f: U \subseteq \mathbb{R}^{n+1} \to \mathbb{R}_n$ can be interpreted as a function of the paravector $\mathbf{x}$. Note also that an element $\mathbf{x}$ will be often denoted as

$$\mathbf{x} = \mathrm{Re}[\mathbf{x}] + \underline{x},$$

to emphasize its real and vector part, respectively.

The theory of slice monogenic functions was first developed in [26] where the authors study a new notion of monogenicity for functions from $\mathbb{R}^{n+1}$ to $\mathbb{R}_n$. It is worth noting, however, that the exposition we propose here offers a significantly improved theory, and reorganizes the ideas of [26] in a new more powerful fashion as in [15], [18], [27], [28], [29], [53].

To introduce the theory of slice monogenic functions, we need some definitions and notation.

Definition 2.2.1. *We will denote by $\mathbb{S}$ the set of unit vectors:*

$$\mathbb{S} = \{\underline{x} = e_1x_1 + \dots + e_nx_n \in \mathbb{R}^{n+1} \mid x_1^2 + \dots + x_n^2 = 1\}.$$

From a geometric point of view, $\mathbb{S}$ is an $(n-1)$-sphere in the Euclidean space of vectors $\mathbb{R}^n$ and if $I \in \mathbb{S}$, then $I^2 = -1$.

The two-dimensional real subspace of $\mathbb{R}^{n+1}$ generated by 1 and I is the plane $\mathbb{R} + I\mathbb{R}$. It will be denoted by $\mathbb{C}_I$, in fact it is isomorphic to the complex plane. Note that the isomorphism between the vector space $\mathbb{C}_I$ and $\mathbb{C}$ is also an algebra isomorphism, thus $\mathbb{C}_I$ will be referred to as a "complex plane".

An element in $\mathbb{C}_I$ will be denoted by $u + Iv$. Conversely, given a paravector $\mathbf{x}$, it will be possible to write it as an element in a suitable complex plane $\mathbb{C}_I$. In fact, either $\mathbf{x}$ is a real number, or we can write it as $\mathbf{x} = \mathrm{Re}[\mathbf{x}] + \frac{\underline{x}}{|\underline{x}|}|\underline{x}|$. Since $\mathrm{Re}[\mathbf{x}]$, $|\underline{x}|$ are real numbers and $\frac{\underline{x}}{|\underline{x}|}$ is a unit vector, we have written the given paravector as $\mathbf{x} = u + I_{\mathbf{x}}v$, with $u = \mathrm{Re}[\mathbf{x}]$, $v = |\underline{x}|$ and $I_{\mathbf{x}} = \frac{\underline{x}}{|\underline{x}|}$.

Definition 2.2.2. *Let $U \subseteq \mathbb{R}^{n+1}$ be an open set and let $f : U \to \mathbb{R}_n$ be a real differentiable function. Let $I \in \mathbb{S}$ and let f_I be the restriction of f to the complex plane $\mathbb{C}_I$ and denote by $u + Iv$ an element on $\mathbb{C}_I$. We say that f is a left slice monogenic (for short* s-monogenic*) function if, for every $I \in \mathbb{S}$, we have*

$$\frac{1}{2}\left(\frac{\partial}{\partial u} + I\frac{\partial}{\partial v}\right) f_I(u + Iv) = \frac{1}{2}\left(\frac{\partial}{\partial u} f_I(u + Iv) + I\frac{\partial}{\partial v} f_I(u + Iv)\right) = 0$$

on $U \cap \mathbb{C}_I$. We will denote by $\mathcal{M}(U)$ the set of left s-monogenic functions on the open set U or by $\mathcal{M}^L(U)$ when confusion may arise. We say that f is a right slice monogenic (for short right s-monogenic*) function if, for every $I \in \mathbb{S}$, we have*

$$\frac{1}{2}\left(\frac{\partial}{\partial u}+\frac{\partial}{\partial v}I\right) f_I(u+Iv)=\frac{1}{2}\left(\frac{\partial}{\partial u}f_I(u+Iv)+\frac{\partial}{\partial v}f_I(u+Iv)I\right)=0,$$

on $U \cap \mathbb{C}_I$. We will denote by $\mathcal{M}^R(U)$ the set of right s-monogenic functions on the open set U.

Remark 2.2.3. The theory of right s-monogenic functions is equivalent to the theory of (left) s-monogenic functions. In the sequel, we will mainly consider s-monogenicity on the left, but we will introduce some basic tools for right s-monogenic functions in order to treat the functional calculus for n-tuples of non-commuting operators.

Definition 2.2.4. *We define the notion of I-derivative by means of the operator:*

$$\partial_I := \frac{1}{2}\left(\frac{\partial}{\partial u}-I\frac{\partial}{\partial v}\right).$$

For consistency, we will denote by $\overline{\partial}_I$ the operator $\frac{1}{2}\left(\frac{\partial}{\partial u}+I\frac{\partial}{\partial v}\right)$.

Using the notation we have just introduced, the condition of left s-monogenicity will be expressed, in short, by

$$\overline{\partial}_I f = 0.$$

Right s-monogenicity will be expressed, with an abuse of notation, by

$$f\overline{\partial}_I = 0.$$

Remark 2.2.5. It is easy to verify that the (left) s-monogenic functions on $U \subseteq \mathbb{R}^{n+1}$ form a right $\mathbb{R}_n$-module. In fact it is trivial that if $f, g \in \mathcal{M}(U)$, then for every $I \in \mathbb{S}$ one has $\overline{\partial}_I f_I = \overline{\partial}_I g_I = 0$, thus $\overline{\partial}_I(f+g)_I = 0$. Moreover, for any $a \in \mathbb{R}_n$ we have $\overline{\partial}_I(f_I a) = (\overline{\partial}_I f)a = 0$. Analogously, the right s-monogenic functions on $U \subseteq \mathbb{R}^{n+1}$ form a left $\mathbb{R}_n$-module.

Definition 2.2.6. *Let U be an open set in $\mathbb{R}^{n+1}$ and let $f : U \to \mathbb{R}_n$ be an s-monogenic function. Its s-derivative ∂_s is defined as*

$$\partial_s(f) = \begin{cases} \partial_I(f)(\mathbf{x}) & \mathbf{x} = u+Iv,\ v \neq 0, \\ \partial_u f(u) & u \in \mathbb{R}. \end{cases} \tag{2.4}$$

Note that the definition of s-derivative is well posed because it is applied only to s-monogenic functions. Moreover, for such functions, it coincides with the partial derivative with respect to the scalar component u, in fact we have:

$$\partial_s(f)(u+Iv) = \partial_I(f_I)(u+Iv) = \partial_u(f_I)(u+Iv). \tag{2.5}$$

Note incidentally that

$$\partial_u(f_I)(u+Iv) = \partial_u(f)(u+Iv).$$

Proposition 2.2.7. *Let U be an open set in $\mathbb{R}^{n+1}$ and let $f: U \to \mathbb{R}_n$ be an s-monogenic function. The s-derivative $\partial_s f$ of f is an s-monogenic function, moreover*

$$\partial_s^m f(u+Iv) = \frac{\partial^m f}{\partial u^m}(u+Iv).$$

Proof. The first part of the statement follows from

$$\overline{\partial}_I(\partial_s f(u+Iv)) = \partial_s(\overline{\partial}_I f(u+Iv)) = 0. \tag{2.6}$$

The second part follows from (2.5). □

We now provide some examples of s-monogenic functions. It is interesting to note that in the classical theory of monogenic functions (see [7], [34]) the monomials, and thus the polynomials, in the paravector variable are not monogenic functions. However polynomials (and also converging power series) in the paravector variable turn out to be s-monogenic functions, provided that the coefficients are written on the right.

Example 2.2.8. The monomials $\mathbf{x}^n a_n$, $a_n \in \mathbb{R}_n$ are s-monogenic, thus also the polynomials $\sum_{n=0}^N \mathbf{x}^n a_n$ are s-monogenic. Note that these polynomials have coefficients written on the right: indeed, polynomials with left coefficients are not, in general, s-monogenic. To avoid confusion, we will call polynomials of the form $\sum_{n=0}^N \mathbf{x}^n a_n$ s-monogenic polynomials. Moreover, as we will see in the sequel, any power series $\sum_{n\geq 0} \mathbf{x}^n a_n$ is s-monogenic in its domain of convergence.

Remark 2.2.9. Note that the complex plane $\mathbb{C} = \mathbb{R}_1$ can be seen both as $\mathbb{R}^2$ and as $\mathbb{R}_1$. It is immediate, from Definition 2.2.2, that the space of holomorphic functions $f: \mathbb{C} \to \mathbb{C}$ coincides with the space of s-monogenic functions from $\mathbb{R}^2$ to $\mathbb{R}_1$. For this reason we will consider the case $n > 1$ (obviously, all the results that we will prove are valid also in the case $n = 1$).

Proposition 2.2.10. *Let $I = I_1 \in \mathbb{S}$. It is possible to choose $I_2, \dots, I_n \in \mathbb{S}$ such that $I_1, \dots, I_n$ form an orthonormal basis for the Clifford algebra $\mathbb{R}_n$ i.e., they satisfy the defining relations $I_r I_s + I_s I_r = -2\delta_{rs}$.*

Proof. First of all, note that since $\underline{x} \wedge \underline{y} = -\underline{y} \wedge \underline{x}$, formula (2.2) gives

$$\underline{x}\,\underline{y} + \underline{y}\,\underline{x} = 2\langle \underline{x}, \underline{y}\rangle.$$

Then it sufficient to select the vectors I_r in a way such that $\langle I_r, I_r\rangle = -1$ and $\langle I_s, I_r\rangle = 0$, for $s = 1, \dots, n$, $r = 2, \dots, n$. Since $I_r = \sum_{\ell=1}^n x_{r\ell} e_\ell$ the two conditions translate into

$$\langle I_r, I_r\rangle = -\sum_{\ell=1}^n x_{r\ell}^2$$

and

$$\langle I_s, I_r \rangle = -\sum_{\ell=1}^{n} x_{s\ell} x_{r\ell}.$$

By identifying each vector I_r with its components $(x_1, \ldots, x_n) \in \mathbb{R}^n$ we conclude using the Gram–Schmidt algorithm. □

A simple and yet extremely important feature of s-monogenic functions is that their restrictions to a complex plane $\mathbb{C}_I$ can be written as a suitable linear combination of 2^{n-1} holomorphic functions, as proved in the following:

Lemma 2.2.11 (Splitting Lemma). *Let $U \subseteq \mathbb{R}^{n+1}$ be an open set. Let $f : U \to \mathbb{R}_n$ be an s-monogenic function. For every $I = I_1 \in \mathbb{S}$ let $I_2, \ldots, I_n$ be a completion to a basis of $\mathbb{R}_n$ satisfying the defining relations $I_r I_s + I_s I_r = -2\delta_{rs}$. Then there exist 2^{n-1} holomorphic functions $F_A : U \cap \mathbb{C}_I \to \mathbb{C}_I$ such that for every $z = u + Iv$,*

$$f_I(z) = \sum_{|A|=0}^{n-1} F_A(z) I_A, \quad I_A = I_{i_1} \ldots I_{i_s},$$

where $A = i_1 \ldots i_s$ is a subset of $\{2, \ldots, n\}$, with $i_1 < \ldots < i_s$, or, when $|A| = 0$, $I_\emptyset = 1$.

Proof. Let $z = u + Iv$. Since f is $\mathbb{R}_n$-valued, there are functions $F_A : U \cap \mathbb{C}_I \to \mathbb{C}_I$ such that

$$f_I(z) = \sum_{|A|=0}^{n-1} F_A(z) I_A = \sum_{|A|=0}^{n-1} (f_A + g_A I) I_A.$$

We now need to show that the functions F_A are holomorphic. Since f is s-monogenic we have that its restriction to $\mathbb{C}_I$ satisfies

$$\left(\frac{\partial}{\partial u} + I\frac{\partial}{\partial v}\right) f_I(u + Iv) = 0$$

and so

$$\sum \left(\frac{\partial}{\partial u} + I\frac{\partial}{\partial v}\right)(f_A + g_A I) I_A$$
$$= \frac{\partial}{\partial u} f_A + I\frac{\partial}{\partial v} f_A + \frac{\partial}{\partial u} g_A I - \frac{\partial}{\partial v} g_A = 0.$$

Since the imaginary units commute with any real-valued function, we obtain the system:

$$\begin{cases} \dfrac{\partial}{\partial u} f_A - \dfrac{\partial}{\partial v} g_A = 0, \\ \dfrac{\partial}{\partial v} f_A + \dfrac{\partial}{\partial u} g_A = 0 \end{cases}$$

for all multi-indices A. Therefore all the functions $F_A = f_A + g_A I$ satisfy the standard Cauchy–Riemann system and so they are holomorphic. □

Example 2.2.12. To clarify our result, we consider explicitly the case of $\mathbb{R}_4$-valued functions. A function $f : U \subseteq \mathbb{R}^5 \to \mathbb{R}_4$ can be written as

$$\begin{aligned} f =& f_0 + f_1 I_1 + f_2 I_2 + f_3 I_3 + f_4 I_4 + f_{12} I_{12} + f_{13} I_{13} + f_{14} I_{14} + f_{23} I_{23} \\ &+ f_{24} I_{24} + f_{34} I_{34} + f_{123} I_{123} + f_{124} I_{124} + f_{134} I_{134} + f_{234} I_{234} + f_{1234} I_{1234} \end{aligned}$$

and grouping as prescribed in the statement of the Lemma, we obtain

$$\begin{aligned} f =& (f_0 + f_1 I_1) + (f_2 + f_{12} I_1) I_2 + (f_3 + f_{13} I_1) I_3 + (f_4 + f_{14} I_1) I_4 \\ &+ (f_{23} + f_{123} I_1) I_{23} + (f_{24} + f_{124} I_1) I_{24} + (f_{34} + f_{134} I_1) I_{34} \\ &+ (f_{234} + f_{1234} I_1) I_{234}. \end{aligned}$$

To develop a meaningful theory of s-monogenic functions we need some additional hypotheses on the open sets on which they are defined. For example, the natural class of open sets in which we can prove the Identity Principle is given by the domains whose intersection with any complex plane $\mathbb{C}_I$ is connected. We introduce these domains in the following definition:

Definition 2.2.13. *Let $U \subseteq \mathbb{R}^{n+1}$ be a domain. We say that U is a* slice domain (s-domain for short) *if $U \cap \mathbb{R}$ is nonempty and if $U \cap \mathbb{C}_I$ is a domain in $\mathbb{C}_I$ for all $I \in \mathbb{S}$.*

In this class of domains it is possible to prove the following Identity Principle:

Theorem 2.2.14 (Identity Principle). *Let U be an s-domain in $\mathbb{R}^{n+1}$. Let $f : U \to \mathbb{R}_n$ be an s-monogenic function, and let Z be its zero set. If there is an imaginary unit I such that $\mathbb{C}_I \cap Z$ has an accumulation point, then $f \equiv 0$ on U.*

Proof. Let us consider the restriction f_I of f to the plane $\mathbb{C}_I$, for $I \in \mathbb{S}$. By the Splitting Lemma we have

$$f_I(z) = \sum_{|A|=0}^{n-1} F_A(z) I_A$$

with $F_A : U \cap \mathbb{C}_I \to \mathbb{C}_I$ holomorphic for every multi-index A and $z = u + Iv$. Since $\mathbb{C}_I \cap Z$ has an accumulation point, we deduce that all the functions F_A vanish identically on $U \cap \mathbb{C}_I$ and thus $f_I = 0$ on $U \cap \mathbb{C}_I$. In particular f_I vanishes in the points of U on the real axis. Any other plane $\mathbb{C}_{I'}$ is such that $f_{I'}$ vanishes on $U \cap \mathbb{R}$ which has an accumulation point. If we apply the Splitting Lemma to $f_{I'}$, we can write $f_{I'} = \sum_{A'} F_{A'} I_{A'}$ and thus its components $F_{A'}$ vanish on $U \cap \mathbb{R}$ and thus they vanish identically on $U \cap \mathbb{C}_{I'}$. This fact implies that also $f_{I'}$ vanish on $\mathbb{C}_{I'}$, thus $f \equiv 0$ on U. □

Analogously to what happens in the complex case, we can prove the following consequence of the Identity Principle.

Corollary 2.2.15. *Let U be an s-domain in $\mathbb{R}^{n+1}$. Let $f, g : U \to \mathbb{R}_n$ be s-monogenic functions. If there is an imaginary unit I such that $f = g$ on a subset of $\mathbb{C}_I$ having an accumulation point, then $f \equiv g$ on U.*

Among the domains in $\mathbb{R}^{n+1}$ there is a special subclass which is useful to provide a Representation Formula for s-monogenic functions. In order to define them, it is useful to suitably denote the $(n-1)$-sphere associated to a paravector.

Let $\mathbf{s} = s_0 + \underline{s} = s_0 + I_{\mathbf{s}}|\underline{s}| \in \mathbb{R}^{n+1}$ be a paravector; we denote by $[\mathbf{s}]$ the set

$$[\mathbf{s}] = \{\mathbf{x} \in \mathbb{R}^{n+1} \mid \mathbf{x} = s_0 + I|\underline{s}|,\ I \in \mathbb{S}\}.$$

The set $[\mathbf{s}]$ is either reduced to a real point or it is the $(n-1)$-sphere defined by $\mathbf{s}$, i.e., the $(n-1)$-dimensional sphere with center at the real point s_0 and radius $|\underline{s}|$.

Remark 2.2.16. Observe that the relation: "$\mathbf{x} \sim \mathbf{s}$ if and only if $x_0 = s_0$ and $|\mathbf{x}| = |\mathbf{s}|$" is an equivalence relation. Given a paravector $\mathbf{s}$, its equivalence class contains only the element $\mathbf{s}$ when $\mathbf{s}$ is a real number, while it contains infinitely many elements when $\mathbf{s}$ is not real and corresponds to the $(n-1)$-dimensional sphere $[\mathbf{s}]$.

Definition 2.2.17. *Let $U \subseteq \mathbb{R}^{n+1}$. We say that U is* axially symmetric *if, for all $\mathbf{s} = u + Iv \in U$, the whole $(n-1)$-sphere $[\mathbf{s}]$ is contained in U.*

Observe that axially symmetric sets are invariant under rotations that fix the real axis.

In order to state the next result we need some notation. Given an element $\mathbf{x} = x_0 + \underline{x} \in \mathbb{R}^{n+1}$ let us set

$$I_{\mathbf{x}} = \begin{cases} \dfrac{\underline{x}}{|\underline{x}|} & \text{if } \underline{x} \neq 0, \\ \text{any element of } \mathbb{S} & \text{otherwise.} \end{cases}$$

We have the following:

Theorem 2.2.18 (Representation Formula). *Let $U \subseteq \mathbb{R}^{n+1}$ be an axially symmetric s-domain and let f be an s-monogenic function on U.*

(1) *For any vector $\mathbf{x} = u + I_{\mathbf{x}}v \in U$ the following formulas hold:*

$$f(\mathbf{x}) = \frac{1}{2}\Big[1 - I_{\mathbf{x}}I\Big]f(u + Iv) + \frac{1}{2}\Big[1 + I_{\mathbf{x}}I\Big]f(u - Iv) \tag{2.7}$$

and

$$f(\mathbf{x}) = \frac{1}{2}\Big[f(u + Iv) + f(u - Iv) + I_{\mathbf{x}}I[f(u - Iv) - f(u + Iv)]\Big]. \tag{2.8}$$

(2) *Moreover, the two quantities*

$$\alpha(u,v) := \frac{1}{2}[f(u+Iv)+f(u-Iv)] \tag{2.9}$$

and

$$\beta(u,v) := I\frac{1}{2}[f(u-Iv)-f(u+Iv)] \tag{2.10}$$

do not depend on $I \in \mathbb{S}$.

Proof. The result is trivial for real paravectors, in fact we have the identity

$$f(u) = \frac{1}{2}\Big[1 - I_{\mathbf{x}}I\Big]f(u) + \frac{1}{2}\Big[1 + I_{\mathbf{x}}I\Big]f(u)$$

for any $I_{\mathbf{x}} \in \mathbb{S}$. If $\mathbf{x} \notin \mathbb{R}$ and if we write $\mathbf{x} = u + I_{\mathbf{x}}v$ we can set

$$\phi(u+I_{\mathbf{x}}v) := \frac{1}{2}\Big[f(u+Iv)+f(u-Iv)+I_{\mathbf{x}}I[f(u-Iv)-f(u+Iv)]\Big],$$

and observe that if $I = I_{\mathbf{x}}$ we have

$$\phi(u+I_{\mathbf{x}}v) = f(\mathbf{x}).$$

Let us show that $\left(\frac{\partial}{\partial u} + I_{\mathbf{x}}\frac{\partial}{\partial v}\right)\phi(u+I_{\mathbf{x}}v) = 0$ for all $\mathbf{x} \in U \cap \mathbb{C}_I$. Indeed we have:

$$\begin{aligned}
&\Big(\frac{\partial}{\partial u} + I_{\mathbf{x}}\frac{\partial}{\partial v}\Big)\phi(u+I_{\mathbf{x}}v)\\
&= \frac{1}{2}\Big[1 - I_{\mathbf{x}}I\Big]\frac{\partial}{\partial u}f(u+Iv) + \frac{1}{2}\Big[1 + I_{\mathbf{x}}I\Big]\frac{\partial}{\partial u}f(u-Iv)\\
&\quad + \frac{1}{2}I_{\mathbf{x}}\Big[1 - I_{\mathbf{x}}I\Big]\frac{\partial}{\partial v}f(u+Iv) + \frac{1}{2}I_{\mathbf{x}}\Big[1 + I_{\mathbf{x}}I\Big]\frac{\partial}{\partial v}f(u-Iv).
\end{aligned}$$

Using the fact that f is s-monogenic, we can write

$$\begin{aligned}
&\Big(\frac{\partial}{\partial u} + I_{\mathbf{x}}\frac{\partial}{\partial v}\Big)\varphi(u+I_{\mathbf{x}}v)\\
&= \frac{1}{2}\Big[1 - I_{\mathbf{x}}I\Big](-I)\frac{\partial}{\partial v}f(u+Iv) + \frac{1}{2}\Big[1 + I_{\mathbf{x}}I\Big]I\frac{\partial}{\partial v}f(u-Iv)\\
&\quad + \frac{1}{2}I_{\mathbf{x}}\Big[1 - I_{\mathbf{x}}I\Big]\frac{\partial}{\partial v}f(u+Iv) + \frac{1}{2}I_{\mathbf{x}}\Big[1 + I_{\mathbf{x}}I\Big]\frac{\partial}{\partial v}f(u-Iv) = 0.
\end{aligned}$$

Since the function ϕ is s-monogenic and $\phi \equiv f$ on $\mathbb{C}_I$, then ϕ coincides with f on U by the Identity Principle. The second part of the proof follows directly from

(2.8). In fact we have

$$\begin{aligned}
&\frac{1}{2}[f(u+Iv)+f(u-Iv)]\\
&=\frac{1}{2}\Big\{\frac{1}{2}\Big[f(u+Jv)+f(u-Jv)\Big]+I\frac{1}{2}\Big[J[f(u-Jv)-f(u+Jv)]\Big]\\
&\quad+\frac{1}{2}\Big[f(u+Jv)+f(u-Jv)\Big]-I\frac{1}{2}\Big[J[f(u-Jv)-f(u+Jv)]\Big]\Big\}\\
&=\frac{1}{2}\Big[f(u+Jv)+f(u-Jv)\Big]
\end{aligned}$$

and so α, and similarly β, depend on u, v only. □

Remark 2.2.19. Note that the operator $\overline{\partial}_I$ is not a constant coefficients differential operator since the imaginary unit I changes with the point $u+Iv$. This shows that f per se does not satisfy a system of constant coefficients differential equations; however, as the next corollary shows, its components α and β do, and they give an s-monogenic function if they satisfy some additional conditions, see [87].

Corollary 2.2.20. *Let $U \subseteq \mathbb{R}^{n+1}$ be an axially symmetric s-domain, and $D \subseteq \mathbb{R}^2$ be such that $u+Iv \in U$ whenever $(u,v) \in D$ and let $f : U \to \mathbb{R}_n$. The function f is an s-monogenic function if and only if there exist two differentiable functions $\alpha, \beta : D \subseteq \mathbb{R}^2 \to \mathbb{R}_n$ satisfying $\alpha(u,v) = \alpha(u,-v)$, $\beta(u,v) = -\beta(u,-v)$ and the Cauchy–Riemann system*

$$\begin{cases} \partial_u\alpha - \partial_v\beta = 0, \\ \partial_u\beta + \partial_v\alpha = 0, \end{cases} \tag{2.11}$$

and such that

$$f(u+Iv) = \alpha(u,v) + I\beta(u,v). \tag{2.12}$$

Proof. If f is s-monogenic, then we can apply Theorem 2.2.18 and we can set $\alpha(u,v)$ and $\beta(u,v)$ as in (2.9) and (2.10). Then $f(u+Iv) = \alpha(u,v)+I\beta(u,v)$, and $\alpha(u,v) = \alpha(u,-v)$, $\beta(u,v) = \beta(u,-v)$ by their definitions. The proof of Theorem 2.2.18 shows that the pair α, β satisfies the Cauchy–Riemann system. The converse is immediate: any function of the form $f(u+Iv) = \alpha(u,v)+I\beta(u,v)$ is well defined on an axially symmetric open set. In fact,

$$f(u-Iv) = \alpha(u,-v) + I\beta(u,-v) = \alpha(u,v) - I\beta(u,v).$$

The fact that α and β satisfy the Cauchy–Riemann system guarantees that f is an s-monogenic function. □

The Representation Formula has several interesting consequences.

Corollary 2.2.21. *Let $U \subseteq \mathbb{R}^{n+1}$ be an axially symmetric s-domain and let $f : U \to \mathbb{R}^{n+1}$ be an s-monogenic function. For any choice of $u, v \in \mathbb{R}$ such that $u+Iv \in U$ there exist $a, b \in \mathbb{R}_n$ such that*

$$f(u+Iv) = a + Ib, \tag{2.13}$$

for all $I \in \mathbb{S}$. In particular, the image of the $(n-1)$-sphere $[u+Iv]$ is the set $\{a+Ib : I \in \mathbb{S}\}$.

Proof. It is a direct application of Theorem 2.2.18. □

Another consequence of the Representation Formula is the fact that any holomorphic map defined on a suitable domain can be uniquely extended to an s-monogenic function:

Lemma 2.2.22 (Extension Lemma). *Let $J \in \mathbb{S}$ and let D be a domain in $\mathbb{C}_J$, symmetric with respect to the real axis and such that $D \cap \mathbb{R} \neq \emptyset$. Let U_D be the axially symmetric s-domain defined by*

$$U_D = \bigcup_{u+Jv \in D,\ I \in \mathbb{S}} (u+Iv).$$

If $f : D \to \mathbb{C}_J$ is holomorphic, then the function $\mathrm{ext}(f) : U_D \to \mathbb{R}_n$ defined by

$$\mathrm{ext}(f)(u+Iv) := \frac{1}{2}\Big[f(u+Jv)+f(u-Jv)\Big]+I\frac{1}{2}\Big[J[f(u-Jv)-f(u+Jv)]\Big] \quad (2.14)$$

is the unique s-monogenic extension of f to U_D.

Similarly, let $J_2, \dots, J_n$ be a completion of J to an orthonormal basis of $\mathbb{R}_n$ and let

$$f : D \to \mathbb{R}_n$$

defined by $f = \sum_{|A|=0}^{n-1} F_A J_A$, $A \subseteq \{2, \dots, n\}$, $F_A : D \to \mathbb{C}_J$ holomorphic. Then, $\overline{\partial}_J f(u+Jv) = 0$ and the function obtained by extending each of its holomorphic components F_A is the unique s-monogenic extension of f to U_D.

Proof. The fact that $\mathrm{ext}(f)$ is s-monogenic follows by the proof of Theorem 2.2.18. When $I = J$ in (2.14) we have that $\mathrm{ext}(f)(u+Jv) = f(u+Jv)$, and hence $\mathrm{ext}(f)$ is the unique extension of f by the Identity Principle. The second part is immediate. □

The second part of Theorem 2.2.18 shows that for every $I, K \in \mathbb{S}$ we have

$$f(u+Iv) = \alpha(u,v) + I\beta(u,v) \qquad \text{and} \qquad f(u+Kv) = \alpha(u,v) + K\beta(u,v).$$

By subtracting the two expressions and assuming that $I \neq K$, we have

$$\alpha(u,v) = (I-K)^{-1}[If(u+Iv) - Kf(u+Kv)]$$

and

$$\beta(u,v) = (I-K)^{-1}[f(u+Iv) - f(u+Kv)].$$

Thus the Representation Formula admits the following generalization:

Theorem 2.2.23 (Representation Formula, II). *Let $U \subseteq \mathbb{R}^{n+1}$ be an axially symmetric s-domain and let f be an s-monogenic function on U. For any vector $u + Jv \in U$ the following formula holds:*

$$f(u+Jv) = (I-K)^{-1}[If(u+Iv) - Kf(u+Kv)] \\ + J(I-K)^{-1}[f(u+Iv) - f(u+Kv)]. \tag{2.15}$$

As a consequence we have that the values of an s-monogenic function f on an axially symmetric set U are uniquely determined by its values on the two half-planes $U \cap \mathbb{C}_J^+$, $U \cap \mathbb{C}_K^+$ through formula (2.15). Moreover we have the following generalization of the extension lemma:

Lemma 2.2.24 (Extension Lemma, II). *Let U be an s-domain in $\mathbb{R}^{n+1}$ and let $f : U \to \mathbb{R}_n$ be an s-monogenic function. Let $\tilde{U}$ be the axially symmetric s-domain defined by*

$$\tilde{U} = \bigcup_{u+Jv \in U,\ I \in \mathbb{S}} (u+Iv)$$

There exists a unique s-monogenic extension of f to the whole $\tilde{U}$.

Proof. By construction, it is immediate that $\tilde{U}$ is an axially symmetric s-domain. Observing that U is an open set, we consider another axially symmetric s-domain W obtained as the union of all the open balls $B(x, r_x) \subset U$ with center at a point on the real axis $x \in U$, i.e.,

$$W = \cup_{x \in U \cap \mathbb{R}} B(x, r_x).$$

The restriction of f to W is an s-monogenic function which can be uniquely extended to a function $\tilde{f}$ defined on a maximal, axially symmetric, s-domain set $U_{\max}$ such that $W \subseteq U_{\max} \subseteq \tilde{U}$. Our goal is now to show that $U_{\max}$ coincides with $\tilde{U}$. Assume the contrary, and suppose that there exists $\mathbf{y} = y_0 + Iy_1 \in \tilde{U} \cap \partial U_{\max}$. Since $\mathbf{y} \in \tilde{U}$, there exists $J \in \mathbb{S}$ such that $y_0 + Jy_1 \in U$ and since U is open, there is an open ball with center at $\mathbf{y}$ contained in U. So there exist $K \in \mathbb{S}$ and $\tilde{\mathbf{y}} = y_0 + Ky_1$ such that the two discs Δ_J and Δ_K of radius ε with center at $\mathbf{y}$ and $\tilde{\mathbf{y}}$ on the plane $\mathbb{C}_J$, $\mathbb{C}_K$, respectively, are contained in U. Let us define

$$\tilde{g}(u+Jv) := (I-K)^{-1}[If(u+Iv) - Kf(u+Kv)] \\ + J(I-K)^{-1}[f(u+Iv) - f(u+Kv)]$$

on the set $D = \{\mathbf{x} = u + Jv \mid (u - y_0)^2 + (v - y_1)^2 < \varepsilon\}$. Then the function $\tilde{g}$ coincides with $\tilde{f}$ on $D \cap U_{\max}$. The function h defined by $h(\mathbf{x}) = \tilde{f}(\mathbf{x})$ for $\mathbf{x} \in U_{\max}$ and $h(\mathbf{x}) = \tilde{g}(\mathbf{x})$ for $\mathbf{x} \in D$ is the s-monogenic extension of f to the axially symmetric open set $D \cup U_{\max}$ contradicting the maximality of $U_{\max}$. This completes the proof. □

2.3 Power series

As we have already observed in the previous section, polynomials in the paravector variable $\mathbf{x}$ are s-monogenic. However, it is no longer true that a polynomial $f(\mathbf{x})$ of the form $f(\mathbf{x}) = (\mathbf{x} - a)^n$, where $a \in \mathbb{R}_n$ is s-monogenic, in general. If $a \in \mathbb{R}$, however, then $f(\mathbf{x})$ is s-monogenic and so are power series centered at a point on the real axis, where they converge. In this section we will provide a detailed study of s-monogenic functions which can be expanded into power series.

Proposition 2.3.1. *If $B = B(0,R) \subseteq \mathbb{R}^{n+1}$ is a ball centered in 0 with radius $R > 0$, then $f : B \to \mathbb{R}_n$ is an s-monogenic function if and only if f has a series expansion of the form*

$$f(\mathbf{x}) = \sum_{m\geq 0} \mathbf{x}^m \frac{1}{m!} \frac{\partial^m f}{\partial u^m}(0) \tag{2.16}$$

converging on B.

Proof. If a function admits a series expansion as in (2.16) it is obviously s-monogenic where the series converges. The converse requires the Splitting Lemma. Consider an element $I = I_1 \in \mathbb{S}$ and the corresponding plane $\mathbb{C}_I$. Let $\Delta \subset \mathbb{C}_I$ be a disc with center in the origin and radius $r < R$ and let us set $z = u + Iv$. The restriction of f to the plane $\mathbb{C}_I$ can be written as $f_I(z) = \sum F_A(z) I_A$. Since every function $F_A(z)$ is holomorphic, it admits an integral representation via the Cauchy formula, i.e.,

$$F_A(z) = \frac{1}{2\pi I} \int_{\partial\Delta(0,r)} \frac{F_A(\zeta)}{\zeta - z}\, d\zeta,$$

for any $z \in \Delta$ and therefore

$$f_I(z) = \sum_{|A|=0}^{n-1} \left(\frac{1}{2\pi I} \int_{\partial\Delta(0,r)} \frac{F_A(\zeta)}{\zeta - z}\, d\zeta \right) I_A.$$

Now observe that ζ and z commute because they lie on the same plane $\mathbb{C}_I$, so we can expand the denominator in each integral in power series, as in the classical case:

$$\begin{aligned} F_A(z) &= \frac{1}{2\pi I} \int_{\partial\Delta(0,r)} \sum_{m\geq 0} \left(\frac{z}{\zeta}\right)^m \frac{F_A(\zeta)}{\zeta}\, d\zeta \\ &= \sum_{m\geq 0} z^m \int_{\partial\Delta(0,r)} \sum_{m\geq 0} \frac{F_A(\zeta)}{\zeta^{m+1}}\, d\zeta \\ &= \sum_{m\geq 0} z^m \frac{1}{m!} \frac{\partial^m F_A}{\partial z^m}(0). \end{aligned}$$

Plugging this expression into $f_I(z) = \sum F_A(z) I_A$ we obtain

$$f_I(z) = \sum_{|A|=0}^{n-1} \sum_{m\geq 0} z^m \frac{1}{m!} \frac{\partial^m F_A}{\partial z^m}(0) I_A = \sum_{m\geq 0} z^m \frac{1}{m!} \frac{\partial^m f}{\partial z^m}(0),$$

and using the definition of s-derivative together with Proposition 2.2.7, we get

$$\sum_{m\geq 0} z^m \frac{1}{m!} \frac{1}{2} \left(\frac{\partial}{\partial u} - I \frac{\partial}{\partial v} \right)^m f(0) = \sum_{m\geq 0} z^m \frac{1}{m!} \frac{\partial^m}{\partial u^m} f(0).$$

Finally observe that the coefficients of the power series do not depend on the choice of the unit I, thus $f_I(z)$ is the restriction to $\mathbb{C}_I$ of the function defined in (2.16) and the statement follows. □

The following two results can be proved as in the complex case:

Proposition 2.3.2. *The s-derivative of a power series*

$$\sum_{n\geq 0} \mathbf{x}^n a_n, \quad a_n \in \mathbb{R}_n$$

equals

$$\sum_{n\geq 0} n\mathbf{x}^{n-1} a_n$$

and has the same radius of convergence of the original series.

Corollary 2.3.3. *Let $f: B \to \mathbb{R}_n$ be an s-monogenic function. Then $f \equiv 0$ on B if and only if $\partial_s^n f(0) = 0$ for all $n \in \mathbb{N}$.*

The next proposition shows that s-monogenic functions whose power series expansion have real coefficients play a privileged role.

Proposition 2.3.4. *The product of two functions $f, g: B(0,R) \to \mathbb{R}_n$ such that the series expansion of f has real coefficients is an s-monogenic function. Moreover, the composition of f with an s-monogenic function $h: B(0,R') \to \mathbb{R}_n$ is an s-monogenic function whenever the composition is defined.*

Proof. Let

$$f(\mathbf{x}) = \sum_{m\geq 0} \mathbf{x}^m a_m,$$

$$g(\mathbf{x}) = \sum_{m\geq 0} \mathbf{x}^m b_m,$$

$$h(\mathbf{x}) = \sum_{m\geq 0} \mathbf{x}^m c_m,$$

be s-monogenic functions with $a_m \in \mathbb{R}$, $b_m, c_m \in \mathbb{R}_n$. Since real coefficients commute with the variable $\mathbf{x}$ we have

$$(fg)(\mathbf{x}) = \sum_{s\geq 0} \mathbf{x}^s (a_0 b_s + a_1 b_{s-1} + \ldots + a_s b_0).$$

Now consider $(h \circ f)(\mathbf{x}) = h(f(\mathbf{x}))$; we have

$$h(f(\mathbf{x})) = \sum_{m\geq 0} (\sum_{r\geq 0} \mathbf{x}^r a_r)^m c_m.$$

Since the coefficients a_r commute with the variables we can group them on the right and the statement follows. □

Corollary 2.3.5. *Let $f : U \to \mathbb{R}_n$ be an s-monogenic function. Then the function $f(\mathbf{x} - y_0)$, $y_0 \in \mathbb{R}$, is an s-monogenic function in the open set $U' = \{\mathbf{x}' = \mathbf{x} - y_0,\ \mathbf{x} \in U\}$.*

Proposition 2.3.6. *Let $B = B(y_0, R) \subseteq \mathbb{R}^{n+1}$ be the ball centered in $y_0 \in \mathbb{R}$ with radius $R > 0$, then $f : B \to \mathbb{R}_n$ is an s-monogenic function if and only if it has a series expansion of the form*

$$f(\mathbf{x}) = \sum_{m\geq 0} (\mathbf{x} - y_0)^m \frac{1}{m!} \frac{\partial^m f}{\partial u^m}(y_0). \tag{2.17}$$

Proof. Consider the transformation of coordinates $\mathbf{z} = \mathbf{x} - y_0$. Since the function $f(\mathbf{z})$ is s-monogenic in a ball centered in the origin with radius $R > 0$, we can apply Proposition 2.3.1. Using the inverse transformation $\mathbf{x} = \mathbf{z} + y_0$, we obtain the statement. □

The result extends to s-domains as follows:

Corollary 2.3.7. *Let f be an s-monogenic function on an s-domain $U \subseteq \mathbb{R}^{n+1}$. Then for any point on the real axis y_0 in U, the function f can be represented in power series*

$$f(\mathbf{x}) = \sum_{n\geq 0} (\mathbf{x} - y_0)^n \frac{1}{n!} \frac{\partial^n f}{\partial u^n}(y_0)$$

on the ball $B(y_0, R)$, where $R = R_{y_0}$ is the largest positive real number such that $B(y_0, R)$ is contained in U.

Proof. Since f is s-monogenic in y_0, then, for every $I \in \mathbb{S}$, f can be expanded in power series on the disc $\Delta_I = B(y_0, R_I)$ of radius R_I on the plane $\mathbb{C}_I$. The radius R turns out to be $\min_{I\in\mathbb{S}} R_I$ which is nonzero because y_0 is an internal point in U. □

Corollary 2.3.8. *Let $f : B(y_0, R) \to \mathbb{R}_n$ be an s-monogenic function. If there exists $I \in \mathbb{S}$ such that $f(\mathbb{C}_I) \subseteq \mathbb{C}_I$, then the series expansion of f,*

$$f(\mathbf{x}) = \sum_{n\geq 0} (\mathbf{x} - y_0)^n \frac{1}{n!} \frac{\partial^n f}{\partial u^n}(y_0),$$

has all its coefficients in $\mathbb{C}_I$. Consequently, if there are two different units $I, J \in \mathbb{S}$ such that $f(\mathbb{C}_I) \subseteq \mathbb{C}_I$ and $f(\mathbb{C}_J) \subseteq \mathbb{C}_J$, then the coefficients are real.

Proof. If $I \in \mathbb{S}$ is such that $f(\mathbb{C}_I) \subseteq \mathbb{C}_I$, then for any real number y_0 we have $f(y_0) = f_I(y_0) \in \mathbb{C}_I$. Therefore $\frac{\partial^n f}{\partial u^n}(y_0) \in \mathbb{C}_I$ for any $n \in \mathbb{N}$, $y_0 \in \mathbb{R}$, and the conclusion follows. The second part is immediate. □

We now introduce a product among s-monogenic polynomials which preserves the s-monogenicity:

Definition 2.3.9. *Let $f(\mathbf{x}) = \sum_{i=0}^{n} \mathbf{x}^i a_i$ and $g(\mathbf{x}) = \sum_{i=0}^{m} \mathbf{x}^i b_i$, for $a_i, b_i \in \mathbb{R}_n$. We define the s-monogenic product of f and g as*

$$f * g(\mathbf{x}) := \sum_{j=0}^{n+m} \mathbf{x}^j c_j$$

with $c_j = \sum_{i+k=j} a_i b_k$. We will denote by f^{*n} the product $f * \ldots * f$, n-times.

This product is computed by taking the coefficients of the polynomials on the right, as in the case in which the variables and the coefficients commute and coincides with the standard product of polynomials with coefficients in a division algebra (see [71]). We adopt this definition also in this setting and we extend it to the case of the product of series. If $f(\mathbf{x}) = \sum_{i\geq 0} \mathbf{x}^i a_i$ and $g(\mathbf{x}) = \sum_{i\geq 0} \mathbf{x}^i b_i$ are s-monogenic series, we define their s-monogenic product as

$$f * g(\mathbf{x}) := \sum_{j\geq 0} \mathbf{x}^j c_j$$

with $c_j = \sum_{i+k=j} a_i b_k$. Note that when the coefficients of a polynomial or a series f are real numbers, the s-monogenic product coincides with the usual product, i.e., $f * g = fg$ (see Proposition 2.3.4). This product will be generalized in the sequel to s-monogenic functions which are not necessarily power series.

We conclude this section by showing that s-monogenic functions are infinitely differentiable:

Proposition 2.3.10. *An s-monogenic function $f : U \to \mathbb{R}_n$ on an axially symmetric s-domain $U \subseteq \mathbb{R}^{n+1}$ is infinitely differentiable on U.*

Proof. The differentiability of f on the real axis follows from Corollary 2.3.7 since for any point of the real axis there is a ball in which the function f can be expressed in power series. To prove differentiability outside the real axis consider formula (2.7) and write $\mathbf{x}$ as $\mathbf{x} = x_0 + \underline{x} = x_0 + \frac{\underline{x}}{|\underline{x}|}\underline{x}$:

$$f(\mathbf{x}) = \frac{1}{2}\Big[f(x_0 + I|\underline{x}|) + f(x_0 - I|\underline{x}|) + \frac{\underline{x}}{|\underline{x}|}I[f(x_0 - I|\underline{x}|) - f(x_0 + I|\underline{x}|]\Big].$$

The function f is s-monogenic and hence, by definition, its restriction f_I to $\mathbb{C}_I$ is infinitely differentiable on $U \cap \mathbb{C}_I$ for any $I \in \mathbb{S}$. It is therefore obvious that f can be obtained as a composition of the functions f_I, x_0, $\underline{x} = \sum_\ell e_\ell x_\ell$, and $|\underline{x}|$, which are all infinitely differentiable outside the real axis with respect to the variables x_ℓ, $\ell = 0, \dots, n$. This concludes the proof. □

2.4 Cauchy integral formula, I

A main result in the theory of s-monogenic functions is an analog of the Cauchy integral formula. We will present two versions of such a Cauchy formula: the one discussed in this section is less general than the second version, but it is enough to prove several properties of s-monogenic functions.

Theorem 2.4.1. *Let $U \subseteq \mathbb{R}^{n+1}$ be an axially symmetric s-domain and let $f : U \to \mathbb{R}_n$ be an s-monogenic function. If $\mathbf{x} \in U$, then*

$$f(\mathbf{x}) = \frac{1}{2\pi}\int_{\partial\Delta_{\mathbf{x}}(a,r)} (\zeta - \mathbf{x})^{-1}\, d\zeta_{I_{\mathbf{x}}} f(\zeta)$$

where $d\zeta_{I_{\mathbf{x}}} := -d\zeta I_{\mathbf{x}}$ and $a \in \mathbb{R}$, $r > 0$ are such that

$$\overline{\Delta_{\mathbf{x}}(a,r)} = \{u + I_{\mathbf{x}}v \mid (u-a)^2 + v^2 \leq r^2\} \subset \mathbb{C}_{I_{\mathbf{x}}}$$

contains $\mathbf{x}$ and is contained in U.

Proof. With no loss of generality, we will assume $a = 0$. Consider the integral

$$\frac{1}{2\pi}\int_{\partial\Delta_{\mathbf{x}}(0,r)} (\xi - \mathbf{x})^{-1} d\xi_{I_{\mathbf{x}}} f(\xi).$$

Set $I_{\mathbf{x}} := I_1$, complete to a basis $I_1, \dots, I_n$ of the Clifford algebra $\mathbb{R}_n$, satisfying the defining relations $I_r I_s + I_s I_r = -2\delta_{rs}$. Using the Splitting Lemma, we can write the restriction of f to $\mathbb{C}_{I_{\mathbf{x}}}$ as $f_{I_{\mathbf{x}}} = \sum_A F_A I_A$. We have

$$\frac{1}{2\pi}\int_{\partial\Delta_{\mathbf{x}}(0,r)} (\xi - \mathbf{x})^{-1} d\xi_{I_{\mathbf{x}}} f_{I_{\mathbf{x}}}(\xi)$$
$$= \frac{1}{2\pi}\int_{\partial\Delta_{\mathbf{x}}(0,r)} (\xi - \mathbf{x})^{-1} d\xi_{I_{\mathbf{x}}} \sum_A F_A(\xi) I_A$$

$$
\begin{aligned}
&= \sum_A \frac{1}{2\pi} \int_{\partial\Delta_{\mathbf{x}}(0,r)} (\xi-\mathbf{x})^{-1} d\xi_{I_{\mathbf{x}}} F_A(\xi) I_A \\
&= \sum_A F_A(\mathbf{x}) I_A \\
&= f(\mathbf{x}).
\end{aligned}
$$

□

Remark 2.4.2. Let $B_1 = B(0,R_1)$, $B_2 = B(0,R_2)$ be two balls centered at the origin and with radii $0 < R_1 < R_2$. The same argument used in the previous proof shows that if a function f is s-monogenic in a neighborhood of the annular domain $B_2 \setminus B_1$, then for any $\mathbf{x} \in B_2 \setminus B_1$, it satisfies

$$
\begin{aligned}
f(\mathbf{x}) = &\frac{1}{2\pi} \int_{\partial(B_2 \cap \mathbb{C}_{I_{\mathbf{x}}})} (\zeta-\mathbf{x})^{-1}\, d\zeta_{I_{\mathbf{x}}} f(\zeta) \\
&- \frac{1}{2\pi} \int_{\partial(B_1 \cap \mathbb{C}_{I_{\mathbf{x}}})} (\zeta-\mathbf{x})^{-1}\, d\zeta_{I_{\mathbf{x}}} f(\zeta).
\end{aligned}
$$

Remark 2.4.3. The function $\mathcal{I}_{\mathbf{y}}(\mathbf{x}) := (\mathbf{x}-\mathbf{y})^{-1}$ corresponding to the Cauchy kernel in Theorem 2.4.1 is not s-monogenic on $\mathbb{R}^{n+1} \setminus \{\mathbf{y}\}$, unless $\mathbf{y} = y_0 \in \mathbb{R}$. In particular, the function

$$
\mathcal{I}_0(\mathbf{x}) = \mathbf{x}^{-1} = \frac{\bar{\mathbf{x}}}{|\mathbf{x}|^2} \tag{2.18}
$$

is s-monogenic in $\mathbb{R}^{n+1} \setminus \{0\}$.

Theorem 2.4.4 (Cauchy formula outside a ball). *Let $B = B(0,R)$ and let $B^c = \mathbb{R}^{n+1} \setminus \overline{B}$. Let $f: B^c \to \mathbb{R}_n$ be an s-monogenic function with $\lim_{\mathbf{x}\to\infty} f(\mathbf{x}) = a$. If $\mathbf{x} \in B^c$, then*

$$
f(\mathbf{x}) = a - \frac{1}{2\pi} \int_{\partial\Delta_{\mathbf{x}}(0,r)} (\zeta-\mathbf{x})^{-1}\, d\zeta_{I_{\mathbf{x}}} f(\zeta)
$$

where $0 < R < r < |\mathbf{x}|$ and the complement of the set $\overline{\Delta_{\mathbf{x}}(0,r)}$ is contained in B^c and contains $\mathbf{x}$.

Proof. The proof is based on the Splitting Lemma and on the analogous result for holomorphic functions of a complex variable. Let $\mathbf{x} \in \mathbb{R}^{n+1} \setminus \overline{B}$ and let $I_{\mathbf{x}}$ be the corresponding imaginary unit. Consider $r' > r > R$, and the discs $\Delta = \Delta_{\mathbf{x}}(0,r)$, $\Delta' = \Delta_{\mathbf{x}}(0,r')$ on the plane $\mathbb{C}_{I_{\mathbf{x}}}$ having radius r and r' respectively and such that $\mathbf{x} \in \Delta'$. Since f is s-monogenic on $\Delta' \setminus \Delta$ we can apply the Cauchy formula to the set $\overline{\Delta'} \setminus \Delta$ to compute $f(\mathbf{x})$. We obtain

$$
\begin{aligned}
f(\mathbf{x}) &= \frac{1}{2\pi} \int_{\partial\Delta' \setminus \partial\Delta} (\xi-\mathbf{x})^{-1} d\xi_{I_{\mathbf{x}}} f(\xi) \\
&= \frac{1}{2\pi} \int_{\partial\Delta'} (\xi-\mathbf{x})^{-1} d\xi_{I_{\mathbf{x}}} f(\xi) \\
&\quad - \frac{1}{2\pi} \int_{\partial\Delta} (\xi-\mathbf{x})^{-1} d\xi_{I_{\mathbf{x}}} f(\xi).
\end{aligned}
$$

Let us set $I_1 := I_{\mathbf{x}}$ and complete to an orthonormal basis $I_1, \ldots, I_n$ of the Clifford algebra $\mathbb{R}_n$. The Splitting Lemma gives $f_{I_{\mathbf{x}}} = \sum_A F_A I_A$ and we can write

$$\begin{aligned} f(\mathbf{x}) &= \frac{1}{2\pi}\int_{\partial\Delta'} (\xi - \mathbf{x})^{-1} d\xi_{I_{\mathbf{x}}} f(\xi) \\ &\quad - \frac{1}{2\pi}\int_{\partial\Delta} (\xi - \mathbf{x})^{-1} d\xi_{I_{\mathbf{x}}} f(\xi) \\ &= \sum_A \frac{1}{2\pi}\int_{\partial\Delta'} (\xi - \mathbf{x})^{-1} d\xi_{I_{\mathbf{x}}} F_A(\xi) I_A \\ &\quad - \sum_A \frac{1}{2\pi}\int_{\partial\Delta} (\xi - \mathbf{x})^{-1} d\xi_{I_{\mathbf{x}}} F_A(\xi) I_A. \end{aligned}$$

Let us now consider a single component F_A at a time. By computing the integral on $\partial\Delta'$ in spherical coordinates, and by letting $r' \to \infty$, we obtain that the integral equals $a_A = \lim_{r'\to\infty} F_A$, and therefore:

$$F_A(\mathbf{x}) = a_A - \frac{1}{2\pi}\int_{\partial\Delta} (\xi - \mathbf{x})^{-1} d\xi_{I_{\mathbf{x}}} F_A(\xi).$$

Taking the sum of the various components multiplied with the corresponding units I_A we get the statement with $a = \sum_A a_A I_A$. □

Theorem 2.4.5 (Cauchy estimates). *Let $U \subseteq \mathbb{R}^{n+1}$ be an axially symmetric s-domain and let $f : U \to \mathbb{R}_n$ be an s-monogenic function. Let $y_0 \in U \cap \mathbb{R}$, $I \in \mathbb{S}$, and $r > 0$ be such that $\Delta_I(y_0, r) = \{(u + Iv) : (u - y_0)^2 + v^2 \leq r^2\}$ is contained in $U \cap \mathbb{C}_I$. If $M_I = \max\{|f(\mathbf{x})| : \mathbf{x} \in \partial\Delta_I(y_0, r)\}$ and if $M = \inf\{M_I : I \in \mathbb{S}\}$, then*

$$\frac{1}{n!}\left|\frac{\partial^n f}{\partial u^n}(y_0)\right| \leq \frac{M}{r^n}, \quad n \geq 0.$$

Proof. For any $I \in \mathbb{S}$, it is possible to write

$$\frac{1}{n!}\frac{\partial^n f}{\partial u^n}(y_0) = \frac{1}{2\pi I}\int_{\partial\Delta_I(y_0,r)} \frac{d\zeta}{(\zeta - y_0)^{n+1}} f(\zeta).$$

Therefore, for any $I \in \mathbb{S}$ we can write the following sequence of inequalities:

$$\begin{aligned} \frac{1}{n!}\left|\frac{\partial^n f}{\partial u^n}(y_0)\right| &\leq \frac{1}{2\pi}\int_{\partial\Delta_I(y_0,r)} \frac{|f(\zeta)|}{r^{n+1}} d\zeta \\ &\leq \frac{1}{2\pi}\int_{\partial\Delta_I(y_0,r)} \frac{M_I}{r^{n+1}} d\zeta = \frac{M_I}{r^n}. \end{aligned}$$

By taking the infimum, for $I \in \mathbb{S}$, of the right-hand side of the inequality we prove the assertion. □

Using the previous result it is immediate to show the following

Theorem 2.4.6 (Liouville). *Let $f:\ \mathbb{R}^{n+1}\to\mathbb{R}_n$ be an entire s-monogenic function. If f is bounded, then f is constant on $\mathbb{R}^{n+1}$.*

Proof. Suppose that $|f|\leq M$ on $\mathbb{R}^{n+1}$. By the previous theorem we have:

$$\frac{1}{n!}\left|\frac{\partial^n f}{\partial u^n}(0)\right|\leq\frac{M}{r^n},\qquad n\geq 0,$$

and by letting $r\to+\infty$ we obtain

$$\frac{\partial^n f}{\partial u^n}(0)=0$$

for any $n>0$, which implies $f(\mathbf{x})=c$, with $c\in\mathbb{R}_n$. □

Corollary 2.4.7. *Let $f:\ \mathbb{R}^{n+1}\to\mathbb{R}_n$ be an entire s-monogenic function. If $\lim_{\mathbf{x}\to\infty}f$ exists, then f is constant on $\mathbb{R}^{n+1}$.*

Theorem 2.4.8. *Let U be an open set in $\mathbb{R}^{n+1}$. If $f:\ U\to\mathbb{R}_n$ is an s-monogenic function, then*

$$\int_{\partial\Delta}d\mathbf{x}f(\mathbf{x})=0$$

for any disc $\Delta\subset U\cap\mathbb{C}_I$ with center in a point on the real axis.

Proof. This result is an easy consequence of the analogous result for holomorphic functions of one complex variable and of the Splitting Lemma. □

Conversely, we have the following result:

Theorem 2.4.9. *Let U be an axially symmetric s-domain and let $f:\ U\to\mathbb{R}_n$ be a real differentiable function. Assume that*

$$\int_{\gamma_I}d\mathbf{x}f(\mathbf{x})=0$$

for any closed, piecewise C^1 curve γ_I contained in $U\cap\mathbb{C}_I$ and homotopic to a point. Then f is an s-monogenic function.

Proof. This is a consequence of the classical Morera's theorem and of the definition of s-monogenic function. □

Proposition 2.4.10. *Let $f:\ B(0,R)\to\mathbb{R}_n$ be the s-monogenic function expressed by the series $\sum\mathbf{x}^m a_m$ converging on B. Then the composition of the functions f and $\mathcal{I}_0=\mathbf{x}^{-1}$ is s-monogenic on $\mathbb{R}^{n+1}\setminus B(0,1/R)$ and it can be expressed by the series $\sum\mathbf{x}^{-m}a_m$ converging on $\mathbb{R}^{n+1}\setminus B(0,1/R)$.*

Proof. Proposition 2.3.4 implies that $f\circ\mathcal{I}_0$ is an s-monogenic function on $\mathbb{R}^{n+1}\setminus B(0,1/R)$. The statement follows from the analogous result for holomorphic functions in one complex variable. □

Theorem 2.4.11 (Laurent series). *Let f be an s-monogenic function in a spherical shell $A = \{\mathbf{x} \in \mathbb{R}^{n+1} \mid R_1 < |\mathbf{x}| < R_2\}$, $0 < R_1 < R_2$. Then f admits the unique Laurent expansion*

$$f(\mathbf{x}) = \sum_{m\geq 0} \mathbf{x}^m a_m + \sum_{m\geq 1} \mathbf{x}^{-m} b_m \tag{2.19}$$

where

$$a_m = \frac{1}{m!}\partial_s^m f(0), \qquad b_m = \frac{1}{2\pi}\int_{\partial(B(0,R_1')\cap\mathbb{C}_{I_\mathbf{x}})} \zeta^{m-1} d\zeta_{I_\mathbf{x}} f(\zeta).$$

The two series in (2.19) converge in the open ball $B(0,R_2)$ and in $\mathbb{R}^{n+1}\setminus\overline{B(0,R_1)}$, respectively.

Proof. Let $\mathbf{x} \in A$, then there exist two positive real numbers R_1', R_2' such that $A' = \{\mathbf{x} \in \mathbb{R}^{n+1} \mid R_1' < |\mathbf{x}| < R_2'\} \subset A$, and $\mathbf{x} \in A'$. Using the Cauchy integral formula, we can write

$$f(\mathbf{x}) = \frac{1}{2\pi}\int_{\partial(A'\cap\mathbb{C}_{I_\mathbf{x}})} (\zeta - \mathbf{x})^{-1}\, d\zeta_{I_\mathbf{x}} f(\zeta) = f_1(\mathbf{x}) + f_2(\mathbf{x})$$

where

$$f_1(\mathbf{x}) = \frac{1}{2\pi}\int_{\partial(B(0,R_2')\cap\mathbb{C}_{I_\mathbf{x}})} (\zeta - \mathbf{x})^{-1}\, d\zeta_{I_\mathbf{x}} f(\zeta)$$

and

$$f_2(\mathbf{x}) = -\frac{1}{2\pi}\int_{\partial(B(0,R_1')\cap\mathbb{C}_{I_\mathbf{x}})} (\zeta - \mathbf{x})^{-1}\, d\zeta_{I_\mathbf{x}} f(\zeta).$$

The first integral is associated to the first series in the Laurent expansion, by Proposition 2.3.1. Let us consider the second integral, set $I_1 = I_\mathbf{x}$ and let us use the Splitting Lemma and write f_2 as $\sum_A F_A I_A$. Now we can reason as in the case of functions in one complex variable, and consider the single components of $f_2(\mathbf{x})$. In $\mathbb{R}^{n+1} \setminus \overline{B(0,R_1')}$, we have

$$\begin{aligned} F_A(\mathbf{x}) &= -\frac{1}{2\pi}\int_{\partial(B(0,R_1')\cap\mathbb{C}_{I_\mathbf{x}})} (\zeta - \mathbf{x})^{-1}\, d\zeta_{I_\mathbf{x}} F_A(\zeta) \\ &= \frac{1}{2\pi}\int_{\partial(B(0,R_1')\cap\mathbb{C}_{I_\mathbf{x}})} \sum_{m\geq 0} \mathbf{x}^{-m-1}\zeta^m d\zeta_{I_\mathbf{x}} F_A(\zeta) \end{aligned}$$

where we have used the fact that on the plane $\mathbb{C}_{I_\mathbf{x}}$ the variables ζ and $\mathbf{x}$ commute. Now, using the uniform convergence of the series we can write

$$F_A(\mathbf{x}) = \sum_{m\geq 0} \mathbf{x}^{-m-1}\frac{1}{2\pi}\int_{\partial(B(0,R_1')\cap\mathbb{C}_{I_\mathbf{x}})} \zeta^m d\zeta_{I_\mathbf{x}} F_A(\zeta) = \sum_{m\geq 0} \mathbf{x}^{-m-1} b_{m+1,A}$$

where

$$b_{m+1,A} := b_{m+1,I_{\mathbf{x}};A} = \frac{1}{2\pi}\int_{\partial(B(0,R_1')\cap\mathbb{C}_{I_{\mathbf{x}}})} \zeta^m d\zeta_{I_{\mathbf{x}}} F_A(\zeta).$$

Finally, we obtain:

$$\tilde{f}_2(\mathbf{x}) = \sum_A F_A(\mathbf{x}) I_A = \sum_{m\geq 0}\sum_A \mathbf{x}^{-m-1} b_{m+1,A} I_A.$$

Note that $\tilde{f}_2(\mathbf{x})$ coincides with $f_2(\mathbf{x})$ on the plane $\mathbb{C}_{I_{\mathbf{x}}}$, thus they coincide everywhere and the coefficients $b_{m+1,A}$ do not depend on the choice of the imaginary unit $I_{\mathbf{x}}$. The statement follows. □

2.5 Zeros of slice monogenic functions

As it is well known, the Fundamental Theorem of Algebra does not hold in $\mathbb{R}_n$ for $n \geq 3$, thus we cannot guarantee that a polynomial in the paravector variable $\mathbf{x}$ has a zero, not even if it is a degree-one polynomial. The following examples are instructive to show what can happen in a Clifford algebra.

Example 2.5.1. Consider the Clifford algebra $\mathbb{R}_n$, $n \geq 2$ and the polynomial $p(\mathbf{x}) = \mathbf{x}e_1 - e_2 \in \mathbb{R}_n[\mathbf{x}]$. The only zero of p is e_1e_2 which does not belong to $\mathbb{R}^{n+1}$.

Example 2.5.2. Consider the Clifford algebra $\mathbb{R}_n$, $n \geq 2$ and the polynomial $p(\mathbf{x}) = \mathbf{x}^2 - \mathbf{x}(e_1e_2 - 2e_1) + 2e_2 \in \mathbb{R}_n[\mathbf{x}]$. It can be easily verified that p vanishes for $\mathbf{x} = -2e_1$ and $\mathbf{x} = -\frac{1}{5}(4e_1 + 3e_1e_2)$. However, only $\mathbf{x} = -2e_1$ is a zero of p in $\mathbb{R}^{n+1}$.

Example 2.5.3. Consider the Clifford algebra $\mathbb{R}_n$, $n \geq 2$ and the polynomial $p(\mathbf{x}) = \mathbf{x}^2 - \mathbf{x}(e_1 + 2e_2) + 2e_1e_2 \in \mathbb{R}_n[\mathbf{x}]$. It can be easily verified that both $\mathbf{x} = e_1$ and $\mathbf{x} = \frac{1}{5}(8e_1 + 6e_2)$ are zeros of p in $\mathbb{R}^{n+1}$.

It is nevertheless interesting to attempt to characterize the set of zeros for those polynomials for which such a set is not empty. Let us start by showing that each $(n-1)$-sphere $[\mathbf{s}]$ is characterized by a second degree equation.

Proposition 2.5.4. *Let $\mathbf{s} = s_0 + \underline{s} \in \mathbb{R}^{n+1}$. Consider the equation*

$$\mathbf{x}^2 - 2\mathrm{Re}[\mathbf{s}]\mathbf{x} + |\mathbf{s}|^2 = 0. \tag{2.20}$$

Then, $\mathbf{x} = x_0 + \underline{x} \in \mathbb{R}^{n+1}$ is a solution if and only if $\mathbf{x} \in [\mathbf{s}]$.

Proof. The result is immediate when $\mathbf{s} = s_0 \in \mathbb{R}$. Let us suppose that $\mathbf{s} \notin \mathbb{R}$. It is immediate that $\mathbf{x} \in [\mathbf{s}]$ is a solution. Conversely, let $\mathbf{x}$ be a solution, i.e., $(x_0 + \underline{x})^2 - 2Re[s](x_0 + \underline{x}) + |\mathbf{s}|^2 = 0$. A direct computation shows that this is possible if and only if $\underline{x} = 0$ or $x_0 = s_0$. The first possibility does not give any solution, while the second gives $|\mathbf{x}| = |\mathbf{s}|$, i.e., the equivalence class of $\mathbf{s}$. □

An obvious consequence of the proposition, which will be useful in the sequel, is that any paravector $\mathbf{s}$ satisfies the identity

$$\mathbf{s}^2 - 2\mathrm{Re}[\mathbf{s}]\mathbf{s} + |\mathbf{s}|^2 = 0. \tag{2.21}$$

As a consequence of the Representation Formula II, we obtain the following immediate result on the zeros of an s-monogenic function:

Proposition 2.5.5. *Let $U \subseteq \mathbb{R}^{n+1}$ be an axially symmetric s-domain and let $f : U \to \mathbb{R}^{n+1}$ be an s-monogenic function. If $f(u+Iv) = f(u+Kv) = 0$ for some $I, K \in \mathbb{S}$, $I \neq K$, then f vanishes on the entire $(n-1)$-sphere $[u+Iv]$.*

In particular, we have

Corollary 2.5.6. *Let $U \subseteq \mathbb{R}^{n+1}$ be an axially symmetric s-domain and let $f : U \to \mathbb{R}^{n+1}$ be an s-monogenic function. If $f(u+Iv) = f(u-Iv) = 0$ for some $I \in \mathbb{S}$, then f vanishes on the entire $(n-1)$-sphere $[u+Iv]$.*

In other words, the zero set of an s-monogenic function having two zeros on a certain $(n-1)$-sphere contains the entire sphere. There are s-monogenic functions whose zero set is made only by the union of isolated $(n-1)$-spheres (in particular, points on the real axis). Among these functions there are power series with real coefficient, as proved in the following:

Proposition 2.5.7. *Let $U \subseteq \mathbb{R}^{n+1}$ be an axially symmetric s-domain and let $f : U \to \mathbb{R}_n$ be an s-monogenic function. If f has a series representation*

$$f(\mathbf{x}) = \sum_{m \geq 0} (\mathbf{x} - y_0)^m a_m$$

with real coefficients a_m, at some point on the real axis $y_0 \in U$, then every real zero is isolated. If $u_0 + v_0 I_0$, for some $I_0 \in \mathbb{S}$, is a nonreal zero, then $u_0 + v_0 I$ is a zero for any $I \in \mathbb{S}$. In particular, if $f \not\equiv 0$, the zero set of f is either empty or it is the union of isolated points (belonging to $\mathbb{R}$) and isolated $(n-1)$-spheres.

Proof. We will first prove that for all $I \in \mathbb{S}$ we have $f(U \cap \mathbb{C}_I) \subseteq U \cap \mathbb{C}_I$. This fact is true in a suitable disc $B \cap \mathbb{C}_I \subset \mathbb{C}_I$ containing y_0, since the series $f(\mathbf{x}) = \sum_{m \geq 0} (\mathbf{x} - y_0)^m a_m$ converging on B, has real coefficients by hypothesis. The Splitting Lemma on the plane $\mathbb{C}_I$ implies that $f_I(u+Iv) = F(u+Iv)$ in that disc on $\mathbb{C}_I$. Therefore, $F_A = 0$ for $A \neq \emptyset$ on $B \cap \mathbb{C}_I$ and by the Identity Principle for holomorphic functions we obtain that all the holomorphic functions F_A are identically zero on $U \cap \mathbb{C}_I$ for $A \neq \emptyset$. Hence $f(U \cap \mathbb{C}_I) \subseteq U \cap \mathbb{C}_I$ for all $I \in \mathbb{S}$ from which it follows that $f(u) \in \mathbb{R}$ for all $u \in U \cap \mathbb{R}$. By the Identity Principle we get that $f(u+I_0 v) \equiv F(u+I_0 v)$ on $U \cap \mathbb{C}_{I_0}$ and, being $F(u)$ real-valued for all $u \in U \cap \mathbb{R}$, we have that $F(u+I_0 v) = \overline{F(u-I_0 v)}$ on $U \cap \mathbb{C}_{I_0}$. Since

$$0 = f(u_0 + I_0 v_0) = F(u_0 + I_0 v_0) = \overline{F(u_0 - I_0 v_0)}$$

it turns out that

$$F(u_0 - I_0 v_0) = f(u_0 - I_0 v_0) = 0.$$

The statement follows from Corollary 2.5.6. The fact that the real zeros and the spheres are isolated follows from the Identity Principle. □

As a consequence, we get a description of the zero set of a polynomial with real coefficients in the paravector variable:

Corollary 2.5.8. *Let p be a polynomial in the paravector variable $\mathbf{x}$ with real coefficients. Then the zero set of p is the union of isolated points (belonging to $\mathbb{R}$) and isolated $(n-1)$-spheres.*

Remark 2.5.9. As we have already pointed out, in the case $n = 1$ the set of s-monogenic functions coincide with the set of holomorphic functions in one complex variable (by identifying $\mathbb{R}^2$ with $\mathbb{C}$). Proposition 2.5.7 corresponds to the well-known result saying that the zeros of a holomorphic function whose series expansion has real coefficients has isolated zeros which are either real or complex conjugates.

To show that any s-monogenic function has zero set consisting of a union of isolated $(n-1)$-spheres (which might be reduced to a point on the real axis) and isolated points, we associate to each s-monogenic function defined on an axially symmetric s-domain U, an auxiliary function defined on U and denoted by f^σ. The function f^σ has two main properties: on one hand it vanishes on the zero set of f, on the other hand, it defines a holomorphic function which takes elements from $U \cap \mathbb{C}_I$ to $\mathbb{C}_I$ for all $I \in \mathbb{S}$.

The idea used to construct the function f^σ is based on the observation that, given a vector with 2^{n-1} complex components w_A, the vector with components $w_A \bar{w}_A$ is zero if and only if $w_A = 0$ for all A. Now note that the Splitting Lemma allows to write the restriction f_I of an s-monogenic function f in terms of a vector of 2^{n-1} holomorphic functions $F_A : U \cap \mathbb{C}_I \to \mathbb{C}_I$ as

$$f_I(z) = \sum_A F_A(z) I_A.$$

Consider the vector with components $F_A(z)\overline{F_A(\bar{z})}$. The components are obviously holomorphic and if $F_A(z_0) = 0$ also $F_A(z_0)\overline{F_A(\bar{z}_0)} = 0$. We then define the function $f_I^\sigma : U \cap \mathbb{C}_I \to \mathbb{C}_I$ by

$$f_I^\sigma(z) = \sum_A F_A(z)\overline{F_A(\bar{z})}.$$

Using the Extension Lemma 2.2.22, we can extend the function f_I^σ to an s-monogenic function defined on U:

Definition 2.5.10. *Let $U \subset \mathbb{R}^{n+1}$ be an axially symmetric s-domain and let $f : U \to \mathbb{R}_n$ be an s-monogenic function. Let $I \in \mathbb{S}$ and let*

$$f_I^\sigma(z) = \sum_A F_A(z)\overline{F_A(\bar{z})}.$$

We define $f^\sigma : U \to \mathbb{R}_n$ by:

$$f^\sigma(\mathbf{x}) := \mathrm{ext}(f_I^\sigma)(\mathbf{x}).$$

We have the following property:

Lemma 2.5.11. *Let $U \subseteq \mathbb{R}^{n+1}$ be an axially symmetric s-domain and let $f : U \to \mathbb{R}_n$ be an s-monogenic function. Then f vanishes identically on U if and only if f^σ vanishes identically on U.*

Proof. When $f \equiv 0$, it is immediate that $f^\sigma \equiv 0$. Conversely, consider the restriction f_I of f to a plane $\mathbb{C}_I$, which, by the Splitting Lemma, can be written as $f_I(z) = \sum_A F_A(z) I_A$ where $F_A : U \cap \mathbb{C}_I \to \mathbb{C}_I$ are holomorphic functions. Then the functions F_A admit series expansion at any point of $U \cap \mathbb{C}_I$. Consider $y_0 \in U \cap \mathbb{C}_I$ belonging to the real axis and the series expansion of F_A at a point y_0:

$$F_A(z) = \sum_{m\geq 0} (z-y_0)^m a_{Am}, \qquad a_{Am} \in \mathbb{C}_I$$

which holds in a suitable disc $\Delta(y_0, R) \subseteq U \cap \mathbb{C}_I$ of radius R and centered in $y_0 \in \mathbb{R}$. Then, on $\Delta(y_0, R)$, we have

$$\overline{F_A(\bar{z})} = \sum_{m\geq 0} (z-y_0)^m \bar{a}_{Am}.$$

Moreover on $\Delta(y_0, R)$ we can write

$$\begin{aligned} f_I^\sigma(z) &= \sum_A F_A(z)\overline{F_A(\bar{z})} \\ &= \sum_A \sum_{m\geq 0} (z-y_0)^m c_{Am} = \sum_{m\geq 0} (z-y_0)^m \Big(\sum_A c_{Am}\Big), \end{aligned}$$

where

$$c_{Am} = \sum_{i=0}^{m} a_{Ai}\bar{a}_{A\ m-i}.$$

Now, if $f^\sigma \equiv 0$, then $f_I^\sigma \equiv 0$. So, in the disc $\Delta(y_0, R)$ we have that $\sum_A c_{A0} = \sum_A |a_{A0}|^2 = 0$ so $a_{A0} = 0$ for all multi-indices A. Now, by induction, assume that $a_{Ai} = 0$ for $i = 0, 1, \ldots, k-1$, $k \geq 1$ for all multi-indices A. Consider the coefficient

$$\sum_A c_{A\,2k} = \sum_A \sum_{i=0}^{2k} a_{Ai}\bar{a}_{A\,2k-i}$$

which is zero because $f_I^\sigma \equiv 0$. By assumption we have $a_{Ai}\bar{a}_{A\ 2k-i} = 0$ when $i = 0, \ldots, k-1$ since $a_{Ai} = 0$ and $a_{Ai}\bar{a}_{A\,2k-i} = 0$ when $i = k+1, \ldots, 2k$ since $\bar{a}_{A\ 2k-i} = 0$. Thus, $\sum_A c_{A\,2k} = \sum_A |a_{Ak}|^2$ is zero if and only if $a_{Ak} = 0$ for all multi-indices A. We conclude that $f_I^\sigma \equiv 0$ in the disc $\Delta(y_0, R) \cap \mathbb{C}_I$ implies that all the coefficients a_{Ai} vanish, thus also f_I vanishes identically on the same disc.

By the Identity Principle f vanishes identically. □

The zero set of f^σ is described in the following result:

Lemma 2.5.12. *Let $U \subseteq \mathbb{R}^{n+1}$ be an axially symmetric s-domain, let $f : U \to \mathbb{R}_n$ be an s-monogenic function, and let $f \not\equiv 0$. If there exists $I \in \mathbb{S}$ for which $f^\sigma(u_0 + Iv_0) = 0$, then $f^\sigma(u_0 + Jv_0) = 0$ for all $J \in \mathbb{S}$. Moreover, the zero set of f^σ consists of isolated $(n-1)$-spheres (which might reduce to points on the real axis).*

Proof. Consider the restriction f_I^σ of f^σ to the plane $\mathbb{C}_I$. We have:

$$\overline{f_I^\sigma(\bar{z})} = \overline{\sum_A F_A(\bar{z})\overline{F_A(z)}} = \sum_A \overline{F_A(\bar{z})}F_A(z) = f_I^\sigma(z),$$

thus $f_I^\sigma(u_0 + Iu_0) = 0$ if and only if $f_I^\sigma(u_0 - Iu_0) = 0$. So, if $f^\sigma(u_0 + Iv_0) = 0$, then, by the Representation Formula, $f^\sigma(u_0 + Ju_0) = 0$ for all $J \in \mathbb{S}$. The second part of the statement follows by the Identity Principle: if the $(n-1)$-spheres of zeros were not isolated, on each plane we would get accumulation points of zeros and thus f^σ would be identically zero by the Identity Principle which contradicts the fact that $f^\sigma \not\equiv 0$ by Lemma 2.5.11. □

Lemma 2.5.13. *Let $U \subseteq \mathbb{R}_n$ be an axially symmetric s-domain and let $f : U \to \mathbb{R}_n$ be an s-monogenic function. If $u + Iv$ is a zero of f, then it is also a zero of f^σ.*

Proof. The restriction of f to the plane $\mathbb{C}_I$ can be written, by the Splitting Lemma, as $f_I(z) = \sum_A F_A(z)I_A$. The condition $f(u + Iv) = 0$ implies that, on the plane $\mathbb{C}_I$ it is also $F_A(u + Iv) = 0$ for all A. Thus $f_I^\sigma(u + Iv) = 0$ and the statement follows. □

We are now in a position to prove the following theorem which describes the zero set of an s-monogenic function defined on an axially symmetric s-domain.

Theorem 2.5.14 (Structure of the Zero Set). *Let $U \subseteq \mathbb{R}^{n+1}$ be an axially symmetric s-domain and let $f : U \to \mathbb{R}_n$ be an s-monogenic function. Suppose that f does not vanish identically. Then if the zero set of f is nonempty, it consists of the union of isolated $(n-1)$-spheres and/or isolated points.*

Proof. Suppose that the zero set of f is nonempty and that f does not vanish identically, thus also f^σ does not vanish identically by Lemma 2.5.11. By Lemma 2.5.13 any zero of f is a zero of f^σ, i.e., denoting by Z_{f^σ} and Z_f the zero set of f^σ and f respectively, we have $Z_f \subseteq Z_{f^\sigma}$. If Z_f contains two points on an $(n-1)$-sphere $[\mathbf{s}]$, then Z_f contains the whole sphere. Indeed, suppose that $u_0 + Iv_0$, $u_0 + Jv_0 \in [\mathbf{s}]$, $I \neq J$, and $f(u_0 + Iv_0) = f(u_0 + Jv_0) = 0$. Then by the Representation Formula we get

$$f(u_0 + Jv_0) = \frac{1}{2}[1 + JI]f(u_0 - Iv_0) = 0.$$

The element $1+JI=(-I+J)I$ is invertible since it is product of two invertible elements, thus $f(u_0-Iv_0)=0$ and the statement follows from Proposition 2.5.6.

When a sphere belongs to Z_f, then it is isolated. Indeed, let $\mathbf{x}_0$ be a point on this sphere. If there were a sequence $\{\mathbf{x}_n\}$ of zeros, $\mathbf{x}_n \notin [u_0+I_0v_0]$, such that $\mathbf{x}_n \to \mathbf{x}_0$, then the corresponding spheres $[\mathbf{x}_n]$ would belong to Z_{f^σ}, which is absurd by Lemma 2.5.12.

Similarly, suppose that Z_f contains a point $\mathbf{x}_0=u_0+Jv_0$, without containing the sphere u_0+Iv_0, $I\in\mathbb{S}$ generated by it. Then we have to show that the point $u_0+I_0v_0$ is isolated. Indeed, if there were a sequence $\{\mathbf{x}_n\}$ of zeros, $\mathbf{x}_n \notin [u_0+Jv_0]$ (otherwise the whole sphere $[u_0+Jv_0]$ would belong to Z_f), such that $\mathbf{x}_n\to\mathbf{x}_0$, then the corresponding spheres would belong to Z_{f^σ} which is absurd by Lemma 2.5.12. □

Remark 2.5.15. The result already obtained in Proposition 2.5.5 can be obtained also as a consequence of the previous theorem. In fact, given a converging power series $\sum_{m\geq 0}\mathbf{x}^m a_m$, $a_m\in\mathbb{R}_n$, if there are two different elements in a given equivalence class $[\mathbf{s}]$, which are solutions to the equation

$$\sum_{m\geq 0}\mathbf{x}^m a_m = 0,$$

then all the elements in the equivalence class are solutions.

We close this section with an immediate corollary of the previous theorem, which yields a nice description of the zero set of a polynomial:

Corollary 2.5.16. *Let $p(\mathbf{x})$ be a polynomial in $\mathbb{R}_n[\mathbf{x}]$, with right coefficients, which does not vanish identically. Then, if the zero set of p is nonempty, it consists of isolated points or isolated $(n-1)$-spheres.*

2.6 The slice monogenic product

It is immediate to see that the product of two s-monogenic functions is not, in general, s-monogenic. Nevertheless, as we indicated in Section 2.3, it is possible to define a product among s-monogenic power series by mimicking the process used to define a product of polynomials in skew fields. We can extend this idea to the case of s-monogenic functions defined on axially symmetric s-domains, to define an s-monogenic product. Let $U\subseteq\mathbb{R}^{n+1}$ be an axially symmetric s-domain and let $f,g:\ U\to\mathbb{R}_n$ be s-monogenic functions. For any $I\in\mathbb{S}$ set $I=I_1$ and consider a completion to a basis $\{I_1,\dots,I_n\}$ of $\mathbb{R}_n$ such that $I_iI_j+I_jI_i=-2\delta_{ij}$. The Splitting

Lemma guarantees the existence of holomorphic functions $F_A, G_A : U \cap \mathbb{C}_I \to \mathbb{C}_I$ such that, for all $z = u + Iv \in U \cap \mathbb{C}_I$,

$$f_I(z) = \sum_A F_A(z) I_A, \qquad g_I(z) = \sum_B G_B(z) I_B,$$

where A, B are subsets of $\{2, \ldots, n\}$ and, by definition, $I_\emptyset = 1$. We define the function $f_I * g_I : U \cap \mathbb{C}_I \to \mathbb{R}_n$ as

$$f_I * g_I(z) = \sum_{|A| \text{even}} (-1)^{\frac{|A|}{2}} F_A(z) G_A(z) + \sum_{|A| \text{odd}} (-1)^{\frac{|A|+1}{2}} F_A(z) \overline{G_A(\bar{z})} \tag{2.22}$$
$$+ \sum_{|A| \text{even}, B \neq A} F_A(z) G_B(z) I_A I_B + \sum_{|A| \text{odd}, B \neq A} F_A(z) \overline{G_B(\bar{z})} I_A I_B.$$

Then $f_I * g_I(z)$ is obviously a holomorphic map on $\mathbb{C}_I$, i.e., $\overline{\partial}_I (f_I * g_I)(z) = 0$, and hence its unique s-monogenic extension to U, according to the Extension Lemma 2.2.22, is given by

$$f * g(\mathbf{x}) := \operatorname{ext}(f_I * g_I)(\mathbf{x}).$$

Definition 2.6.1. *Let $U \subseteq \mathbb{R}^{n+1}$ be an axially symmetric s-domain and let $f, g : U \to \mathbb{R}_n$ be s-monogenic functions. The function*

$$f * g(\mathbf{x}) = \operatorname{ext}(f_I * g_I)(\mathbf{x})$$

defined as the extension of (2.22) is called the s-monogenic product of f and g. This product is also called the $*$-product of f and g.

Remark 2.6.2. It is immediate to verify that the $*$-product is associative, distributive but, in general, not commutative.

The following example shows the dramatic difference between polynomials in a division algebra and polynomials in a Clifford algebra. Even a simple result such as $\deg(p_1 * p_2) = \deg(p_1) + \deg(p_2)$ fails (we can only conclude that $\deg(p_1 * p_2) \leq \deg(p_1) + \deg(p_2)$) and it is impossible to deduce the zeros of the product from the zeros of the factors. This is in stark contrast with the case of polynomials in division algebras, where it is possible to obtain explicit formulas to deduce the zeros of $f * g$ from the zeros of f and g (see, e.g., [71]).

Example 2.6.3. Consider the two polynomials $p_1(\mathbf{x}) = 1 + \mathbf{x}(1 - e_1e_2e_3)$ and $p_2(\mathbf{x}) = 1 + \mathbf{x}(1 + e_1e_2e_3) \in \mathbb{R}_3[\mathbf{x}]$. None of them has roots in $\mathbb{R}^4$ because $(1 \pm e_1e_2e_3)$ are zero divisors. Their product $p_1 * p_2(\mathbf{x}) = (1 + \mathbf{x}(1 - e_1e_2e_3)) * (1 + \mathbf{x}(1 + e_1e_2e_3)) = 1 + 2\mathbf{x}$ is a degree-one polynomial and has the real number $-1/2$ as its root.

The s-monogenic product is however an important tool to obtain s-monogenic functions. In particular, it allows us to define the inverse of an s-monogenic function with respect to the $*$-product. As we have already mentioned, not all the Clifford numbers admit an inverse with respect to the product in the Clifford algebra

$\mathbb{R}_n$. Those Clifford numbers $a \in \mathbb{R}_n$ for which $a\bar{a}$ is a real nonzero number admit inverse $a^{-1} = \bar{a}(a\bar{a})^{-1}$. In particular the existence of the inverse can be guaranteed for all nonzero vectors. Similarly, for s-monogenic functions we can guarantee the existence of an inverse with respect to the $*$-product, if we suitably restrict their codomains. To introduce the notion of inverse we need some preliminary definitions.

Let $U \subseteq \mathbb{R}^{n+1}$ be an axially symmetric s-domain and let $f : U \to \mathbb{R}_n$ be an s-monogenic function. Let us consider the restriction $f_I(z)$ of f to the plane $\mathbb{C}_I$ and it usual representation (given by the Splitting Lemma)

$$f_I(z) = \sum_A F_A(z) I_A.$$

Let us define the function $f_I^c : U \cap \mathbb{C}_I \to \mathbb{C}_I$ as

$$\begin{aligned} f_I^c(z) &:= \sum_A F_A^c(z) I_A \\ &= \sum_{|A|\equiv 0} \overline{F_A(\bar{z})} I_A - \sum_{|A|\equiv 1} F_A(z) I_A - \sum_{|A|\equiv 2} \overline{F_A(\bar{z})} I_A + \sum_{|A|\equiv 3} F_A(z) I_A, \end{aligned} \tag{2.23}$$

where the equivalence $\equiv$ is intended as $\equiv$ (mod4), i.e., the congruence modulo 4. Since any function F_A is obviously holomorphic it can be uniquely extended to an s-monogenic function on U, according to the Extension Lemma 2.2.22. Thus we can give the following definition:

Definition 2.6.4. *Let $U \subseteq \mathbb{R}^{n+1}$ be an axially symmetric s-domain and let $f : U \to \mathbb{R}_n$ be an s-monogenic function. The function*

$$f^c(\mathbf{x}) = \text{ext}(f_I^c)(\mathbf{x})$$

is called the s-monogenic conjugate of f.

This definition of conjugate behaves, for power series and thus for polynomials, as the conjugation on the coefficients as proven in the next result:

Proposition 2.6.5. *Let $f : B(y_0, R) \to \mathbb{R}_n$ be an s-monogenic function on an open ball in $\mathbb{R}^{n+1}$ centered at a point on the real axis y_0. If*

$$f(\mathbf{x}) = \sum_{m\geq 0} (\mathbf{x} - y_0)^m a_m,$$

then, for $a_m \in \mathbb{R}_n$, we have

$$f^c(\mathbf{x}) = \sum_{m\geq 0} (\mathbf{x} - y_0)^m \bar{a}_m.$$

Proof. We will suppose without loss of generality that $y_0 = 0$. By Corollary 2.3.7, given any $I \in \mathbb{S}$, the coefficients of the power series expansion of f can be obtained as the coefficients of the power series of f_I. By the Splitting Lemma with respect to an orthonormal completion of I to a basis of $\mathbb{R}_n$, for all $z = u + Iv \in B(0, R) \cap \mathbb{C}_I$ we have

$$f_I(z) = \sum_A F_A(z) I_A = \sum_A \sum_{m \geq 0} z^m \frac{1}{m!} \Big(\frac{\partial^m F_A}{\partial u^m}(0) \Big) I_A = \sum_{m \geq 0} z^m \frac{1}{m!} \partial_s^m f(0)$$

and hence the relation

$$f_I^c(z) = \sum_{|A| \equiv 0} \overline{F_A(\bar{z})} I_A - \sum_{|A| \equiv 1} F_A(z) I_A - \sum_{|A| \equiv 2} \overline{F_A(\bar{z})} I_A + \sum_{|A| \equiv 3} F_A(z) I_A \qquad (2.24)$$

$$= \sum_{m \geq 0} \frac{z^m}{m!} \Big(\sum_{|A| \equiv 0} \frac{\partial^m \overline{F_A}}{\partial u^m}(0) - \sum_{|A| \equiv 1} \frac{\partial^m F_A}{\partial u^m}(0) - \sum_{|A| \equiv 2} \frac{\partial^m \overline{F_A}}{\partial u^m}(0) + \sum_{|A| \equiv 3} \frac{\partial^m F_A}{\partial u^m}(0) \Big) I_A \qquad (2.25)$$

$$= \sum_{m \geq 0} z^m \frac{1}{m!} \overline{\partial_s^m f(0)}, \qquad (2.26)$$

where the equivalence $\equiv$ is intended as the congruence modulo 4, proves the assertion. □

Using the notion of $*$-multiplication of s-monogenic functions, it is possible to associate to any s-monogenic function f its "symmetrization" or "normal form", denoted by f^s. We will show that all the zeros of f^s are $(n-1)$-spheres (possibly reduced to a point on the real axis) and that if $\mathbf{x}$ is a zero of f (isolated or not), then the $(n-1)$-sphere $[\mathbf{x}]$ is a zero of f^s.

Let $U \subseteq \mathbb{R}^{n+1}$ be an axially symmetric s-domain and let $f : U \to \mathbb{R}_n$ be an s-monogenic function. As usual, using the Splitting Lemma we can write

$$f_I(z) = \sum_A F_A(z) I_A;$$

here we will use the notation $[f_I]_0$ to denote the "scalar" part of the function f_I, i.e., the part whose coefficient in the Splitting Lemma is $I_\emptyset = 1$. With this notation, we define the function $f^s : U \cap \mathbb{C}_I \to \mathbb{C}_I$ as

$$\begin{aligned} f_I^s &:= [f_I * f_I^c]_0 \qquad (2.27) \\ &= \Big[\Big(\sum_B F_B(z) I_B\Big)\Big(\sum_{|A| \equiv 0} \overline{F_A(\bar{z})} I_A - \sum_{|A| \equiv 1} F_A(z) I_A \\ &\quad - \sum_{|A| \equiv 2} \overline{F_A(\bar{z})} I_A + \sum_{|A| \equiv 3} F_A(z) I_A \Big) \Big]_0 . \end{aligned}$$

We have

$$\begin{aligned} f_I * f_I^c = & \sum_{|B| \text{even}, |A| \equiv 0} F_B(z)\overline{F_A(\bar{z})} I_B I_A - \sum_{|B| \text{even}, |A| \equiv 1} F_B(z) F_A(z) I_B I_A \\ & - \sum_{|B| \text{even}, |A| \equiv 2} F_B(z)\overline{F_A(\bar{z})} I_B I_A + \sum_{|B| \text{even}, |A| \equiv 3} F_B(z) F_A(z) I_B I_A \\ & + \sum_{|B| \text{odd}, |A| \equiv 0} F_B(z) F_A(z) I_B I_A - \sum_{|B| \text{odd}, |A| \equiv 1} F_B(z)\overline{F_A(\bar{z})} I_B I_A \\ & - \sum_{|B| \text{odd}, |A| \equiv 2} F_B(z) F_A(z) I_B I_A + \sum_{|B| \text{odd}, |A| \equiv 3} F_B(z)\overline{F_A(\bar{z})} I_B I_A. \end{aligned}$$

The terms from which the scalar part arises are the ones with $A = B$, i.e.,

$$\begin{aligned} [f_I * f_I^c]_0 = & \sum_{|A| \equiv 0} F_A(z)\overline{F_A(\bar{z})} I_A^2 - \sum_{|A| \equiv 2} F_A(z)\overline{F_A(\bar{z})} I_A^2 \\ & - \sum_{|A| \equiv 1} F_A(z)\overline{F_A(\bar{z})} I_A^2 + \sum_{|A| \equiv 3} F_A(z)\overline{F_A(\bar{z})} I_A^2 = \sum_A F_A(z)\overline{F_A(\bar{z})}. \end{aligned}$$

Then f_I^s is obviously holomorphic and hence its unique s-monogenic extension to U defined by

$$f^s(\mathbf{x}) := \text{ext}(f_I^s)(\mathbf{x})$$

is s-monogenic.

Definition 2.6.6. *Let $U \subseteq \mathbb{R}^{n+1}$ be an axially symmetric s-domain and let $f : U \to \mathbb{R}_n$ be an s-monogenic function. The function*

$$f^s(\mathbf{x}) = \text{ext}(f_I^s)(\mathbf{x})$$

defined by the extension of $f_I^s = [f_I * f_I^c]_0$ from $U \cap \mathbb{C}_I$ to the whole U is called the symmetrization of f.

Remark 2.6.7. Notice that formula (2.27) yields that, for all $I \in \mathbb{S}$, $f^s(U \cap \mathbb{C}_I) \subseteq \mathbb{C}_I$.

Remark 2.6.8. Note that the function f^σ introduced in Definition 2.5.10 to study the zero set of an s-monogenic function coincides with f^s for all s-monogenic functions f.

It is now easy to verify the following facts.

Proposition 2.6.9. *Let $U \subseteq \mathbb{R}^{n+1}$ be an axially symmetric s-domain and let $f, g \in \mathcal{M}(U)$. Then*

$$f^s g = f^s * g = g * f^s.$$

Moreover, if Z_{f^s} is the zero set of f^s, then

$$(f^s)^{-1} g = (f^s)^{-1} * g = g * (f^s)^{-1} \quad \text{on } U \setminus Z_{f^s}.$$

Proof. Since $f^s(U\cap\mathbb{C}_I)\subseteq\mathbb{C}_I$, the series expansion of f_I^s in a small ball with center at a real point has real coefficients so, in that ball, we have $f_I^s g_I = f_I^s * g_I = g_I * f_I^s$. By the Identity Principle $f_I^s * g_I = f_I^s g_I = g_I * f_I^s$ on $U\cap\mathbb{C}_I$ and so, by the Extension Lemma, $f^s * g = g * f^s$. Reasoning in the same way with the function $(f^s)^{-1}$, whose restriction to $\mathbb{C}_I$ takes $(U\setminus Z_{f^s})\cap\mathbb{C}_I$ to $\mathbb{C}_I$, we get the final part of the statement. □

Definition 2.6.10. *Let $U\subseteq\mathbb{R}^{n+1}$ be an axially symmetric s-domain. Let $f: U\to\mathbb{R}_n$ be an s-monogenic function such, that for some $I\in\mathbb{S}$ its restriction f_I to the complex plane $\mathbb{C}_I$ satisfies the condition*

$$f_I * f_I^c \ \ \textit{has values in } \mathbb{C}_I.$$

We define the function:

$$f^{-*} := \mathrm{ext}((f_I^s)^{-1} f_I^c)$$

where $f_I^s = [f_I * f_I^c]_0 = f_I * f_I^c$, and we will call it s-monogenic inverse of the function f.

The next proposition shows that the function f^{-*} is the inverse of f with respect to the $*$-product:

Proposition 2.6.11. *Let $U\subseteq\mathbb{R}^{n+1}$ be an axially symmetric s-domain. Let $f: U\to\mathbb{R}_n$ be an s-monogenic function such that for some $I\in\mathbb{S}$ we have $f_I * f_I^c$ has values in $\mathbb{C}_I$. Then on $U\setminus Z_{f^s}$ we have:*

$$f^{-*} * f = f * f^{-*} = 1.$$

Proof. To prove the statement it is sufficient to show that on the plane $\mathbb{C}_I$ we have:

$$f_I * (f_I^s)^{-1} f_I^c = (f_I^s)^{-1} f_I^c * f_I = 1.$$

Using associativity and Proposition 2.6.9, we easily compute:

$$f_I * ((f_I^s)^{-1} * f_I^c) = (f_I^s)^{-1} * f_I * f_I^c = f_I^{s\,-1} * (f_I^s) = 1,$$

and

$$((f_I^s)^{-1} * f_I^c) * f_I = (f_I^s)^{-1} * f_I^c * f_I = (f_I^s)^{-1} * f_I^s = 1.$$

The result now follows from the Extension Lemma 2.2.22. □

Example 2.6.12. Consider the function $f(\mathbf{x}) = \mathbf{x}-\mathbf{s}$ defined on $\mathbb{R}^{n+1}$. As it is well known, the inverse $(\mathbf{x}-\mathbf{s})^{-1}$ is not an s-monogenic function, unless $\mathbf{s}\in\mathbb{R}$. Since the function

$$(f_I * f_I^c)(z) = (z-\mathbf{s})(z-\overline{\mathbf{s}}) = z^2 - 2\mathrm{Re}[\mathbf{s}]z + |\mathbf{s}|^2$$

has real coefficients and thus has values in $\mathbb{C}_I$, we can consider the s-monogenic inverse of f. According to Definition 2.6.10, f^{-*} is defined for $\mathbf{x}\notin[\mathbf{s}]$, and it is the function

$$f^{-*}(\mathbf{x}) = (\mathbf{x}^2 - 2\mathrm{Re}[\mathbf{s}]\mathbf{x} + |\mathbf{s}|^2)^{-1}(\mathbf{x}-\overline{\mathbf{s}}).$$

As we will see in the next section, the expression $(\mathbf{x}^2 - 2\mathrm{Re}[\mathbf{s}]\mathbf{x} + |\mathbf{s}|^2)^{-1}(\mathbf{x} - \overline{\mathbf{s}})$ cannot be simplified, unless $\mathbf{s} \in \mathbb{R}$ and in this case it coincides with $(\mathbf{x} - \mathbf{s})^{-1}$, i.e., the standard inverse of f.

Example 2.6.13. The notions of s-monogenic inverse and s-monogenic multiplication allow us to introduce s-monogenic quotients (left and right) of s-monogenic functions. Let $f, g : U \subseteq \mathbb{R}^{n+1} \to \mathbb{R}_n$ be two s-monogenic functions. On $U \setminus Z_{g^s}$ we can define the functions

$$g^{-*} * f \qquad \text{and} \qquad f * g^{-*}.$$

Let us consider the function $g^{-*} * f$ (the other case can be treated in a similar way): by definition it is the extension of

$$g_I^{-*} * f_I = (g_I^s)^{-1} g_I^c * f_I,$$

which is an $\mathbb{R}_n$-valued function satisfying

$$\overline{\partial}_I((g_I^s)^{-1} g_I^c * f_I) = 0$$

and such that $Z_{g^s} \cap \mathbb{C}_I$ consists of isolated points.

2.7 Slice monogenic Cauchy kernel

We begin this section with the following crucial definition, which is the starting point to find a Cauchy formula with s-monogenic kernel.

Definition 2.7.1. *Let* $\mathbf{x}$, $\mathbf{s} \in \mathbb{R}^{n+1}$. *We call*

$$S^{-1}(\mathbf{s}, \mathbf{x}) := \sum_{n \geq 0} \mathbf{x}^n \mathbf{s}^{-1-n}$$

the noncommutative Cauchy kernel series.

Remark 2.7.2. The noncommutative Cauchy kernel series is convergent for $|\mathbf{x}| < |\mathbf{s}|$.

Theorem 2.7.3. *Let* $\mathbf{x}$, $\mathbf{s} \in \mathbb{R}^{n+1}$ *be such that* $\mathbf{x}\mathbf{s} \neq \mathbf{s}\mathbf{x}$. *Then, the function*

$$S(\mathbf{s}, \mathbf{x}) = -(\mathbf{x} - \overline{\mathbf{s}})^{-1}(\mathbf{x}^2 - 2\mathrm{Re}[\mathbf{s}]\mathbf{x} + |\mathbf{s}|^2),$$

is the inverse of the noncommutative Cauchy kernel series.

Proof. Let us verify that

$$-(\mathbf{x} - \overline{\mathbf{s}})^{-1}(\mathbf{x}^2 - 2\mathrm{Re}[\mathbf{s}]\mathbf{x} + |\mathbf{s}|^2) \sum_{n \geq 0} \mathbf{x}^n \mathbf{s}^{-1-n} = 1.$$

We therefore obtain

$$(-|\mathbf{s}|^2 - \mathbf{x}^2 + 2\mathrm{Re}[\mathbf{s}]\mathbf{x}) \sum_{n\geq 0} \mathbf{x}^n \mathbf{s}^{-1-n} = \mathbf{s} + \mathbf{x} - 2\ \mathrm{Re}[\mathbf{s}]. \tag{2.28}$$

Observing that $-|\mathbf{s}|^2 - \mathbf{x}^2 + 2\mathrm{Re}[\mathbf{s}]\mathbf{x}$ commutes with $\mathbf{x}^n$ we can rewrite this last equation as

$$\sum_{n\geq 0} \mathbf{x}^n(-|\mathbf{s}|^2 - \mathbf{x}^2 + 2\mathrm{Re}[\mathbf{s}]\mathbf{x})\mathbf{s}^{-1-n} = \mathbf{s} + \mathbf{x} - 2\ \mathrm{Re}[\mathbf{s}].$$

Now the left-hand side can be written as

$$\begin{aligned}
&\sum_{n\geq 0} \mathbf{x}^n(-|\mathbf{s}|^2 - \mathbf{x}^2 + 2\mathrm{Re}[\mathbf{s}]\mathbf{x})s^{-1-n}\\
&= (-|\mathbf{s}|^2 - \mathbf{x}^2 + 2\mathrm{Re}[\mathbf{s}]\mathbf{x})\mathbf{s}^{-1} + \mathbf{x}^1(-|\mathbf{s}|^2 - \mathbf{x}^2 + 2\mathrm{Re}[\mathbf{s}]\mathbf{x})\mathbf{s}^{-2}\\
&\quad + \mathbf{x}^2(-|\mathbf{s}|^2 - \mathbf{x}^2 + 2\mathrm{Re}[\mathbf{s}]\mathbf{x})\mathbf{s}^{-3} + \dots\\
&= -\Big(|\mathbf{s}|^2\mathbf{s}^{-1} + \mathbf{x}(-2\mathrm{Re}[\mathbf{s}]\mathbf{s} + |\mathbf{s}|^2)\mathbf{s}^{-2} + \mathbf{x}^2(\mathbf{s}^2 - 2\mathrm{Re}[\mathbf{s}]\mathbf{s} + |\mathbf{s}|^2)\mathbf{s}^{-3}\\
&\quad + \mathbf{x}^3(\mathbf{s}^2 - 2\mathrm{Re}[\mathbf{s}]\mathbf{s} + |\mathbf{s}|^2)\mathbf{s}^{-4} + \dots\Big).
\end{aligned}$$

Using the identity (2.20)

$$\mathbf{s}^2 - 2\mathrm{Re}[\mathbf{s}]\mathbf{s} + |\mathbf{s}|^2 = 0$$

we get

$$\begin{aligned}
&\sum_{n\geq 0} \mathbf{x}^n(-|\mathbf{s}|^2 - \mathbf{x}^2 + 2\mathrm{Re}[\mathbf{s}]\mathbf{x})\mathbf{s}^{-1-n} = -|\mathbf{s}|^2\mathbf{s}^{-1} + \mathbf{x}\mathbf{s}^2\mathbf{s}^{-2}\\
&= -|\mathbf{s}|^2\mathbf{s}^{-1} + \mathbf{x} = -\bar{\mathbf{s}}\mathbf{s}\mathbf{s}^{-1} + \mathbf{x} = -\bar{\mathbf{s}} + \mathbf{x} = \mathbf{s} - 2\ \mathrm{Re}[\mathbf{s}] + \mathbf{x}
\end{aligned}$$

which equals the right-hand side of (2.28). □

When $\mathbf{x}$, $\mathbf{s}$ commute, the function $S(\mathbf{s},\mathbf{x})$ becomes

$$S(\mathbf{s},\mathbf{x}) = -(\mathbf{x}-\bar{\mathbf{s}})^{-1}(\mathbf{x}^2 - 2\mathbf{x}\mathrm{Re}[\mathbf{s}] + |\mathbf{s}|^2) = -(\mathbf{x}-\bar{\mathbf{s}})^{-1}(\mathbf{x}-\bar{\mathbf{s}})(\mathbf{x}-\mathbf{s}) = \mathbf{s}-\mathbf{x}$$

which is, trivially, the inverse of the standard sum of the Cauchy kernel series $S^{-1}(\mathbf{s},\mathbf{x}) = \sum_{n\geq 0} \mathbf{x}^n\mathbf{s}^{-1-n} = (\mathbf{s}-\mathbf{x})^{-1}$.

As a direct consequence of this observation and of the previous result, we can explicitly write the sum of the noncommutative Cauchy kernel series:

Theorem 2.7.4. *Let* $\mathbf{x}$, $\mathbf{s} \in \mathbb{R}^{n+1}$ *be such that* $\mathbf{x}\mathbf{s} \neq \mathbf{s}\mathbf{x}$. *Then*

$$\sum_{n\geq 0} \mathbf{x}^n\mathbf{s}^{-1-n} = -(\mathbf{x}^2 - 2\mathrm{Re}[\mathbf{s}]\mathbf{x} + |\mathbf{s}|^2)^{-1}(\mathbf{x}-\bar{\mathbf{s}}),$$

for $|\mathbf{x}| < |\mathbf{s}|$. *If* $\mathbf{xs} = \mathbf{sx}$, *then*

$$\sum_{n\geq 0} \mathbf{x}^n \mathbf{s}^{-1-n} = (\mathbf{s} - \mathbf{x})^{-1},$$

for $|\mathbf{x}| < |\mathbf{s}|$.

Definition 2.7.5. *We will call the expression*

$$S^{-1}(\mathbf{s}, \mathbf{x}) = -(\mathbf{x}^2 - 2\mathrm{Re}[\mathbf{s}]\mathbf{x} + |\mathbf{s}|^2)^{-1}(\mathbf{x} - \overline{\mathbf{s}}), \tag{2.29}$$

defined for $\mathbf{x}^2 - 2\mathrm{Re}[\mathbf{s}]\mathbf{x} + |\mathbf{s}|^2 \neq 0$, *the noncommutative Cauchy kernel.*

Remark 2.7.6. With an abuse of notation we have used the same symbol $S^{-1}(\mathbf{s}, \mathbf{x})$ to denote the noncommutative Cauchy kernel series and the noncommutative Cauchy kernel. This notation will not create confusion in the following since from the context it will be clear which object we are considering.

Note that the noncommutative Cauchy kernel is defined on a set which is larger than the set $\{(\mathbf{x}, \mathbf{s}) \ : \ |\mathbf{x}| < |\mathbf{s}|\}$ where the noncommutative Cauchy kernel series is convergent.

Remark 2.7.7. We now observe that the expression

$$(\mathbf{x}^2 - 2\mathrm{Re}[\mathbf{s}]\mathbf{x} + |\mathbf{s}|^2)^{-1}(\mathbf{x} - \overline{\mathbf{s}})$$

involves an inverse which does not exist if we set $\mathbf{x} = \overline{\mathbf{s}}$; indeed, in this case we have

$$\overline{\mathbf{s}}^2 - 2\mathrm{Re}[\mathbf{s}]\overline{\mathbf{s}} + |\mathbf{s}|^2 = 0.$$

One may wonder if the factor $(\mathbf{x} - \overline{\mathbf{s}})$ can be simplified. The next theorem shows that this is not possible and the function

$$(\mathbf{x}^2 - 2\mathrm{Re}[\mathbf{s}]\mathbf{x} + |\mathbf{s}|^2)^{-1}(\mathbf{x} - \overline{\mathbf{s}})$$

cannot be extended to a continuous function in $\mathbf{x} = \overline{\mathbf{s}}$.

Theorem 2.7.8. *Let* $S^{-1}(\mathbf{s}, \mathbf{x})$ *be the noncommutative Cauchy kernel and let* $\mathbf{xs} \neq \mathbf{sx}$. *Then* $S^{-1}(\mathbf{s}, \mathbf{x})$ *is irreducible and* $\lim_{\mathbf{x}\to\overline{\mathbf{s}}} S^{-1}(\mathbf{s}, \mathbf{x})$ *does not exist.*

Proof. We prove that we cannot find a degree-one polynomial $Q(\mathbf{x})$ such that

$$\mathbf{x}^2 - 2\mathrm{Re}[\mathbf{s}]\mathbf{x} + |\mathbf{s}|^2 = (\mathbf{s} + \mathbf{x} - 2\mathrm{Re}[\mathbf{s}])Q(\mathbf{x}).$$

The existence of $Q(\mathbf{x})$ would allow the simplification

$$S^{-1}(\mathbf{s}, \mathbf{x}) = Q^{-1}(\mathbf{x})(\mathbf{s} + \mathbf{x} - 2\mathrm{Re}[\mathbf{s}])^{-1}(\mathbf{s} + \mathbf{x} - 2\mathrm{Re}[\mathbf{s}]) = Q^{-1}(\mathbf{x}).$$

We proceed as follows: first of all note that $Q(\mathbf{x})$ has to be a monic polynomial of degree one, so we set

$$Q(\mathbf{x}) = \mathbf{x} - \mathbf{r}$$

where $\mathbf{r} = r_0 + \sum_{j=1}^{n} r_j e_j$. The equality

$$(\mathbf{s} + \mathbf{x} - 2\ \mathrm{Re}[\mathbf{s}])(\mathbf{x} - \mathbf{r}) = \mathbf{x}^2 - 2\mathrm{Re}[\mathbf{s}]\mathbf{x} + |\mathbf{s}|^2$$

gives

$$\mathbf{s}\mathbf{x} - \mathbf{s}\mathbf{r} - \mathbf{x}\mathbf{r} + 2\mathrm{Re}[\mathbf{s}]\mathbf{r} - |\mathbf{s}|^2 = 0.$$

Solving for $\mathbf{r}$, we get

$$\mathbf{r} = (\mathbf{s} + \mathbf{x} - 2\ \mathrm{Re}[\mathbf{s}])^{-1}(\mathbf{s}\mathbf{x} - |\mathbf{s}|^2),$$

which depends on $\mathbf{x}$. Let us now prove that the limit does not exist. Let $\mathfrak{e} = \varepsilon_0 + \sum_{j=1}^{n} \varepsilon_j e_j$, and consider

$$\begin{aligned}
S^{-1}(\mathbf{s}, \overline{\mathbf{s}} + \mathfrak{e}) &= ((\overline{\mathbf{s}} + \mathfrak{e})^2 - 2(\overline{\mathbf{s}} + \mathfrak{e})\mathrm{Re}[\mathbf{s}] + |\mathbf{s}|^2)^{-1}\mathfrak{e} \\
&= ((\overline{\mathbf{s}} + \mathfrak{e})^2 - 2(\overline{\mathbf{s}} + \mathfrak{e})\mathrm{Re}[\mathbf{s}] + |\mathbf{s}|^2)^{-1}\mathfrak{e} \\
&= (\overline{\mathbf{s}}\mathfrak{e} + \mathfrak{e}\overline{\mathbf{s}} + \mathfrak{e}^2 - 2\mathfrak{e}\mathrm{Re}[\mathbf{s}])^{-1}\mathfrak{e} \\
&= (\mathfrak{e}^{-1}(\overline{\mathbf{s}}\mathfrak{e} + \mathfrak{e}\overline{\mathbf{s}} + \mathfrak{e}^2 - 2\mathfrak{e}\mathrm{Re}[\mathbf{s}]))^{-1} \\
&= (\mathfrak{e}^{-1}\overline{\mathbf{s}}\mathfrak{e} + \overline{\mathbf{s}} + \mathfrak{e} - 2\mathrm{Re}[\mathbf{s}]))^{-1}.
\end{aligned}$$

If we now let $\mathfrak{e} \to \mathbf{0}$, we obtain that the term $\mathfrak{e}^{-1}\overline{\mathbf{s}}\mathfrak{e}$ does not have a limit because the element

$$\mathfrak{e}^{-1}\overline{\mathbf{s}}\mathfrak{e} = \frac{\overline{\mathfrak{e}}}{|\mathfrak{e}|^2}\overline{\mathbf{s}}\mathfrak{e}$$

has scalar components of the type $\dfrac{\varepsilon_i \varepsilon_j s_\ell}{|\mathfrak{e}|^2}$ with $i, j, \ell \in \{0, 1, 2, 3\}$, which do not have limit. □

Proposition 2.7.9. *The function $S^{-1}(\mathbf{s}, \mathbf{x})$ is left s-monogenic in the variable $\mathbf{x}$ and right s-monogenic in the variable $\mathbf{s}$ in its domain of definition.*

Proof. The proof follows by direct computations. Consider any $I \in \mathbb{S}$ and set $\mathbf{x} = u + Iv$. We have:

$$\begin{aligned}
&\frac{\partial}{\partial u} S^{-1}(\mathbf{s}, u + Iv) \\
&\quad = ((u + Iv)^2 - 2\mathrm{Re}[\mathbf{s}](u + Iv) + |\mathbf{s}|^2)^{-2}(2u + 2Iv - 2\mathrm{Re}[\mathbf{s}])(u + Iv - \overline{\mathbf{s}}) \\
&\qquad\qquad - ((u + Iv)^2 - 2\mathrm{Re}[\mathbf{s}](u + Iv) + |\mathbf{s}|^2)^{-1},
\end{aligned}$$

$$\begin{aligned}
&\frac{\partial}{\partial v} S^{-1}(\mathbf{s}, u + Iv) \\
&\quad = ((u + Iv)^2 - 2\mathrm{Re}[\mathbf{s}](u + Iv) + |\mathbf{s}|^2)^{-2}(2uI - 2v - 2\mathrm{Re}[\mathbf{s}]I)(u + Iv - \overline{\mathbf{s}}) \\
&\qquad\qquad - ((u + Iv)^2 - 2\mathrm{Re}[\mathbf{s}](u + Iv) + |\mathbf{s}|^2)^{-1}I,
\end{aligned}$$

so we obtain:

$$\begin{aligned}
&\frac{\partial}{\partial u}S^{-1}(\mathbf{s},u+Iv)+I\frac{\partial}{\partial v}S^{-1}(\mathbf{s},u+Iv)\\
&=((u+Iv)^2-2\mathrm{Re}[\mathbf{s}](u+Iv)+|\mathbf{s}|^2)^{-2}(2u+2Iv-2\mathrm{Re}[\mathbf{s}])(u+Iv-\bar{\mathbf{s}})\\
&\quad-((u+Iv)^2-2\mathrm{Re}[\mathbf{s}](u+Iv)+|\mathbf{s}|^2)^{-1}\\
&\quad+((u+Iv)^2-2\mathrm{Re}[\mathbf{s}](u+Iv)+|\mathbf{s}|^2)^{-2}(-2u-2vI+2\mathrm{Re}[\mathbf{s}])(u+Iv-\bar{\mathbf{s}})\\
&\quad+((u+Iv)^2-2\mathrm{Re}[\mathbf{s}](u+Iv)+|\mathbf{s}|^2)^{-1}=0.
\end{aligned}$$

Let us now set $\mathbf{s}=u+Iv$. Then $S^{-1}(u+Iv,\mathbf{x})=F(u,v,\mathbf{x})(\mathbf{x}-u+Iv)$ where $F(u,v,\mathbf{x})$ is a function involving $\mathbf{x}$, the real variables u,v but not the imaginary unit I. Then we have:

$$\begin{aligned}
\frac{\partial}{\partial u}S^{-1}(u+Iv,\mathbf{x})=(\mathbf{x}^2-2\mathbf{x}u+u^2+v^2)^{-2}(-2\mathbf{x}+2u)(\mathbf{x}-u+Iv)&\\
+(\mathbf{x}^2-2\mathbf{x}u+u^2+v^2)^{-1}&,
\end{aligned}$$

$$\frac{\partial}{\partial v}S^{-1}(u+Iv,\mathbf{x})=(\mathbf{x}^2-2\mathbf{x}u+u^2+v^2)^{-2}2v(\mathbf{x}-u+Iv)-(\mathbf{x}^2-2\mathbf{x}u+u^2+v^2)^{-1}I.$$

It follows that

$$\begin{aligned}
&\frac{\partial}{\partial u}S^{-1}(u+Iv,\mathbf{x})+\frac{\partial}{\partial y}S^{-1}(u+Iv,\mathbf{x})I\\
&=(\mathbf{x}^2-2\mathbf{x}u+u^2+v^2)^{-2}(-2\mathbf{x}+2u)(\mathbf{x}-u+Iv)-(\mathbf{x}^2-2\mathbf{x}u+u^2+v^2)^{-1}\\
&\quad+(\mathbf{x}^2-2\mathbf{x}u+u^2+v^2)^{-2}2v(\mathbf{x}-u+Iv)I-(\mathbf{x}^2-2\mathbf{x}u+u^2+v^2)^{-1}\\
&=2(\mathbf{x}^2-2\mathbf{x}u+u^2+v^2)^{-2}(\mathbf{x}^2-2\mathbf{x}u+u^2+v^2)-2(\mathbf{x}^2-2\mathbf{x}u+u^2+v^2)^{-1}\\
&=0.
\end{aligned}$$

□

This result is obviously trivial when $S^{-1}(\mathbf{s},\mathbf{x})$ coincides with the Cauchy kernel series. However, as we have pointed out after Definition 2.29, the function $S^{-1}(\mathbf{s},\mathbf{x})$ is defined on a set which is larger than the domain of convergence of the series and therefore the direct argument in the preceding proof is necessary.

We now state some equalities which are important to prove further properties of the Cauchy kernel function.

Proposition 2.7.10. *Let* $\mathbf{x}$, $\mathbf{s}\in\mathbb{R}^{n+1}$ *be such that* $\mathbf{x}\neq\bar{\mathbf{s}}$. *Then the following identity holds:*

$$(\mathbf{x}-\bar{\mathbf{s}})^{-1}\mathbf{s}(\mathbf{x}-\bar{\mathbf{s}})-\mathbf{x}=-(\mathbf{s}-\bar{\mathbf{x}})\mathbf{x}(\mathbf{s}-\bar{\mathbf{x}})^{-1}+\mathbf{s},$$

or, equivalently,

$$-(\mathbf{x}-\bar{\mathbf{s}})^{-1}(\mathbf{x}^2-2\mathbf{x}\mathrm{Re}[\mathbf{s}]+|\mathbf{s}|^2)=(\mathbf{s}^2-2\mathrm{Re}[\mathbf{x}]\mathbf{s}+|\mathbf{x}|^2)(\mathbf{s}-\bar{\mathbf{x}})^{-1};\tag{2.30}$$

finally, if $\mathbf{x}\notin[\mathbf{s}]$ *we have*

$$-(\mathbf{x}^2-2\mathrm{Re}[\mathbf{s}]\mathbf{x}+|\mathbf{s}|^2)^{-1}(\mathbf{x}-\bar{\mathbf{s}})=(\mathbf{s}-\bar{\mathbf{x}})(\mathbf{s}^2-2\mathrm{Re}[\mathbf{x}]\mathbf{s}+|\mathbf{x}|^2)^{-1}.\tag{2.31}$$

Proof. One may prove the identities by direct computations. Let us prove (2.31). To show that the formula is an identity, we multiply by $(\mathbf{x}^2 - 2\mathrm{Re}[\mathbf{s}]\mathbf{x} + |\mathbf{s}|^2)$ on the left and by $(\mathbf{s}^2 - \mathrm{Re}[\mathbf{x}]\mathbf{s} + |\mathbf{x}|^2)$ on the right. We obtain:

$$\mathbf{x}^2\mathbf{s} - 2\mathrm{Re}[\mathbf{s}]\mathbf{x}\mathbf{s} + 2\mathrm{Re}[\mathbf{s}]|\mathbf{x}|^2 - \overline{\mathbf{x}}|\mathbf{s}|^2 = -\mathbf{x}\mathbf{s}^2 + 2\mathrm{Re}[\mathbf{x}]\mathbf{x}\mathbf{s} - 2\mathrm{Re}[\mathbf{x}]|s|^2 + \overline{\mathbf{s}}|\mathbf{x}|^2$$

which becomes

$$(\mathbf{x}^2 - \mathrm{Re}[\mathbf{x}] + |\mathbf{x}|^2)\mathbf{s} = -\mathbf{x}(\mathbf{s}^2 - \mathrm{Re}[\mathbf{s}] + |\mathbf{s}|^2)$$

that is an identity by (2.21). Note that (2.31) holds for $\mathbf{x} \notin [\mathbf{s}]$, which is equivalent to $\mathbf{s} \notin [\mathbf{x}]$. The identity (2.30) can be proven by taking the inverse of (2.31) and it holds for $\mathbf{x} \neq \overline{\mathbf{s}}$. Easy computations show the validity of the remaining identity. □

We now consider the function $S^{-1}(\mathbf{s}, \mathbf{x}) = S^{-1}_{\mathbf{s}}(\mathbf{x})$ as a function of $\mathbf{x}$. Clearly, its singularities are the entire $(n-1)$-sphere $[\mathbf{s}]$ which reduces to the point $\{\mathbf{s}\}$ when $\mathbf{s}$ is real. The next result analyzes in detail the singularities of $S^{-1}_{\mathbf{s}}(\mathbf{x})$ on each plane $\mathbb{C}_I$ when $\mathbf{s} \notin \mathbb{R}$.

Proposition 2.7.11. *Let* $\mathbf{s} \in \mathbb{R}^{n+1} \backslash \mathbb{R}$. *If* $I \neq I_{\mathbf{s}}$, *then the function* $S^{-1}(\mathbf{s}, \mathbf{x}) = S^{-1}_{\mathbf{s}}(\mathbf{x})$ *has two singularities* $\mathrm{Re}[\mathbf{s}] \pm I|\underline{s}|$ *on the plane* $\mathbb{C}_I$. *On the plane* $\mathbb{C}_{I_{\mathbf{s}}}$, *the restriction of* $S^{-1}_{\mathbf{s}}(\mathbf{x})$, *i.e.,* $(\mathbf{x} - \mathbf{s})^{-1}$, *has only one singularity at the point* $\mathbf{s}$.

Proof. Suppose $\mathbf{s} \in \mathbb{R}^{n+1} \backslash \mathbb{R}$ and consider $S^{-1}_{\mathbf{s}}(\mathbf{x}) = (\mathbf{s}^2 - 2\mathrm{Re}[\mathbf{x}]\mathbf{s} + |\mathbf{x}|^2)^{-1}(\mathbf{s} - \overline{\mathbf{x}})$. The singularities of $S^{-1}_{\mathbf{s}}(\mathbf{x})$ corresponds to the roots of $\mathbf{s}^2 - 2\mathrm{Re}[\mathbf{x}]\mathbf{s} + |\mathbf{x}|^2 = 0$. This equation can be written by splitting real and imaginary parts as

$$\begin{aligned} &\mathrm{Re}[\mathbf{s}]^2 - |\underline{s}|^2 - 2\mathrm{Re}[\mathbf{s}]\mathrm{Re}[\mathbf{x}] + |\mathbf{x}|^2 = 0, \\ &(\mathrm{Re}[\mathbf{s}] - \mathrm{Re}[\mathbf{x}])\underline{s} = 0. \end{aligned}$$

The assumption $\underline{s} \neq 0$ implies $\mathrm{Re}[\mathbf{x}] = \mathrm{Re}[\mathbf{s}]$ and so $|\mathbf{x}| = |\mathbf{s}|$, i.e., the roots correspond to the $(n-1)$-sphere $[\mathbf{s}]$. Consider now the plane $\mathbb{C}_I$. When $I \neq I_{\mathbf{s}}$, $\mathbb{C}_I$ intersect the $(n-1)$-sphere $[\mathbf{s}]$ in $\mathrm{Re}[\mathbf{s}] \pm I|\underline{s}|$ while, when $I = I_{\mathbf{s}}$, $\mathbf{x}$ and $\mathbf{s}$ commute, so

$$S^{-1}_{\mathbf{s}}(\mathbf{x}) = -(\mathbf{x} - \mathbf{s})^{-1}(\mathbf{x} - \overline{\mathbf{s}})^{-1}(\mathbf{x} - \overline{\mathbf{s}}) = -(\mathbf{x} - \mathbf{s})^{-1}$$

and $\mathbf{x}$ is the only singularity of the restriction of $S^{-1}_{\mathbf{s}}(\mathbf{x})$ to the plane $\mathbb{C}_{I_{\mathbf{s}}}$. □

Remark 2.7.12. The previous proposition states that the restriction of $S^{-1}(\mathbf{s}, \mathbf{x})$ to the plane $\mathbb{C}_{I_{\mathbf{s}}}$, has a removable singularity at the point $\mathbf{x} = \overline{\mathbf{s}}$. However, equality (2.31) and the proof of Theorem 2.7.8 show that the function $S^{-1}(\mathbf{s}, \mathbf{x})$ still has a singularity at the point $\mathbf{x} = \overline{\mathbf{s}}$.

The kernel $S^{-1}(\mathbf{s}, \mathbf{x})$ is a left s-monogenic function in $\mathbf{x}$ and a right s-monogenic function in $\mathbf{s}$ so, in principle, it cannot be used in both the Cauchy formulas for left and for right s-monogenic functions. Thus one has to establish which kernel has to be used for a Cauchy formula for right s-monogenic functions. Note that the series expansion of a kernel which is right (resp. left) s-monogenic in the variable $\mathbf{x}$ (resp. $\mathbf{s}$) is of the following form

Definition 2.7.13. *Let* $\mathbf{x}$, $\mathbf{s} \in \mathbb{R}^{n+1}$. *We call*

$$S_R^{-1}(\mathbf{s},\mathbf{x}) := \sum_{n\geq 0} \mathbf{s}^{-n-1}\mathbf{x}^n, \tag{2.32}$$

the right noncommutative Cauchy kernel series.

Remark 2.7.14. *The right noncommutative Cauchy kernel series is convergent for* $|\mathbf{x}| < |\mathbf{s}|$.

We have the following:

Proposition 2.7.15. *The sum of the series* (2.32) *is given by the function*

$$S_R^{-1}(\mathbf{s},\mathbf{x}) = -(\mathbf{x}-\bar{\mathbf{s}})(\mathbf{x}^2 - 2\mathrm{Re}[\mathbf{s}]\mathbf{x} + |\mathbf{s}|^2)^{-1}, \tag{2.33}$$

which is defined for $\mathbf{x} \notin [\mathbf{s}]$. *Moreover,* $S_R^{-1}(\mathbf{s},\mathbf{x})$ *is right (resp. left) s-monogenic in the variable* $\mathbf{x}$ *(resp.* $\mathbf{s}$*).*

Proof. It follows the same lines of the proof of Theorem 2.7.4. We just sketch some of the computations. The statement is proved if we show that, for $|\mathbf{x}| < |\mathbf{s}|$, we have

$$\Big(\sum_{n\geq 0} \mathbf{s}^{-n-1}\mathbf{x}^n\Big)(\mathbf{x}^2 - 2\mathrm{Re}[\mathbf{s}]\mathbf{x} + |\mathbf{s}|^2) = -(\mathbf{x}-\bar{\mathbf{s}}). \tag{2.34}$$

By computing the product at the left-hand side of (2.34), we obtain:

$$\begin{aligned}
&\mathbf{s}^{-1}\mathbf{x}^2 - 2\mathbf{s}^{-1}\mathrm{Re}[\mathbf{s}]\mathbf{x} + \mathbf{s}^{-1}|\mathbf{s}|^2 + \mathbf{s}^{-2}\mathbf{x}^3 - 2\mathbf{s}^{-2}\mathrm{Re}[\mathbf{s}]\mathbf{x}^2 + \mathbf{s}^{-2}\mathbf{x}|\mathbf{s}|^2 + \dots\\
&= -2\mathbf{s}^{-1}\mathrm{Re}[\mathbf{s}]\mathbf{x} + \mathbf{s}^{-1}|\mathbf{s}|^2 + \mathbf{s}^{-2}\mathbf{x}|\mathbf{s}|^2 + \sum_{n\geq 2} \mathbf{s}^{-(n+1)}(\mathbf{s}^2 - 2\mathrm{Re}[\mathbf{s}]\mathbf{s} + |\mathbf{s}|^2)\mathbf{x}^n\\
&= -2\mathbf{s}^{-1}\mathrm{Re}[\mathbf{s}]\mathbf{x} + \mathbf{s}^{-1}|\mathbf{s}|^2 + \mathbf{s}^{-2}\mathbf{x}|\mathbf{s}|^2\\
&= \mathbf{s}^{-2}(-2\mathrm{Re}[\mathbf{s}] + \bar{\mathbf{s}})\mathbf{s}\mathbf{x} + \mathbf{s}^{-1}\mathbf{s}\bar{\mathbf{s}} = -\mathbf{x} + \bar{\mathbf{s}}.
\end{aligned}$$

The fact that function $S_R^{-1}(\mathbf{s},\mathbf{x})$, which is defined for $\mathbf{x} \notin [\mathbf{s}]$, is left s-monogenic in the variable $\mathbf{s}$ and right s-monogenic in the variable $\mathbf{x}$ can be proved by a direct computation. This concludes the proof. □

Definition 2.7.16. *We will call the expression*

$$S_R^{-1}(\mathbf{s},\mathbf{x}) = -(\mathbf{x}-\bar{\mathbf{s}})(\mathbf{x}^2 - 2\mathrm{Re}[\mathbf{s}]\mathbf{x} + |\mathbf{s}|^2)^{-1}, \tag{2.35}$$

defined for $\mathbf{x}^2 - 2\mathbf{x}\mathrm{Re}[\mathbf{s}] + |\mathbf{s}|^2 \neq 0$, *the right noncommutative Cauchy kernel.*

Remark 2.7.17. Analogous considerations as in Remarks 2.7.6 and 2.7.7 and in Theorem 2.7.8 can be done for the right noncommutative Cauchy kernel $S_R^{-1}(\mathbf{s},\mathbf{x})$.

Proposition 2.7.18. *Suppose that* $\mathbf{x}$ *and* $\mathbf{s} \in \mathbb{R}^{n+1}$ *are such that* $\mathbf{x} \notin [\mathbf{s}]$. *The following identity holds:*

$$S_R^{-1}(\mathbf{s},\mathbf{x}) = (\mathbf{s}^2 - 2\mathrm{Re}[\mathbf{x}]\mathbf{s} + |\mathbf{x}|^2)^{-1}(\mathbf{s}-\bar{\mathbf{x}}) = -(\mathbf{x}-\bar{\mathbf{s}})(\mathbf{x}^2 - 2\mathrm{Re}[\mathbf{s}]\mathbf{x} + |\mathbf{s}|^2)^{-1}. \tag{2.36}$$

Proof. One may prove the identity by direct computations (compare with the proof of Proposition 2.7.10). □

Remark 2.7.19. The identities (2.31) and (2.36) can be proved not only by direct computation but also in a longer way which can be of some interest. We sketch the lines of this alternative proof. Consider the function $f(\mathbf{x}) = \mathbf{s} - \mathbf{x}$. It is such that $f_I * f_I^c$ has values in $\mathbb{C}_I$ thus it admits an s-monogenic inverse (see Example 2.6.12). One may construct its s-monogenic inverse with respect to the two variables $\mathbf{x}$ and $\mathbf{s}$ on the left and on the right. If one constructs, e.g., the left inverse with respect to $\mathbf{x}$, see Definition 2.6.10, one gets

$$(\mathbf{x}^2 - 2\mathrm{Re}[\mathbf{s}]\mathbf{x} + |\mathbf{s}|^2)^{-1}(\bar{\mathbf{s}} - \mathbf{x}).$$

By direct computation it follows that this function is right s-monogenic with respect to $\mathbf{s}$, thus it must coincide, by the Identity Principle, with the right s-monogenic inverse of $(\mathbf{s} - \mathbf{x})$ with respect to $\mathbf{s}$, i.e.,

$$(\mathbf{s} - \bar{\mathbf{x}})(\mathbf{s}^2 - 2\mathrm{Re}[\mathbf{x}]\mathbf{s} + |\mathbf{x}|^2)^{-1}$$

thus relation (2.31) holds. Note that we have not provided the construction of the right s-monogenic inverse of a function f, but it is not difficult to check that, when it exists, it coincides with the extension of the function $f_I^c(f_I * f_I^c)^{-1}$. Similarly, one can construct the left s-monogenic inverse of $\mathbf{s} - \mathbf{x}$ with respect to $\mathbf{s}$, then one shows that it is right s-monogenic with respect to $\mathbf{x}$ and so it follows that it must coincide with the right s-monogenic inverse with respect to $\mathbf{x}$, thus equality (2.36) holds.

By comparing the Cauchy kernel functions $S^{-1}(\mathbf{s}, \mathbf{x})$ and $S_R^{-1}(\mathbf{s}, \mathbf{x})$, we conclude that the two functions are different, thus the kernel to be used for the Cauchy formula for right s-monogenic functions is not the kernel $S^{-1}(\mathbf{s}, \mathbf{x})$ used for left s-monogenic functions. However we have the following relation.

Proposition 2.7.20. *Let $x, s \in \mathbb{R}^{n+1}$. The following identity holds:*

$$S^{-1}(\mathbf{x}, \mathbf{s}) = -S_R^{-1}(\mathbf{s}, \mathbf{x}), \quad \textit{for} \quad \mathbf{x} \notin [\mathbf{s}].$$

Proof. The identities (2.31) and (2.36) show that by exchanging the role of the variables $\mathbf{x}$ and $\mathbf{s}$ we get $S^{-1}(\mathbf{x}, \mathbf{s}) = -S_R^{-1}(\mathbf{s}, \mathbf{x})$. □

2.8 Cauchy integral formula, II

In this section we prove a Cauchy formula for an s-monogenic function with s-monogenic kernel which is more general than the one proved in Section 2.4. In fact, the formula does not depend on the plane in which the integration path is chosen.

Let us recall the well-known Stokes' theorem in the complex plane (see for example [2]).

Theorem 2.8.1. *Let C be a bounded open set in $\mathbb{C}$ such that its boundary ∂C is a finite union of continuously differentiable Jordan curves. If $f \in \mathcal{C}^1(\overline{C})$, then*

$$\int_{\partial C} f dz = \int_C df \wedge dz = 2i \int_C \frac{\partial f}{\partial \bar{z}} dx \wedge dy.$$

When considering $\mathbb{R}_n$-valued functions, the Stokes' theorem can be rephrased as follows:

Lemma 2.8.2. *Let D_I be a bounded open set on a plane $\mathbb{C}_I$ such that its boundary ∂D_I is a finite union of continuously differentiable Jordan curves. Let $f, g \in \mathcal{C}^1(\overline{D}_I)$ be $\mathbb{R}_n$-valued functions. Then*

$$\int_{\partial D_I} g(\mathbf{s}) d\mathbf{s}_I f(\mathbf{s}) = 2 \int_{D_I} ((g(\mathbf{s})\bar{\partial}_I) f(\mathbf{s}) + g(\mathbf{s})(\bar{\partial}_I f(\mathbf{s}))) d\sigma$$

where $\mathbf{s} = u + Iv$ is the variable on $\mathbb{C}_I$, $d\mathbf{s}_I = -Id\mathbf{s}$, $d\sigma = du \wedge dv$.

Proof. Let us choose $n-1$ imaginary units $I_2, \dots, I_n$ such that $I, I_2, \dots, I_n$ form an orthonormal basis of $\mathbb{R}_n$ satisfying the defining relations $I_r I_s + I_s I_r = -2\delta_{rs}$. Then it is possible to write

$$f(\mathbf{s}) = \sum_{|A|=0}^{n-1} F_A(\mathbf{s}) I_A,$$

$$g(\mathbf{s}) = \sum_{|A|=0}^{n-1} I_A G_A(\mathbf{s}),$$

where $\mathbf{s} \in \mathbb{C}_I$, $I_A = I_{i_1} \dots I_{i_s}$, $A = i_1 \dots i_s$ is a subset of $\{2, \dots, n\}$ and $F_A(\mathbf{s})$, $G_A(\mathbf{s})$ have values in the complex plane $\mathbb{C}_I$. We have

$$\begin{aligned}\int_{\partial D_I} g(\mathbf{s}) d\mathbf{s}_I f(\mathbf{s}) &= \int_{\partial D_I} \Big(\sum_{|A|=0}^{n-1} I_A G_A(\mathbf{s}) \Big) d\mathbf{s}_I \Big(\sum_{|B|=0}^{n-1} F_B(\mathbf{s}) I_B \Big) \\ &= \sum_{|A|=0,|B|=0}^{n-1} I_A \Big(\int_{\partial D_I} G_A(\mathbf{s}) d\mathbf{s}_I F_B(\mathbf{s}) \Big) I_B.\end{aligned}$$

We now use the usual Stokes' theorem in the complex plane $\mathbb{C}_I$ and we write

$$\int_{\partial D_I} g(\mathbf{s}) d\mathbf{s}_I f(\mathbf{s}) = \sum_{|A|=0,|B|=0}^{n-1} I_A \Big(\int_{D_I} \frac{\partial}{\partial \bar{\mathbf{s}}} (G_A(\mathbf{s}) F_B(\mathbf{s})) d\bar{\mathbf{s}} \wedge d\mathbf{s}_I \Big) I_B$$

$$= 2 \sum_{|A|=0,|B|=0}^{n-1} I_A \Big(\int_{D_I} (\partial_u + I\partial_v)(G_A(\mathbf{s}) F_B(\mathbf{s})) d\sigma \Big) I_B;$$

we recall that I commutes with F_A and G_B which have values in $\mathbb{C}_I$, and that $d\sigma$ is real, thus we obtain

$$\begin{aligned}\int_{\partial D_I} g(\mathbf{s})d\mathbf{s}_I f(\mathbf{s}) &= 2\sum_{|A|,|B|=0}^{n-1}\Big(\int_{D_I} I_A(\partial_u(G_A)+\partial_v(G_A)I)F_B I_B d\sigma \\ &\quad + \int_{D_I} I_A G_A(\partial_u F_B + I\partial_v F_B)I_B d\sigma\Big) \\ &= 2\int_{D_I}\sum_{|A|,|B|=0}^{n-1} I_A(G_A\overline{\partial}_I)F_B I_B d\sigma \\ &\quad + 2\int_{D_I}\sum_{|A|,|B|=0}^{n-1} I_A G_A(\overline{\partial}_I F_B)I_B d\sigma \\ &= 2\int_{D_I}((g(\mathbf{s})\bar{\partial}_I)f(\mathbf{s}) + g(\mathbf{s})(\bar{\partial}_I f(\mathbf{s})))d\sigma\end{aligned}$$

and we get the statement. □

An immediate consequence of the above lemma is the following:

Corollary 2.8.3. *Let f and g be left s-monogenic and right s-monogenic functions, respectively, defined on an open set U. For any $I \in \mathbb{S}$ and any open bounded set D_I in $U \cap \mathbb{C}_I$ whose boundary is a finite union of continuously differentiable Jordan curves, we have*

$$\int_{\partial D_I} g(\mathbf{s})d\mathbf{s}_I f(\mathbf{s}) = 0.$$

Theorem 2.8.4 (The Cauchy formula with s-monogenic kernel). *Let $U \subset \mathbb{R}^{n+1}$ be an axially symmetric s-domain. Suppose that $\partial(U \cap \mathbb{C}_I)$ is a finite union of continuously differentiable Jordan curves for every $I \in \mathbb{S}$. Set $d\mathbf{s}_I = -d\mathbf{s}I$ for $I \in \mathbb{S}$. If f is a (left) s-monogenic function on a set that contains $\overline{U}$, then*

$$f(\mathbf{x}) = \frac{1}{2\pi}\int_{\partial(U\cap\mathbb{C}_I)} S^{-1}(\mathbf{s},\mathbf{x})d\mathbf{s}_I f(\mathbf{s}) \tag{2.37}$$

where $S^{-1}(\mathbf{s},\mathbf{x})$ is defined in (2.29) *and the integral does not depend on U and on the imaginary unit $I \in \mathbb{S}$.*

If f is a right s-monogenic function on a set that contains $\overline{U}$, then

$$f(\mathbf{x}) = \frac{1}{2\pi}\int_{\partial(U\cap\mathbb{C}_I)} f(\mathbf{s})d\mathbf{s}_I S_R^{-1}(\mathbf{s},\mathbf{x}) \tag{2.38}$$

$$= -\frac{1}{2\pi}\int_{\partial(U\cap\mathbb{C}_I)} f(\mathbf{s})d\mathbf{s}_I S^{-1}(\mathbf{x},\mathbf{s}) \tag{2.39}$$

where $S_R^{-1}(\mathbf{s},\mathbf{x})$ is defined in (2.35) *and the integral* (2.38) *does not depend on the choice of the imaginary unit $I \in \mathbb{S}$ and on U.*

Proof. First of all, the integral at the right-hand side of (2.37) does not depend on the open set U: this follows from the fact that $S^{-1}(\mathbf{s},\mathbf{x})$ is right s-monogenic in s, and Corollary 2.8.3. Let us show that the integral (2.37) does not depend on the choice of the imaginary unit $I \in \mathbb{S}$. The zeros of the function $\mathbf{x}^2 - 2s_0\mathbf{x} + |\mathbf{s}|^2 = 0$ consist either of a real point $\mathbf{x}$ or a 2-sphere $[\mathbf{x}]$. On $\mathbb{C}_{I_\mathbf{x}}$ we find only the point $\mathbf{x}$ as a singularity and the result follows from the Cauchy formula on the plane $\mathbb{C}_{I_\mathbf{x}}$. When the singularity is a real number, the integral reduces again to the Cauchy integral of complex analysis. If the zero is not real, on any complex plane $\mathbb{C}_I$ we find the two zeros $s_{1,2} = x_0 \pm I|\underline{x}|$. In this case, we calculate the residues in the points s_1 and s_2 on the plane $\mathbb{C}_I$ for $I \neq I_\mathbf{x}$. Let us start with s_1 by setting the positions

$$\begin{aligned} \mathbf{s} &= x_0 + I|\underline{x}| + \varepsilon e^{I\theta}, \\ s_0 &= x_0 + \varepsilon\cos\theta, \\ \bar{\mathbf{s}} &= x_0 - I|\underline{x}| + \varepsilon e^{-I\theta}, \\ d\mathbf{s}_I &= -[\varepsilon I e^{I\theta}]Id\theta = \varepsilon e^{I\theta}d\theta, \end{aligned}$$

and

$$|\mathbf{s}|^2 = x_0^2 + 2x_0\varepsilon\cos\theta + \varepsilon^2 + |\underline{x}|^2 + 2\varepsilon\sin\theta|\underline{x}|.$$

We have

$$\begin{aligned} 2\pi I_1^\varepsilon = \int_0^{2\pi} &-(-2\mathbf{x}\varepsilon\cos\theta + 2x_0\varepsilon\cos\theta + \varepsilon^2 + 2\varepsilon\sin\theta|\underline{x}|)^{-1} \\ &\cdot(\mathbf{x} - [x_0 - I|\underline{x}| + \varepsilon e^{-I\theta}])\varepsilon e^{I\theta}d\theta f(x_0 + I|\underline{x}| + \varepsilon e^{I\theta}), \end{aligned}$$

and for $\varepsilon \to 0$ we get

$$\begin{aligned} &2\pi I_1^0 \\ &= \int_0^{2\pi} (2\mathbf{x}\cos\theta - 2x_0\cos\theta - 2\sin\theta|\underline{x}|)^{-1}(\underline{x} + I|\underline{x}|)e^{I\theta}d\theta f(x_0 + I|\underline{x}|) \\ &= \frac{1}{2}\int_0^{2\pi} (\underline{x}\cos\theta - \sin\theta|\underline{x}|)^{-1}(\underline{x} + I|\underline{x}|)e^{I\theta}d\theta f(x_0 + I|\underline{x}|) \\ &= -\frac{1}{2|\underline{x}|^2}\int_0^{2\pi} (\underline{x}\cos\theta + \sin\theta|\underline{x}|)(\underline{x} + I|\underline{x}|)[\cos\theta + I\sin\theta]d\theta f(x_0 + I|\underline{x}|) \\ &= -\frac{1}{2|\underline{x}|^2}\int_0^{2\pi} [(\underline{x})^2\cos\theta + \sin\theta|\underline{x}|\underline{x} + \underline{x}I|\underline{x}|\cos\theta \\ &\quad + \sin\theta|\underline{x}|^2 I][\cos\theta + I\sin\theta]d\theta f(x_0 + I|\underline{x}|). \end{aligned}$$

With some calculations we obtain

$$\begin{aligned} 2\pi I_1^0 &= -\frac{1}{2|\underline{x}|^2}\int_0^{2\pi} \Big[(\underline{x})^2 + \underline{x}I|\underline{x}|\cos^2\theta + \sin^2\theta|\underline{x}|\underline{x}I\Big]d\theta f(x_0 + I|\underline{x}|) \\ &= -\frac{1}{2|\underline{x}|^2}\Big[2\pi(\underline{x})^2 + \pi\underline{x}I|\underline{x}| + \pi|\underline{x}|\underline{x}I\Big]f(x_0 + I|\underline{x}|) \end{aligned}$$

$$= \frac{\pi}{|\underline{x}|}\Big[|\underline{x}| - \underline{x}I\Big] f(x_0 + I|\underline{x}|).$$

Recalling that $\underline{x}/|\underline{x}| = I_{\mathbf{x}}$ we get the first residue

$$I_1^0 = \frac{1}{2}\Big[1 - I_{\mathbf{x}}I\Big] f(x_0 + I|\underline{x}|).$$

With analogous calculations we prove that the residue in s_2 is

$$I_2^0 = \frac{1}{2}\Big[1 + I_{\mathbf{x}}I\Big] f(x_0 - I|\underline{x}|).$$

So by the residues theorem we get

$$\frac{1}{2\pi}\int_{\partial(U\cap\mathbb{C}_I)} S^{-1}(\mathbf{s},\mathbf{x})d\mathbf{s}_I f(\mathbf{s}) = I_1^0 + I_2^0.$$

The statement now follows from the Representation Formula. Formula (2.38) can be deduced with similar arguments while formula (2.39) is a consequence of Proposition 2.7.20. □

We conclude this section with the formula for the derivatives of an s-monogenic function using the s-monogenic Cauchy kernel.

Theorem 2.8.5 (Derivatives using the s-monogenic Cauchy kernel). *Let $U \subset \mathbb{R}^{n+1}$ be an axially symmetric s-domain. Suppose that $\partial(U\cap\mathbb{C}_I)$ is a finite union of continuously differentiable Jordan curves for every $I\in\mathbb{S}$. Set $d\mathbf{s}_I = -d\mathbf{s}I$ for $I\in\mathbb{S}$. Let f be an s-monogenic function on an open set that contains $\overline{U}$ and set $\mathbf{x} = x_0 + \underline{x}$, $\mathbf{s} = s_0 + \underline{s}$. Then*

$$\begin{aligned}\partial_{x_0}^n f(\mathbf{x}) &= \frac{n!}{2\pi}\int_{\partial(U\cap\mathbb{C}_I)} (\mathbf{x}^2 - 2s_0\mathbf{x} + |\mathbf{s}|^2)^{-n-1}(\mathbf{x}-\overline{\mathbf{s}})^{*(n+1)} d\mathbf{s}_I f(\mathbf{s})\\ &= \frac{n!}{2\pi}\int_{\partial(U\cap\mathbb{C}_I)} [S^{-1}(\mathbf{s},\mathbf{x})(\mathbf{x}-\overline{\mathbf{s}})^{-1}]^{n+1}(\mathbf{x}-\overline{\mathbf{s}})^{*(n+1)} d\mathbf{s}_I f(\mathbf{s}) \qquad (2.40)\end{aligned}$$

where

$$(\mathbf{x}-\overline{\mathbf{s}})^{*n} = \sum_{k=0}^{n} \frac{n!}{(n-k)!k!}\mathbf{x}^{n-k}\overline{\mathbf{s}}^k, \qquad (2.41)$$

and $S^{-1}(\mathbf{s},\mathbf{x})$ is defined in (2.29). Moreover, the integral does not depend on U and on the imaginary unit $I\in\mathbb{S}$.

Proof. First of all, we recall that the s-derivative defined in (2.4) coincides, for s-monogenic functions, with the partial derivative with respect to the scalar coordinate x_0. To compute $\partial_{x_0}^n f(\mathbf{x})$, we can compute the derivative of the integrand, since f and its derivatives with respect to x_0 are continuous functions on $\partial(U\cap\mathbb{C}_I)$. Thus we get

$$\partial_{x_0}^n f(\mathbf{x}) = \frac{1}{2\pi}\int_{\partial(U\cap\mathbb{C}_I)} \partial_{x_0}^n [S^{-1}(\mathbf{s},\mathbf{x})] d\mathbf{s}_I f(\mathbf{s}).$$

To prove the statement, it is sufficient to compute $\partial^n_{x_0}[S^{-1}(\mathbf{s},\mathbf{x})]$ by recurrence. Consider the derivative of $\partial_{x_0}S^{-1}(\mathbf{s},\mathbf{x})$:

$$\begin{aligned}
&\partial_{x_0}S^{-1}(\mathbf{s},\mathbf{x})\\
&= -(\mathbf{x}^2-2s_0\mathbf{x}+|\mathbf{s}|^2)^{-2}(2\mathbf{x}-2s_0)(\mathbf{x}-\bar{\mathbf{s}})-(\mathbf{x}^2-2s_0\mathbf{x}+|\mathbf{s}|^2)^{-1}\\
&= (\mathbf{x}^2-2s_0\mathbf{x}+|\mathbf{s}|^2)^{-2}[2\mathbf{x}^2-2\mathbf{x}\bar{\mathbf{s}}-2s_0\mathbf{x}+2s_0\bar{\mathbf{s}}-\mathbf{x}^2+2s_0\mathbf{x}-|\mathbf{s}|^2]\\
&= (\mathbf{x}^2-2s_0\mathbf{x}+|\mathbf{s}|^2)^{-2}[\mathbf{x}^2-2\mathbf{x}\bar{\mathbf{s}}+\bar{\mathbf{s}}^2]=(\mathbf{x}^2-2s_0\mathbf{x}+|\mathbf{s}|^2)^{-2}(\mathbf{x}-\bar{\mathbf{s}})^{*2}.
\end{aligned}$$

We now assume

$$\partial^n_{x_0}S^{-1}(\mathbf{s},\mathbf{x})=(-1)^{n+1}n!(\mathbf{x}^2-2s_0\mathbf{x}+|\mathbf{s}|^2)^{-(n+1)}(\mathbf{x}-\bar{\mathbf{s}})^{*(n+1)},$$

and we compute $\partial^{n+1}_{x_0}S^{-1}(\mathbf{s},\mathbf{x})$. We have

$$\begin{aligned}
\partial^{n+1}_{x_0}S^{-1}(\mathbf{s},\mathbf{x})&=\partial_{x_0}[(-1)^{n+1}n!(\mathbf{x}^2-2s_0\mathbf{x}+|\mathbf{s}|^2)^{-(n+1)}(\mathbf{x}-\bar{\mathbf{s}})^{*(n+1)}]\\
&=(-1)^{n+2}(n+1)!(\mathbf{x}^2-2s_0\mathbf{x}+|\mathbf{s}|^2)^{-(n+2)}(2\mathbf{x}-2s_0)(\mathbf{x}-\bar{\mathbf{s}})^{*(n+1)}\\
&\quad+(-1)^{n+1}(n+1)!(\mathbf{x}^2-2s_0\mathbf{x}+|\mathbf{s}|^2)^{-(n+1)}(\mathbf{x}-\bar{\mathbf{s}})^{*n}\\
&=(-1)^{n+2}(n+1)!(\mathbf{x}^2-2s_0\mathbf{x}+|\mathbf{s}|^2)^{-(n+2)}[(2\mathbf{x}-2s_0)(\mathbf{x}-\bar{\mathbf{s}})\\
&\quad-(\mathbf{x}^2-2s_0\mathbf{x}+|\mathbf{s}|^2)]*(\mathbf{x}-\bar{\mathbf{s}})^{*n};
\end{aligned}$$

here we have used the fact that the s-monogenic product coincides with the usual one when the coefficients are real numbers, so

$$\begin{aligned}
&\partial^{n+1}_{x_0}S^{-1}(\mathbf{s},\mathbf{x})\\
&=(-1)^{n+2}(n+1)!(\mathbf{x}^2-2s_0\mathbf{x}+|\mathbf{s}|^2)^{-(n+2)}[\mathbf{x}^2-2\mathbf{x}\bar{\mathbf{s}}+\bar{\mathbf{s}}^2]*(\mathbf{x}-\bar{\mathbf{s}})^{*n}.
\end{aligned}$$

We get the last equality in (2.40) by recalling that

$$S^{-1}(\mathbf{s},\mathbf{x})(\mathbf{x}-\bar{\mathbf{s}})^{-1}=(\mathbf{x}^2-2s_0\mathbf{x}+|\mathbf{s}|^2)^{-1}.$$ □

Theorem 2.8.6 (Cauchy formula II outside an axially symmetric s-domain). *Let $U\subset\mathbb{R}^{n+1}$ be a bounded axially symmetric s-domain and assume that $U^c=\mathbb{R}^{n+1}\setminus\overline{U}$ is connected. Let $f:U^c\to\mathbb{R}_n$ be a left s-monogenic function with $\lim_{\mathbf{x}\to\infty}f(x)=a$. If $\mathbf{x}\in U^c$, then*

$$f(\mathbf{x})=a-\frac{1}{2\pi}\int_{\partial(V\cap\mathbb{C}_I)}S^{-1}(\mathbf{s},\mathbf{x})\,d\mathbf{s}_I f(\mathbf{s}),$$

where V is an axially symmetric s-domain containing U such that $\partial(V\cap\mathbb{C}_I)$ is a union of a finite number of continuously differentiable Jordan curves for every $I\in\mathbb{S}$. Moreover, the integral does not depend on V and on the imaginary unit $I\in\mathbb{S}$.

Proof. Let $\mathbf{x} \in U^c$. Then there exists $r > 0$ and a real point α such that the ball $B = B(\alpha, r)$ satisfies $B \supset U$ and $\mathbf{x} \in B$. Let V be an axially symmetric s-domain containing U such that $\partial(V \cap \mathbb{C}_I)$ is a union of a finite number of continuously differentiable Jordan curves for every $I \in \mathbb{S}$. Then f is s-monogenic on $B \setminus V$ and we can apply the Cauchy formula to compute $f(\mathbf{x})$. We obtain

$$\begin{aligned} f(\mathbf{x}) &= \frac{1}{2\pi} \int_{\partial((B\setminus V)\cap\mathbb{C}_I)} S^{-1}(\mathbf{s}, \mathbf{x}) d\mathbf{s}_I f(\mathbf{s}) \\ &= \frac{1}{2\pi} \int_{\partial(B\cap\mathbb{C}_I)} S^{-1}(\mathbf{s}, \mathbf{x}) d\mathbf{s}_I f(\mathbf{s}) - \frac{1}{2\pi} \int_{\partial(V\cap\mathbb{C}_I)} S^{-1}(\mathbf{s}, \mathbf{x}) d\mathbf{s}_I f(\mathbf{s}). \end{aligned}$$

By setting the positions

$$\mathbf{s} = \alpha + re^{I\theta}$$

we can compute the integral on $\partial(B \cap \mathbb{C}_I)$ in the standard way, and letting $r \to \infty$ we obtain that the integral equals $a = \lim_{r\to\infty} f$, therefore,

$$f(\mathbf{x}) = a - \frac{1}{2\pi} \int_{\partial(V\cap\mathbb{C}_I)} S^{-1}(\mathbf{s}, \mathbf{x}) d\mathbf{s}_I f(\mathbf{s}).$$

The integral does not depend on V and on the imaginary unit $I \in \mathbb{S}$, thanks to the Cauchy formula on bounded axially symmetric s-domains. □

We finally obtain a version of the Borel-Pompeiu formula.

Theorem 2.8.7 (Borel-Pompeiu formula). *Let $U \subset \mathbb{R}^{n+1}$ be an axially symmetric open bounded set such that $\partial(U \cap \mathbb{C}_I)$ is a union of a finite number of continuously differentiable Jordan curves for every $I \in \mathbb{S}$. Let $f : \overline{U} \to \mathbb{R}^{n+1}$ be a function of class C^1 and set $d\mathbf{s}_I = -Id\mathbf{s}$. For every $\mathbf{x} \in U$, $\mathbf{x} = u + I_{\mathbf{x}}v$ and $I \in \mathbb{S}$, we have*

$$\begin{aligned} &\frac{1}{2}\Big[1 - I_{\mathbf{x}}I\Big] f(u + Iv) + \frac{1}{2}\Big[1 + I_{\mathbf{x}}I\Big] f(u - Iv) \\ &= \frac{1}{2\pi}\Big(\int_{\partial(U\cap\mathbb{C}_I)} S^{-1}(\mathbf{s}, \mathbf{x}) d\mathbf{s}_I f(\mathbf{s}) + \int_{U\cap\mathbb{C}_I} S^{-1}(\mathbf{s}, \mathbf{x}) \overline{\partial}_I f(\mathbf{s}) d\mathbf{s}_I \wedge d\bar{\mathbf{s}} \Big). \end{aligned} \tag{2.42}$$

In particular, when $I = I_{\mathbf{x}}$ we have

$$\begin{aligned} f(\mathbf{x}) = \frac{1}{2\pi}\Big(&\int_{\partial(U\cap\mathbb{C}_{I_{\mathbf{x}}})} S^{-1}(\mathbf{s}, \mathbf{x}) d\mathbf{s}_{I_{\mathbf{x}}} f(\mathbf{s}) \\ &+ \int_{U\cap\mathbb{C}_{I_{\mathbf{x}}}} S^{-1}(\mathbf{s}, \mathbf{x}) \overline{\partial}_{I_{\mathbf{x}}} f(s) d\mathbf{s}_{I_{\mathbf{x}}} \wedge d\bar{\mathbf{s}} \Big). \end{aligned} \tag{2.43}$$

Proof. Let us set $\mathbf{x} = u + I_{\mathbf{x}}v$ and let us define

$$U_\varepsilon = \{\mathbf{s} = u' + I_s v' \in U \mid |(u + Iv) - (u' + Iv')| > \varepsilon \ \ \forall I \in \mathbb{S}\}$$

where ε is a positive number less than the distance from the $(n-1)$-sphere $u + \mathbb{S}v$

defined by $\mathbf{x}$ to the complement of U. The zeros of the function $\mathbf{x}^2 - 2\mathrm{Re}[\mathbf{s}]\mathbf{x} + |\mathbf{s}|^2 = 0$ consist either of a point on the real axis, or an $(n-1)$-sphere $u + Iv$. On $\mathbb{C}_{I_\mathbf{x}}$ we find only the point $\mathbf{x}$ as a singularity and the result follows from the Pompeiu formula on the complex plane $\mathbb{C}_{I_\mathbf{x}}$. When the singularity is a real number, S^{-1} is the standard Cauchy kernel and again the statement follows from the Pompeiu formula on the complex plane $\mathbb{C}_I$ for every $I \in \mathbb{S}$. If the zero is not real, on any complex plane $\mathbb{C}_I$ we find two zeros $\mathbf{s}_{1,2} = x_0 \pm I|\underline{x}|$. Thus $\partial U_\varepsilon = \partial U - \partial B_1 - \partial B_2$ where ∂B_i is the boundary of ball B_i with center $\mathbf{s}_i$ and radius ε.

From Lemma 2.8.2 applied to the functions $S^{-1}(\mathbf{s},\mathbf{x})$, $f(\mathbf{s})$ and since $S^{-1}(\mathbf{s},\mathbf{x})$ is right s-monogenic in the variable $\mathbf{s}$, we obtain

$$\frac{1}{2\pi}\int_{U_\varepsilon \cap \mathbb{C}_I} S^{-1}(\mathbf{s},\mathbf{x})\overline{\partial}_I f(\mathbf{s}) d\mathbf{s}_I \wedge d\bar{\mathbf{s}} + \frac{1}{2\pi}\int_{\partial(U\cap\mathbb{C}_I)} S^{-1}(\mathbf{s},\mathbf{x}) d\mathbf{s}_I f(\mathbf{s})$$
$$= \mathfrak{I}_1^\varepsilon(\mathbf{x}) + \mathfrak{I}_2^\varepsilon(\mathbf{x})$$

where

$$\mathfrak{I}_1^\varepsilon(\mathbf{x}) := \frac{1}{2\pi}\int_{\partial(B_1\cap\mathbb{C}_I)} S^{-1}(s,\mathbf{x}) d\mathbf{s}_I f(\mathbf{s}),$$
$$\mathfrak{I}_2^\varepsilon(\mathbf{x}) := \frac{1}{2\pi}\int_{\partial(B_2\cap\mathbb{C}_I)} S^{-1}(\mathbf{s},\mathbf{x}) d\mathbf{s}_I f(\mathbf{s}).$$

With similar computations as in the proof of Theorem 2.8.4, by letting $\varepsilon \to 0$ and after some computations we get

$$\mathfrak{I}_1^0(\mathbf{x}) = \frac{1}{2}\Big[1 - I_\mathbf{x} I\Big] f(x_0 + I|\underline{x}|).$$

Similarly, the integral related to $\mathbf{s}_2$ turns out to be

$$\mathfrak{I}_2^0(\mathbf{x}) = \frac{1}{2}\Big[1 + I_\mathbf{x} I\Big] f(x_0 - I|\underline{x}|).$$

So we get

$$\mathfrak{I}_1^0(\mathbf{x}) + \mathfrak{I}_2^0(\mathbf{x}) = \frac{1}{2}\Big[1 - I_q I\Big] f(x_0 + I|\underline{x}|) + \frac{1}{2}\Big[1 + I_\mathbf{x} I\Big] f(x_0 - I|\underline{x}|),$$

and this concludes the proof. □

Remark 2.8.8. Note that formula (2.43) is not surprising and in fact is the exact analog of the Borel-Pompeiu formula in the complex case. Formula (2.42) on the other hand, highlights a new phenomenon: given a point $\mathbf{x}$ and an imaginary unit $I \in \mathbb{S}$ there are exactly two points in $\mathbb{C}_I$ on the same sphere of $\mathbf{x}$ and formula (2.42) shows how to obtain an integral representation of f at those points.

Remark 2.8.9. The Cauchy formula in Theorem 2.8.4 follows as an immediate consequence of the Borel-Pompeiu formula and of the Representation Formula.

2.9 Duality Theorems

In this section we prove the algebraic isomorphism between the $\mathbb{R}_n$-module of functionals acting on $\mathcal{G}(K) := \operatorname{ind\,lim}_U \mathcal{M}^R(U)$ where K is a connected axially symmetric compact set such that its intersection with every complex plane $\mathbb{C}_I$ remains connected, and the $\mathbb{R}_n$-module of s-monogenic functions defined in the complement of K and vanishing at infinity. The results we obtain are the analogs, in this setting, of those obtained by Köthe in [67] and generalized by Grothendieck, see [57].

Consider the set $\mathcal{C}^\infty(U, \mathbb{R}_n)$ of infinitely differentiable functions defined on an open set $U \subseteq \mathbb{R}^{n+1}$ with values in $\mathbb{R}_n$. This set is an $\mathbb{R}_n$-bimodule with respect to the standard sum of functions and multiplication of a function by a Clifford number. To endow $\mathcal{C}^\infty(U, \mathbb{R}_n)$ with a locally convex topology, we follow [7] and consider an increasing sequence of compact sets $\{K_j\}_{j\in\mathbb{N}}$, $K_j \subset \mathbb{R}^{n+1}$, such that

$$K_0 \Subset K_1 \Subset \dots, \qquad U = \cup_{j=0}^{\infty} K_j,$$

and we introduce the family of seminorms $\{p_{j,r},\ j, r \in \mathbb{N}\}$ defined by

$$p_{j,r}(f) := \sup_{|\alpha| \le r} \sup_{\mathbf{x}\in K_j} |\partial^\alpha f(\mathbf{x})|, \quad f \in \mathcal{C}^\infty(U, \mathbb{R}_n),$$

where

$$\partial^\alpha = \frac{\partial^{\alpha_0}}{\partial x_0^{\alpha_0}} \cdots \frac{\partial^{\alpha_n}}{\partial x_n^{\alpha_n}}, \qquad |\alpha| = \sum_{i=0}^{n} \alpha_i.$$

This topology coincides with the product topology $\prod_A \mathcal{C}^\infty(U, \mathbb{R})$ where A is a multi-index which can be identified with an element in the power set of $\{1, \dots, n\}$. Thus we have the following result:

Theorem 2.9.1. *The set $\mathcal{C}^\infty(U, \mathbb{R}_n)$ is a Fréchet $\mathbb{R}_n$-bimodule.*

Proposition 2.9.2. *Let U be an open set in $\mathbb{R}^{n+1}$. The sets $\mathcal{M}^R(U)$ (resp. $\mathcal{M}^L(U)$) are Fréchet left (resp. right) $\mathbb{R}_n$-modules with respect to the topology of uniform convergence over compact sets.*

Proof. The set $\mathcal{C}^\infty(U)$ with the topology of uniform convergence on compact sets is a Fréchet bimodule. The sets $\mathcal{M}^R(U)$ and $\mathcal{M}^L(U)$ are closed submodules of $\mathcal{C}^\infty(U)$. Indeed, if we choose a sequence $\{f_m\}_{m\in\mathbb{N}} \subset \mathcal{M}^R(U)$, then, by definition, for every $I \in \mathbb{S}$ we have that the function f_m satisfies $\overline{\partial}_I f_{m,I}(u+Iv) = 0$ on $U \cap \mathbb{C}_I$. Let f be the limit function of $\{f_m\}_{m\in\mathbb{N}}$ in $\mathcal{C}^\infty(U)$. The restriction of $f \in \mathcal{C}^\infty(U)$ to a plane $\mathbb{C}_I$ is the limit of the restrictions $f_{m,I}$ thus, by the uniform convergence of the derivatives of $\{f_{m,I}\}$, it satisfies $\overline{\partial}_I f_I(u + Iv) = 0$ on $U \cap \mathbb{C}_I$. This proves that $\mathcal{M}^R(U)$ is a Fréchet module with the topology induced by the topology of $\mathcal{C}^\infty(U)$. The same argument applies to $\mathcal{M}^L(U)$. □

Remark 2.9.3. The same argument used in the proof shows also that $\mathcal{M}^R(U)$ and $\mathcal{M}^L(U)$ are Montel modules, since they are closed submodules of $\mathcal{C}^\infty(U)$ which is a Montel $\mathbb{R}_n$-bimodule.

Definition 2.9.4. *Let $K \subset \mathbb{R}^{n+1}$ be a compact set. We define a set of germs of functions defined by*

$$\mathcal{G}(K) := \operatorname{ind} \lim_{U \text{open} \supset K} \mathcal{M}^R(U).$$

In the sequel, we will use the same letter φ both to denote an element $\varphi \in \mathcal{G}(K)$ and an s-monogenic extension of φ to some neighborhood $U \subseteq \mathbb{R}^{n+1}$ of K. Because of Proposition 2.9.2, $\mathcal{G}(K)$ is a limit of Fréchet $\mathbb{R}_n$-modules, and it is naturally endowed with an LF–topology: a seminorm on $\mathcal{G}(K)$ is every seminorm that is continuous on every $\mathcal{M}^R(U)$. Even though $\mathcal{G}(K)$ is not a Fréchet $\mathbb{R}_n$-module itself, it is possible to characterize its topology in terms of convergence of sequences as in the following result:

Proposition 2.9.5. *Let $K \subset \mathbb{R}^{n+1}$ be a compact set. A sequence $\{\varphi_j\}$ of germs in $\mathcal{G}(K)$ converges to a germ $\varphi \in \mathcal{G}(K)$, if $\varphi_j(\mathbf{x})$ converges uniformly to $\varphi(\mathbf{x})$ in a neighborhood $U \subset \mathbb{R}^{n+1}$ of K.*

Proof. It is a consequence of the definition of inductive limit topology of $\mathcal{G}(K)$. □

Definition 2.9.6. *We call a connected compact set K such that $K \cap \mathbb{R} \neq \emptyset$ and its intersection $K \cap \mathbb{C}_I$ is connected for all $I \in \mathbb{S}$ an s-compact set.*

Let K be an s-compact set in $\overline{\mathbb{R}}^{n+1} := \mathbb{R}^{n+1} \cup \{\infty\}$.

We denote by $\mathcal{M}^L_\infty(\overline{\mathbb{R}}^{n+1} \setminus K)$ the right $\mathbb{R}_n$-module of left s-monogenic functions on $\overline{\mathbb{R}}^{n+1} \setminus K$ which vanish at infinity.

Theorem 2.9.7. *Let K be an axially symmetric s-compact set in $\mathbb{R}^{n+1}$. There is an $\mathbb{R}_n$-module isomorphism*

$$(\mathcal{G}(K))' \cong \mathcal{M}^L_\infty(\overline{\mathbb{R}}^{n+1} \setminus K)$$

where $(\mathcal{G}(K))'$ is the set of left $\mathbb{R}_n$–linear continuous functionals on $\mathcal{G}(K)$.

Proof. Let us define a map $T : \mathcal{M}^L_\infty(\overline{\mathbb{R}}^{n+1} \setminus K) \to (\mathcal{G}(K))'$. For any function $f \in \mathcal{M}^L_\infty(\overline{\mathbb{R}}^{n+1} \setminus K)$ we construct a functional $\mu = \mu_f$. Let $g \in \mathcal{G}(K)$ and let us denote by the same symbol g also its s-monogenic extension to an axially symmetric s-domain $U \supset K$. Let us fix an element $I \in \mathbb{S}$ and define

$$\langle \mu_f, g \rangle := \int_{\partial(U \cap \mathbb{C}_I)} g(\mathbf{s}) d\mathbf{s}_I f(\mathbf{s}). \tag{2.44}$$

We have to show that the definition does not depend on the choice of U and on the extension g. If we replace U by another axially symmetric s-domain V containing K we have

$$\begin{aligned} &\int_{\partial(U \cap \mathbb{C}_I)} g(\mathbf{s}) d\mathbf{s}_I f(\mathbf{s}) - \int_{\partial(V \cap \mathbb{C}_I)} g(\mathbf{s}) d\mathbf{s}_I f(\mathbf{s}) \\ &= \int_{\partial((U \setminus V) \cap \mathbb{C}_I)} g(\mathbf{s}) d\mathbf{s}_I f(\mathbf{s}) = 0 \end{aligned}$$

by Lemma 2.8.2; indeed f, g are s-monogenic functions on the left and on the right, respectively, on $U \setminus V$. If we replace g by another extension, the value of integral (2.44) is not affected since all the extensions of g coincide on small open sets containing K. The map μ_f is left $\mathbb{R}_n$-linear and continuous on $\mathcal{G}(K)$ by its definition. Thus the map T defined by

$$T(f) = \mu_f,$$

is well defined and right $\mathbb{R}_n$-linear. Let us now show that there is a map T' which is the inverse of T. Let us consider any $\mu \in (\mathcal{G}(K))'$, and define the function

$$\mathfrak{F}(\mathbf{x}) := -\frac{1}{2\pi} \langle \mu, S^{-1}(\mathbf{s}, \mathbf{x}) \rangle. \tag{2.45}$$

Note that μ acts on the variable s and $S^{-1}(\mathbf{s}, \mathbf{x})$ is right s-monogenic with respect to it. Since μ is a linear functional, we have

$$\begin{aligned} \bar{\partial}_I \mathfrak{F}_I(\mathbf{x}) &= -\frac{1}{2\pi} \bar{\partial}_I \langle \mu, S^{-1}(\mathbf{s}, u + Iv) \rangle \\ &= -\frac{1}{2\pi} \langle \mu, \bar{\partial}_I S^{-1}(\mathbf{s}, u + Iv) \rangle = 0, \qquad \forall I \in \mathbb{S}. \end{aligned}$$

Thus the function $\mathfrak{F}(\mathbf{x})$ is left s-monogenic for $\mathbf{x} \notin [\mathbf{s}]$, $\mathbf{s} \in K$ so, by the hypothesis on K, it is s-monogenic on the complement of K and vanishes at infinity, i.e., $\mathfrak{F} \in \mathcal{M}^L_\infty(\overline{\mathbb{R}}^{n+1} \setminus K)$. Define now

$$T' : (\mathcal{G}(K))' \to \mathcal{M}^L_\infty(\overline{\mathbb{R}}^{n+1} \setminus K), \qquad T'(\mu) = \mathfrak{F}.$$

The map T' is well defined and right $\mathbb{R}_n$-linear. Let us show that T' is a right inverse of T, i.e., that $T \cdot T' = \mathrm{id}_{(\mathcal{G}(K))'}$. Let $\mu \in (\mathcal{G}(K))'$ and consider $T'(\mu) = \mathfrak{F}$.

The functional $T(T'(\mu))$ acts on right s-monogenic functions as follows:

$$\begin{aligned}\langle T(T'(\mu)), g\rangle = \langle T(\mathfrak{F}), g\rangle &= \int_{\partial(U\cap\mathbb{C}_I)} g(\mathbf{x})d\mathbf{x}_I\mathfrak{F}(\mathbf{x})\\ &= -\frac{1}{2\pi}\int_{\partial(U\cap\mathbb{C}_I)} g(\mathbf{x})d\mathbf{x}_I\langle\mu, S^{-1}(\mathbf{s},\mathbf{x})\rangle\\ &= \langle\mu, -\frac{1}{2\pi}\int_{\partial(U\cap\mathbb{C}_I)} g(\mathbf{x})d\mathbf{x}_I S^{-1}(\mathbf{s},\mathbf{x})\rangle\\ &= \langle\mu, \frac{1}{2\pi}\int_{\partial(U\cap\mathbb{C}_I)} g(\mathbf{x})d\mathbf{x}_I S_R^{-1}(\mathbf{x},\mathbf{s})\rangle = \langle\mu, g\rangle,\end{aligned}$$

so we get $T(T'(\mu)) = \mu$. Let us now show that T' is a left inverse of T, i.e., that $T'\cdot T = \mathrm{id}_{\mathcal{M}^L_\infty(\overline{\mathbb{R}}^{n+1}\backslash K)}$. Consider $f \in \mathcal{M}^L_\infty(\overline{\mathbb{R}}^{n+1}\backslash K)$, the functional $T(f) = \mu_f$ defined in (2.44) and $T'(\mu_f)$. By Theorem 2.8.6 and the fact that f vanishes at infinity, we have

$$\begin{aligned}T'(T(f)) = T'(\mu_f) &= -\frac{1}{2\pi}\langle\mu_f, S^{-1}(\mathbf{s},\mathbf{x})\rangle\\ &= -\frac{1}{2\pi}\int_{\partial(U\cap\mathbb{C}_I)} S^{-1}(\mathbf{s},\mathbf{x})d\mathbf{s}_I f(\mathbf{s}) = f(\mathbf{x}),\end{aligned}$$

that is $T'(T(f)) = f$. This concludes the proof. □

In analogy with the complex case, we give the following definition.

Definition 2.9.8. *The function*

$$\mathfrak{F}(\mathbf{x}) := -\frac{1}{2\pi}\langle\mu, S^{-1}(\mathbf{s},\mathbf{x})\rangle$$

is called the Fantappié indicatrix of the functional $\mu \in (\mathcal{G}(K))'$.

One could be tempted to dualize Theorem 2.9.7 by simply taking the dual of the sets in its statement. Since

$$(\mathcal{G}(K))' \cong \mathcal{M}^L_\infty(\overline{\mathbb{R}}^{n+1}\backslash K),$$

one could take the dual on both sides and obtain

$$(\mathcal{G}(K))'' \cong (\mathcal{M}^L_\infty(\overline{\mathbb{R}}^{n+1}\backslash K))',$$

and attempt to conclude that

$$\mathcal{G}(K) \cong (\mathcal{M}^L_\infty(\overline{\mathbb{R}}^{n+1}\backslash K))'$$

by using some reflexivity property of $\mathcal{G}$. This approach, however, is premature, and at this stage we need to give a direct proof of such an isomorphism. In the next section, we will show how to make such an attempt rigorous.

Theorem 2.9.9. *Let K be an axially symmetric s-compact set in $\mathbb{R}^{n+1}$. Then there is an $\mathbb{R}_n$-module isomorphism*

$$(\mathcal{M}^L_\infty(\overline{\mathbb{R}}^{n+1}\setminus K))' \cong \mathcal{G}(K).$$

Proof. Fix any $g \in \mathcal{G}(K)$ and consider, for every $f \in \mathcal{M}^L_\infty(\overline{\mathbb{R}}^{n+1}\setminus K)$, the integral

$$\langle \phi_g, f\rangle := \int_{\partial(U\cap\mathbb{C}_I)} g(\mathbf{s})d\mathbf{s}_I f(\mathbf{s}) \tag{2.46}$$

where U denotes an axially symmetric s-domain containing K and where we have fixed $I \in \mathbb{S}$. For any $g \in \mathcal{G}(K)$, the integral (2.46) defines a continuous right linear $\mathbb{R}_n$–functional ϕ_g on $\mathcal{M}^L_\infty(\overline{\mathbb{R}}^{n+1}\setminus K)$. Therefore we have a map

$$\mathcal{T} : \mathcal{G}(K) \longrightarrow (\mathcal{M}^L_\infty(\overline{\mathbb{R}}^{n+1}\setminus K))',$$

defined by setting $\mathcal{T}(g) = \phi_g$ for any fixed $g \in \mathcal{G}(K)$. The map is injective: if $g_1 \neq g_2$, then the functionals ϕ_{g_1}, ϕ_{g_2} (defined by g_1 and g_2) are different. Indeed, let $\mathbf{x} \in K$ and consider the action of the two functionals on the function $S_R^{-1}(\mathbf{s},\mathbf{x}) = S_{R,\mathbf{x}}^{-1}(\mathbf{s}) \in \mathcal{M}^L_\infty(\overline{\mathbb{R}}^{n+1}\setminus K)$; then we have

$$\frac{1}{2\pi}\langle \phi_{g_1}, S_R^{-1}(\mathbf{s},\mathbf{x})\rangle = \frac{1}{2\pi}\int_{\partial(U\cap\mathbb{C}_I)} g_1(\mathbf{s})d\mathbf{s}_I S_R^{-1}(\mathbf{s},\mathbf{x}) = g_1(\mathbf{x}),$$

and

$$\frac{1}{2\pi}\langle \phi_{g_2}, S_R^{-1}(\mathbf{s},\mathbf{x})\rangle = \frac{1}{2\pi}\int_{\partial(U\cap\mathbb{C}_I)} g_2(\mathbf{s})d\mathbf{s}_I S_R^{-1}(\mathbf{s},\mathbf{x}) = g_2(\mathbf{x}),$$

hence we have a one-to-one mapping:

$$\mathcal{T} : \mathcal{G}(K) \longrightarrow (\mathcal{M}^L_\infty(\overline{\mathbb{R}}^{n+1}\setminus K))'.$$

To conclude the proof it is sufficient to show that $\mathcal{T}$ admits a right inverse. Let

$$\phi : \mathcal{M}^L_\infty(\overline{\mathbb{R}}^{n+1}\setminus K) \to \mathbb{R}_n$$

be a continuous right $\mathbb{R}_n$–linear map, acting continuously on $\mathcal{M}^L_\infty(\overline{\mathbb{R}}^{n+1}\setminus K)$ with its natural topology. It allows us to define

$$\psi(\mathbf{x}) = \frac{1}{2\pi}\langle \phi, S_R^{-1}(\mathbf{s},\mathbf{x})\rangle, \tag{2.47}$$

where the functional ϕ acts on the variable $\mathbf{s}$. The function $\psi(\mathbf{x})$ is right s-monogenic, as one can check directly, hence $\psi \in \mathcal{G}(K)$. Let

$$\mathcal{T}' : \ (\mathcal{M}^L_\infty(\overline{\mathbb{R}}^{n+1}\setminus K))' \to \mathcal{G}(K)$$

be the map defined by

$$\mathcal{T}'(\phi) = \frac{1}{2\pi}\langle \phi, S_R^{-1}(\mathbf{s}, \mathbf{x})\rangle = \psi(\mathbf{x}).$$

Now we have to show that $\mathcal{T} \cdot \mathcal{T}' = \mathrm{id}_{(\mathcal{M}_\infty^L(\overline{\mathbb{R}}^{n+1}\setminus K))'}$. Since

$$\begin{aligned}\langle \mathcal{T}(\mathcal{T}'(\phi)), f\rangle &= \langle \mathcal{T}(\psi), f\rangle = \frac{1}{2\pi}\langle \mathcal{T}\Big(\langle \phi, S_R^{-1}(\mathbf{s}, \mathbf{x})\rangle\Big), f\rangle \\ &= \frac{1}{2\pi}\int_{\partial(U\cap\mathbb{C}_I)} \langle \phi, S_R^{-1}(\mathbf{s}, \mathbf{x})\rangle d\mathbf{x}_I f(\mathbf{x}) \\ &= \langle \phi, -\frac{1}{2\pi}\int_{\partial(U\cap\mathbb{C}_I)} S^{-1}(\mathbf{x}, \mathbf{s}) d\mathbf{x}_I f(\mathbf{x})\rangle,\end{aligned}$$

by Theorem 2.8.6 we get

$$\langle \phi, -\frac{1}{2\pi}\int_{\partial(U\cap\mathbb{C}_I)} S^{-1}(\mathbf{x}, \mathbf{s}) d\mathbf{x}_I f(\mathbf{x})\rangle = \langle \phi, f\rangle,$$

which concludes the proof. □

Corollary 2.9.10. *Let $\overline{B} = \overline{B(0, r)}$ be the closed ball in $\mathbb{R}^{n+1}$ centered at the origin and with radius $r > 0$. The dual of $\mathcal{M}_\infty^L(\overline{\mathbb{R}}^{n+1}\setminus B)$ is the set of all right s-monogenic functions defined in a neighborhood of $\overline{B}$.*

2.10 Topological Duality Theorems

In the previous sections we have proved the two $\mathbb{R}_n$-modules isomorphisms:

$$(\mathcal{G}(K))' \cong \mathcal{M}_\infty^L(\overline{\mathbb{R}}^{n+1}\setminus K)$$

and

$$\mathcal{G}(K) \cong (\mathcal{M}_\infty^L(\overline{\mathbb{R}}^{n+1}\setminus K))'.$$

We now want to show that those isomorphisms are actually topological isomorphisms. To this end we need to introduce a special class of infinite-order differential operators which is of independent interest. We recall that for s-monogenic functions, the s-derivatives coincide with the partial derivative ∂_u with respect to the scalar part u of a paravector, so

$$F(\partial_s) = F(\partial_u) = \sum_{m\geq 0} \partial_u^m a_m.$$

Proposition 2.10.1. *Let $F(\partial_s)$ be defined as above and let f be a left s-monogenic function in an axially symmetric s-domain $U \subseteq \mathbb{R}^{n+1}$. The function $F(\partial_s)f$ is a left s-monogenic function in U if and only if*

$$\lim_{m\to+\infty} \sqrt[m]{|a_m| m!} = 0. \tag{2.48}$$

Proof. Suppose that condition (2.48) holds, choose $I \in \mathbb{S}$, and consider the restriction of f to the plane $\mathbb{C}_I$. By the Splitting Lemma, f_I can be written as $f_I(z) = \sum_A F_A(z) I_A$, $z = u + Iv$ and each holomorphic function F_A can be expanded into a power series at a point $z_0 \in \mathbb{C}_I$. Thus $f_I(z)$ can be expanded into a power series with center at z_0 and, by the usual Cauchy estimates on the plane $\mathbb{C}_I$, we also deduce that

$$\frac{1}{m!}\left|\frac{\partial^m f}{\partial u^m}(z_0)\right| \leq \frac{M}{\delta^m}, \quad m \geq 0, \quad \text{for } |\mathbf{x} - z_0| \leq \delta.$$

Since $|a_m m!| < \varepsilon$ for all $m \in \mathbb{N}$, we deduce that the series $\sum_m \partial_s^m f(\mathbf{x}) a_m$ converges locally uniformly on $\mathbb{C}_I$. It is immediate to verify that

$$\overline{\partial}_I[F(\partial_s) f_I(z)] = \overline{\partial}_I[\sum_{m\geq 0} \partial_u^m f_I(z) a_m] = \sum_{m\geq 0} \partial_u^m \overline{\partial}_I f_I(z) a_m = 0,$$

and since the choice of I is arbitrary we get that $\sum_{m\geq 0} \partial_u^m f(\mathbf{x}) a_m$ is an s-monogenic function.

Conversely, suppose by an absurdity that $\sum_{m\geq 0} \partial_s^m f(\mathbf{x}) a_m$ is s-monogenic but (2.48) does not hold. The result follows as in the complex case, see [63], Lemma 1.8.1. Indeed, suppose we negate (2.48). Then for some $\varepsilon > 0$ there is a subsequence a_{k_j} such that

$$\sqrt[k_j]{|a_{k_j}| k_j!} \geq 2\varepsilon \qquad \text{for all } k_j, \qquad k_j \to +\infty.$$

We now apply $F(\partial_s)$ to the s-monogenic function $(\mathbf{x} - y_0)^{-1}$, with $y_0 \in U \cap \mathbb{R}$, and we obtain

$$F(\partial_s)(\mathbf{x} - y_0)^{-1} = \sum_{k\geq 0} \frac{a_k(-1)^k k!}{(\mathbf{x} - y_0 - \varepsilon)^{k+1}} = \sum_{k\geq 0} F_k(\mathbf{x}). \tag{2.49}$$

Consider $|\mathbf{x} - y_0| \leq \varepsilon$, and assume, by taking if necessary a subsequence $\mathbf{x}_j$, that $\mathbf{x}_j \to y_0$. Then we get

$$|F_k(\mathbf{x}_j)| \geq \frac{(2\varepsilon)^{k_j}}{|\mathbf{x}_j - y_0 - \varepsilon|^{k_j+1}} \geq \frac{1}{2\varepsilon},$$

thus for $|\mathbf{x} - y_0| \leq \varepsilon$ the series (2.49) does not converge locally uniformly which contradicts the hypothesis. □

Proposition 2.10.2. *Let $U \subseteq \mathbb{R}^{n+1}$ be an axially symmetric s-domain. An operator of the type $F(\partial_{\mathbf{s}})$ acts continuously on $\mathcal{M}^L(U)$, for any U.*

Proof. If $f \in \mathcal{M}^L(U)$ we know that the estimate for $F(\partial_{\mathbf{s}})f$ depends only on the maximum norm of f, so continuity follows. □

Theorem 2.10.3. *Let $K \subseteq \mathbb{R}^{n+1}$ be an axially symmetric s-compact set. The sequence $\{g_k\}$ converges to $g \in \mathcal{G}(K)$ if and only if the sequence $\{F(\partial_{\mathbf{s}}) g_k(\mathbf{x})\}$ converges pointwise on K for all $F(\partial_{\mathbf{s}})$.*

Proof. Let g_k be defined in an axially symmetric s-domain U containing K and let $g_{k,I} = \sum_A F_{k,A}(z)I_A$ be the restriction of g_k to a plane $\mathbb{C}_I$ obtained using the Splitting Lemma 2.2.11. The convergence of g_k to g in the topology of $\mathcal{M}^R(U)$ is equivalent to the convergence, for every multi-index A, of $\{F_{k,A}\}$ to some function F_A which is holomorphic in $U \cap \mathbb{C}_I$. Theorem 4.1.10 in [63] shows that the convergence of $F_{k,A}$ is equivalent, for every A, to the pointwise convergence of $\{F(\partial_s)F_{k,A}\}$ for every $F(\partial_s)$. This in turn is equivalent to the convergence of $g_{k,I}$ on the plane $\mathbb{C}_I$. We conclude the proof by applying the Representation Formula (2.7). □

Theorem 2.10.4. *Let $K \subset \mathbb{R}^{n+1}$ be an axially symmetric s-compact set. The isomorphism*

$$(\mathcal{M}^L_\infty(\overline{\mathbb{R}}^{n+1}\backslash K))' \cong \mathcal{G}(K).$$

is topological.

Proof. If $g_k \to g$ in $\mathcal{G}(K)$ it means that $g_k \to g$ uniformly in a neighborhood of K. With respect to the duality defined by (2.46), we have $\langle \phi_{g_k}, f\rangle \to \langle \phi_g, f\rangle$ uniformly when f varies in a bounded subset of $\mathcal{M}^L_\infty(\overline{\mathbb{R}}^{n+1}\backslash K)$, thus $\phi_{g_k} \to \phi_g$.

Conversely, suppose that $\phi_k \to \phi$ in $(\mathcal{M}^L_\infty(\overline{\mathbb{R}}^{n+1}\backslash K))'$, with its natural topology. Then the functions

$$g_k(\mathbf{x}) = \frac{1}{2\pi}\langle \phi_k, S_R^{-1}(\mathbf{s},\mathbf{x})\rangle$$

defined by (2.47) are right s-monogenic in a neighborhood U of K which can be chosen to be an axially symmetric s-domain. Now we have to show that the sequence $\{g_k\}$ converges uniformly in some suitable neighborhood of K. By Theorem 2.10.3 it is enough to prove that $\{F(\partial_u)g_k\}$ converges pointwise for all infinite-order differential operators $F(\partial_u)$ satisfying condition (2.48). From the continuity of ϕ_k, fixing any $\mathbf{x} \in K$, we have

$$\begin{aligned} F(\partial_u)g_k(\mathbf{x}) = \frac{1}{2\pi}\langle \phi_k, F(\partial_u)S_R^{-1}(\mathbf{s},\mathbf{x})\rangle &\to \frac{1}{2\pi}\langle \phi, F(\partial_u)S_R^{-1}(\mathbf{s},\mathbf{x})\rangle \\ &= F(\partial_u)\langle \phi, \frac{1}{2\pi}S_R^{-1}(\mathbf{s},\mathbf{x})\rangle \end{aligned}$$

and the statement follows by setting $g(\mathbf{x}) = \frac{1}{2\pi}\langle \phi, S_R^{-1}(\mathbf{s},\mathbf{x})\rangle$. □

Corollary 2.10.5. *Let $K \subset \mathbb{R}^{n+1}$ be an axially symmetric s-compact set. The isomorphism*

$$(\mathcal{G}(K))' \cong \mathcal{M}^L_\infty(\overline{\mathbb{R}}^{n+1}\backslash K)$$

is topological.

Proof. We have pointed out that $\mathcal{M}^L_\infty(\overline{\mathbb{R}}^{n+1}\backslash K)$ is a Montel module thus it is reflexive. So, by Theorem 2.10.4, the dual of $\mathcal{G}(K)$ is $\mathcal{M}^L_\infty(\overline{\mathbb{R}}^{n+1}\backslash K)$ itself. □

We conclude this section by looking at a very special case of a compact set, namely $K = \{0\}$. Recall that s-monogenic functions outside the origin can be represented by a Laurent-type series of the form

$$f(\mathbf{x}) = \sum_{m\geq 0} \mathbf{x}^m a_m + \sum_{m\geq 1} \mathbf{x}^{-m} b_m \tag{2.50}$$

converging in a spherical shell

$$A = \{\mathbf{x} \in \mathbb{R}^{n+1} \mid R_1 < |\mathbf{x}| < R_2\}, \qquad 0 < R_1 < R_2.$$

This formula contains two series: one with positive powers of the variable, and one with negative powers of the variable. It is clear that, in order for the Laurent series to give a function which vanishes at infinity, the portion with positive powers must vanish. Thus, we can say that s-monogenic functions outside the origin, which vanish at infinity, are represented by Laurent series where only negative powers of the variable appear. An additional condition is the consequence of the fact that we are requiring the Laurent series to converge everywhere. For this to be true, we need to ask that the series has radius of convergence equal to infinity, and this yields, once again, condition (2.48). We can therefore state the following result:

Corollary 2.10.6. *The $\mathbb{R}_n$-module $(\mathcal{G}(\{0\}))'$ is isomorphic to the $\mathbb{R}_n$-module of infinite-order differential operators acting on s-monogenic functions.*

2.11 Notes

Note 2.11.1. On the kernel $S^{-1}(\mathbf{s}, \mathbf{x})$. Unlike the case of regular or monogenic functions which are defined as the elements of the kernel of first-order differential operators (the Cauchy Fueter operator for the case of regular functions and the Dirac operator for the case of monogenic functions), it is not possible to consider s-regular and s-monogenic functions as solutions of a globally defined operator. Specifically, these functions are defined as those functions whose restrictions to a family of planes satisfy a family of first-order operators on those planes. The Cauchy kernel that we have constructed is, on each of those planes, the fundamental solution for the relevant operator; this justifies our choice of nomenclature, even though strictly speaking this is somewhat of an abuse of notation because the kernel is not the solution on $\mathbb{R}^4$ or $\mathbb{R}^{n+1}$ of a globally-defined operator.

In fact, the fundamental solution to the equation

$$\frac{1}{2}\left(\frac{\partial}{\partial u} + I\frac{\partial}{\partial v}\right) f_I(u + Iv) = \delta(u + Iv), \quad I \in \mathbb{S}, \tag{2.51}$$

on the plane $\mathbb{C}_I$, where $\delta(u + Iv)$ is the Dirac delta distribution, is (see [59])

$$f_I(u + Iv) = \frac{1}{\pi}\frac{1}{u + Iv}, \quad I \in \mathbb{S}. \tag{2.52}$$

By the Extension Lemma, we uniquely extend the function in (2.52) to the entire space $\mathbb{R}^{n+1} \setminus \{0\}$ to get

$$f(x) = \frac{1}{\pi}\frac{1}{\mathbf{x}} = -\frac{1}{\pi}\, S^{-1}(0, \mathbf{x}).$$

If the delta distribution is not centered at the origin but at a point α on the real axis, the solution becomes

$$f(\mathbf{x}-\alpha) = \frac{1}{\pi}\,\frac{1}{\mathbf{x}-\alpha} = -\frac{1}{\pi}\, S^{-1}(\alpha, \mathbf{x}).$$

If α is not real, then the function $(\mathbf{x}-\alpha)^{-1}$ is not s-monogenic, thus we have to consider its s-monogenic inverse $(\mathbf{x}-\alpha)^{-*}$ which is precisely

$$(\mathbf{x}-\alpha)^{-*} = -S^{-1}(\alpha, \mathbf{x}).$$

Let us now consider another feature of the Cauchy kernel series. Let $\mathbf{x}$, $\mathbf{s} \in \mathbb{R}^{n+1}$ such that $\mathbf{xs} \neq \mathbf{sx}$ and denote by $S(\mathbf{s}, \mathbf{x})$ the inverse of the noncommutative Cauchy kernel series $S^{-1}(\mathbf{s}, \mathbf{x})$. Our next goal is to show that the function $S(\mathbf{s}, \mathbf{x})$, satisfying the equation

$$S^2(\mathbf{s}, \mathbf{x}) + S(\mathbf{s}, \mathbf{x})\mathbf{x} - \mathbf{s}\, S(\mathbf{s}, \mathbf{x}) = 0 \tag{2.53}$$

is the inverse of the noncommutative Cauchy kernel series.

Lemma 2.11.2. *Let* $\mathbf{x}$, $\mathbf{s} \in \mathbb{R}^{n+1}$. *Then* $S(\mathbf{s}, \mathbf{x}) := \mathbf{s} - \mathbf{x}$ *is a solution of equation (2.53) if and only if* $\mathbf{s}\mathbf{x} = \mathbf{x}\mathbf{s}$.

In general, when $\mathbf{s}$, $\mathbf{x}$ do not commute, the equation (2.53) has another nontrivial solution:

Theorem 2.11.3. *Let* $\mathbf{x}$, $\mathbf{s} \in \mathbb{R}^{n+1}$ *be such that* $\mathbf{xs} \neq \mathbf{sx}$. *The equation* (2.53) *has the nontrivial solution*

$$S(\mathbf{s}, \mathbf{x}) = -(\mathbf{x} - \overline{\mathbf{s}})^{-1}(\mathbf{x}^2 - 2\mathrm{Re}[\mathbf{s}]\mathbf{x} + |\mathbf{s}|^2).$$

Proof. Let us plug $-(\mathbf{x} - \overline{\mathbf{s}})^{-1}(\mathbf{x}^2 - 2\mathrm{Re}[\mathbf{s}]\mathbf{x} + |\mathbf{s}|^2)$ into (2.53) and show that

$$\begin{aligned}
&(\mathbf{x} - \overline{\mathbf{s}})^{-1}(\mathbf{x}^2 - 2\mathrm{Re}[\mathbf{s}]\mathbf{x} + |\mathbf{s}|^2)(\mathbf{x} - \overline{\mathbf{s}})^{-1}(\mathbf{x}^2 - 2\mathrm{Re}[\mathbf{s}]\mathbf{x} + |\mathbf{s}|^2)\\
&- (\mathbf{x} - \overline{\mathbf{s}})^{-1}(\mathbf{x}^2 - 2\mathrm{Re}[\mathbf{s}]\mathbf{x} + |\mathbf{s}|^2)\mathbf{x}\\
&+ \mathbf{s}(\mathbf{x} - \overline{\mathbf{s}})^{-1}(\mathbf{x}^2 - 2\mathrm{Re}[\mathbf{s}]\mathbf{x} + |\mathbf{s}|^2) = 0
\end{aligned}$$

is an identity. We multiply on the left by $(\mathbf{x} - \overline{\mathbf{s}})$ and we get

$$\begin{aligned}
&(\mathbf{x}^2 - 2\mathrm{Re}[\mathbf{s}]\mathbf{x} + |\mathbf{s}|^2)(\mathbf{x} - \overline{\mathbf{s}})^{-1}(\mathbf{x}^2 - 2\mathrm{Re}[\mathbf{s}]\mathbf{x} + |\mathbf{s}|^2)\\
&- (\mathbf{x}^2 - 2\mathrm{Re}[\mathbf{s}]\mathbf{x} + |\mathbf{s}|^2)\mathbf{x}\\
&+ (\mathbf{x} - \overline{\mathbf{s}})\mathbf{s}(\mathbf{x} - \overline{\mathbf{s}})^{-1}(\mathbf{x}^2 - 2\mathrm{Re}[\mathbf{s}]\mathbf{x} + |\mathbf{s}|^2) = 0.
\end{aligned} \tag{2.54}$$

We observe that $\mathbf{x}$ and $(\mathbf{x}^2 - 2\mathrm{Re}[\mathbf{s}]\mathbf{x} + |\mathbf{s}|^2)$ commute and that the element

$$\mathbf{u} := (\mathbf{x}^2 - 2\mathrm{Re}[\mathbf{s}]\mathbf{x} + |\mathbf{s}|^2)$$

is invertible where it is nonzero. Indeed

$$\begin{aligned}\mathbf{u}\bar{\mathbf{u}} &= (\mathbf{x}^2 - 2\mathrm{Re}[\mathbf{s}]\mathbf{x} + |\mathbf{s}|^2)(\bar{\mathbf{x}}^2 - 2\mathrm{Re}[\mathbf{s}]\bar{\mathbf{x}} + |\mathbf{s}|^2)\\ &= |\mathbf{x}|^4 - 2\mathbf{x}|\mathbf{x}|^2\mathrm{Re}[\mathbf{s}] + \mathbf{x}^2|\mathbf{s}|^2 - 2\bar{\mathbf{x}}|\mathbf{x}|^2\mathrm{Re}[\mathbf{s}] + 4|\mathbf{x}|^2\mathrm{Re}[\mathbf{s}]^2\\ &\quad - 2\mathbf{x}\mathrm{Re}[\mathbf{s}]|\mathbf{s}|^2 + \bar{\mathbf{x}}^2|\mathbf{s}|^2 - 2\bar{\mathbf{x}}\mathrm{Re}[\mathbf{s}]|\mathbf{s}|^2 + |\mathbf{s}|^4\\ &= |\mathbf{x}|^4 - 2\mathrm{Re}[\mathbf{x}]|\mathbf{x}|^2\mathrm{Re}[\mathbf{s}]\\ &\quad + (\mathrm{Re}[\mathbf{s}]^2 - |\underline{s}|^2)|\mathbf{s}|^2 + 4|\mathbf{x}|^2\mathrm{Re}[\mathbf{s}]^2 - 2\mathrm{Re}[\mathbf{x}]\mathrm{Re}[\mathbf{s}]|\mathbf{s}|^2 + |\mathbf{s}|^4\end{aligned}$$

therefore $\mathbf{u}\bar{\mathbf{u}} \in \mathbb{R}$, thus the inverse of $\mathbf{u}$ is $\bar{\mathbf{u}}/|\mathbf{u}|^2$. By multiplying equality (2.54) by $\mathbf{u}^{-1}$ on the right, we obtain:

$$(\mathbf{x}^2 - 2\mathrm{Re}[\mathbf{s}]\mathbf{x} + |\mathbf{s}|^2)(\mathbf{x} - \bar{\mathbf{s}})^{-1} - \mathbf{x} + (\mathbf{x} - \bar{\mathbf{s}})\mathbf{s}(\mathbf{x} - \bar{\mathbf{s}})^{-1} = 0.$$

We multiply by $\mathbf{x} - \bar{\mathbf{s}}$ on the right and we get the identity

$$-2\mathrm{Re}[\mathbf{s}]\mathbf{x} + \mathbf{x}\bar{\mathbf{s}} + \mathbf{x}\mathbf{s} = 0. \qquad \square$$

Note 2.11.4. Historical notes and further readings. The study of s-monogenic functions is a relatively new field of research: they were introduced in 2007 in [26] (published two years later), in an effort to generalize the notion of slice regularity (see [48], [49]) to the setting of Clifford algebras. Further properties of s-monogenic functions which are collected in this book are treated in [18], [27], [28], [29]. The Runge theorem is proved in [30] for a slightly different class of functions that, however, coincide with the class of s-monogenic functions over axially symmetric s-domains.

The most studied and well-known generalization of holomorphic functions to the Clifford algebras setting is Clifford analysis, intended as the study of functions in the kernel of the Dirac operator. It is nowadays a widely developed topic which the reader can approach in the classical references [7] and [34]. More recent books, which address in a less detailed way the topic of monogenic functions but give some insights to further developments of the theory, are [23] and [31]. Finally, a very friendly introduction to classical complex analysis and its higher-dimensional generalizations containing also historical remarks is given in the textbook [58]. Clifford analysis is a very rich and well-developed theory which, however, does not allow one to treat power series in the paravector variable and for this reason other theories have been introduced. With no claim of completeness, we mention for example the hyperholomorphic functions studied by Eriksson and Leutwiler in [74], [38], [39], [40] and Cliffordian holomorphic functions introduced by Laville and Ramadanoff [72], [73]. Slice monogenic functions admit power series expansion in terms of the paravector variable, at least on discs centered at points on the real axis, and this property will allow us to deal with a functional calculus for n-tuples of linear operators (see the next chapter).

It is worth noticing, however, that the theory of s-monogenic functions is not, strictly speaking, a generalization of the theory of holomorphic functions of a complex variable: holomorphic functions, can be obtained as s-monogenic functions for $n = 1$, see Remark 2.2.9, but given an s-monogenic function there is no possibility to restrict its domain or codomain in order to obtain a holomorphic function.

The s-monogenic functions, as well as s-regular functions in one quaternionic variable, have several forerunners in the literature. Fueter in his paper [43], but see also [33], [98], considered the problem of constructing regular functions (in the sense of Cauchy–Fueter) starting from holomorphic functions. Thus he introduced functions of the form

$$f(q) = \alpha(q_0, |\mathrm{Im}(q)|) + \frac{\mathrm{Im}(q)}{|\mathrm{Im}(q)|}\beta(q_0, |\mathrm{Im}(q)|) \tag{2.55}$$

where α, β are defined on the upper complex plane $\mathbb{C}^+$, have real values and α, β satisfy the Cauchy–Riemann system. The function Δf, now called the Fueter transform of f, is Cauchy–Fueter regular. Note that, in light of this result, the function $\sum \Delta q^n a_n$ is (Cauchy–Fueter) regular in q where it converges. This approach was generalized to functions of a paravector variable: it is sufficient to rewrite (2.55) by replacing the quaternion q by a paravector $\mathbf{x} \in \mathbb{R}^{n+1}$ and $\mathrm{Im}(q)$ by the vector $\underline{x}$. If n is odd, it is possible to show that $\Delta^{(n-1)/2} f$ is a monogenic function in the sense of [7]. This result, known as Fueter's mapping theorem, has been proved by Sce in [94] and then generalized by Qian, see [87], when n is an even number. Later on, Fueter's theorem was generalized to the case in which a function f as above is multiplied by a monogenic homogeneous polynomial of degree k, see [68], [83], [96] and to the case in which the function f is defined on an open set U, not necessarily chosen in the upper complex plane, see [88]. This last result is important because in this case a function of the form (2.55), with q replaced by $\mathbf{x}$, is s-monogenic in the sense of our definition, even though we are allowed to consider α and β with values in the Clifford algebra $\mathbb{R}_n$. Fueter's mapping theorem allows us to construct monogenic functions starting from s-monogenic functions, moreover it allows us to show that the class of monogenic functions which comes from s-monogenic ones corresponds to the axially monogenic functions (see [24]).

The class of functions (2.55), whose importance for Fueter's mapping theorem is clear, is also known in the literature as the class of radially holomorphic functions, see for example [58]. They are also related to the so-called standard intrinsic functions studied by Rinehart and then by Cullen, see [89], [32] respectively. These studies were the starting point for a deep generalization carried out by Ghiloni and Perotti in their paper [53]. In this paper, the authors study functions with values in a real alternative algebra A which are slice functions, i.e., they are of the form $f(u, v) = \alpha(u, v) + I\beta(u, v)$ where $\alpha(u, -v) = \alpha(u, v)$ and $\beta(u, -v) = -\beta(u, v)$, I is an element chosen in a suitable subset of the algebra such that $I^2 = -1$, (u, v)

are real numbers which correspond to the "real part" and to the modulus of the "imaginary part" of a variable chosen in a suitable subset of the algebra A. Then by requiring that the pair of functions (α, β) satisfy the Cauchy–Riemann system, one obtains the so-called slice regular functions according to [53] (compare with Corollary 2.2.20). We do not enter into the details of this interesting construction: it is sufficient to observe that the treatment is general enough to include, when we consider open sets that are axially symmetric and which properly intersect the real axis, the case of s-monogenic and s-regular functions treated in this book.

Finally, we point out that the study of zeros of polynomials of a paravector variable, which we started in our work as a byproduct of the study of s-monogenic functions, has been the topic of the researches of Qian and Yang, see [104]. Moreover, polynomials with coefficients in a Clifford algebra can also be treated with the techniques developed by Ghiloni and Perotti, see the aforementioned papers and [54].

Chapter 3

Functional calculus for n-tuples of operators

The goal of this chapter is to construct a functional calculus for n-tuples of not necessarily commuting operators on a Banach space V over the real numbers. We start by introducing the basic notions which will allow us, given an n-tuple of linear operators acting on V, to construct a new operator acting on a suitable module over a real Clifford algebra. The idea to use a Clifford algebra approach is not new and goes back to Coifman and Murray, see [80] and also to the works of McIntosh, Pryde and Jefferies (see [62] and the references therein).

Let then V be a Banach space over $\mathbb{R}$ with norm $\|\cdot\|_V$. It is possible to embed V into a wider set V_n which possesses the structure of a Clifford module and to endow V with an operation of multiplication by elements of $\mathbb{R}_n$ which gives a two-sided module over $\mathbb{R}_n$.

Specifically, by V_n we denote the two-sided Banach module over $\mathbb{R}_n$ corresponding to $V \otimes \mathbb{R}_n$. An element in V_n is of the type

$$\sum_A v_A \otimes e_A$$

$A = i_1 \dots i_r$, $i_\ell \in \{1, 2, \dots, n\}$, $i_1 < \dots < i_r$ is a multi-index, $v_A \in V$, and e_A is a basis element in the Clifford algebra $\mathbb{R}_n$. The multiplications of an element $v \in V_n$ with a scalar $a \in \mathbb{R}_n$ are defined as

$$va = \sum_A v_A \otimes (e_A a)$$

and

$$av = \sum_A v_A \otimes (a e_A).$$

We will write $\sum_A v_A e_A$ instead of $\sum_A v_A \otimes e_A$ and we define

$$\|v\|_{V_n} = \sum_A \|v_A\|_V.$$

Remark 3.0.1. A two-sided module V_n over $\mathbb{R}_n$ is called a Banach module over $\mathbb{R}_n$, if there exists a constant $C \geq 1$ such that

$$\|va\|_{V_n} \leq C\|v\|_{V_n}|a| \quad \text{and} \quad \|av\|_{V_n} \leq C|a|\|v\|_{V_n}$$

for all $v \in V_n$ and $a \in \mathbb{R}_n$. In general, the constant C can be chosen equal to 2^n. The importance of Banach modules is well known and for a thorough discussion we refer the reader to [5].

We denote by $\mathcal{B}(V)$ the space of all bounded $\mathbb{R}$-homomorphisms from the Banach space V into itself endowed with the natural norm denoted by $\|\cdot\|_{\mathcal{B}(V)}$. Let $T_A \in \mathcal{B}(V)$. We define the operator

$$T = \sum_A T_A e_A$$

and its action on the generic element of V_n,

$$v = \sum_B v_B e_B,$$

as

$$T(v) = \sum_{A,B} T_A(v_B) e_A e_B.$$

The operator T is a right-module homomorphism which is a bounded linear map on V_n. The set of all such bounded operators is denoted by $\mathcal{B}_n(V_n)$. We define a norm in $\mathcal{B}_n(V_n)$ by setting

$$\|T\|_{\mathcal{B}_n(V_n)} = \sum_A \|T_A\|_{\mathcal{B}(V)}.$$

It can be proved that

$$\|TS\|_{\mathcal{B}_n(V_n)} \leq \|T\|_{\mathcal{B}_n(V_n)} \|S\|_{\mathcal{B}_n(V_n)}.$$

From now on we will omit the subscripts V_n and $\mathcal{B}_n(V_n)$ when dealing with the norms. The context will clarify which norm we will be using.

3.1 The S-resolvent operator and the S-spectrum

Given an n-tuple $(T_1, \dots, T_n)$ of not necessarily commuting operators, we can construct the operator in $\mathcal{B}_n(V_n)$ given by

$$\sum_{j=1}^n e_j T_j, \quad T_j \in \mathcal{B}(V). \tag{3.1}$$

As we will show in the sequel, when $n = 1$, i.e., in the case of a single operator, this approach is consistent with the standard Riesz–Dunford calculus. However, our methods allow us to treat a more general situation and therefore we can also consider an operator of the form

$$T = T_0 + \sum_{j=1}^{n} e_j T_j, \qquad T_\mu \in \mathcal{B}(V), \qquad \mu = 0, 1, \dots, n, \tag{3.2}$$

which is slightly more general than the form (3.1).

Warning. We will develop our theory for these more general operators even though, when dealing with n-tuples of operators, we always consider them as in the form (3.1) or, in other words, in the form (3.2) with $T_0 \equiv 0$. With an abuse of language we will always refer to them as n-tuples of operators. Note that our construction embeds an n-tuple of operators into $\mathcal{B}_n(V_n)$ as a right linear operator acting on V_n.

The set of bounded operators of the form (3.1) or (3.2) will be denoted by $\mathcal{B}_n^1(V_n)$ or $\mathcal{B}_n^{0,1}(V_n)$, respectively. We obviously have the inclusions

$$\mathcal{B}_n^1(V_n) \subset \mathcal{B}_n^{0,1}(V_n) \subset \mathcal{B}_n(V_n).$$

We now give a definition and a theorem which are of crucial importance to construct a functional calculus for n-tuples of noncommuting operators using the theory of s-monogenic functions. In this section we consider left s-monogenic functions, but it is possible to develop this theory using right s-monogenic functions.

Definition 3.1.1 (**S-resolvent operator series**). *Let $T \in \mathcal{B}_n^{0,1}(V_n)$ and $\mathbf{s} \in \mathbb{R}^{n+1}$. We define the Cauchy kernel operator series or S-resolvent operator series as*

$$S^{-1}(\mathbf{s}, T) := \sum_{n \geq 0} T^n \mathbf{s}^{-1-n} \tag{3.3}$$

for $\|T\| < |\mathbf{s}|$.

Remark 3.1.2. Note that $\|T^n s^{-1-n}\| < C\|T\|^n |\mathbf{s}|^{-1-n}$ since $\mathbf{s}$ is a paravector (compare with the definition of a Banach module). Thus we have

$$\begin{aligned} \|\sum_{n\geq 0} T^n \mathbf{s}^{-n-1}\| &\leq \sum_{n\geq 0} \|T^n \mathbf{s}^{-n-1}\| \\ &\leq C \sum_{n\geq 0} \|T^n\| |\mathbf{s}^{-n-1}| \leq C \sum_{n\geq 0} \|T^n\| |\mathbf{s}|^{-n-1}. \end{aligned}$$

Theorem 3.1.3. *Let $T \in \mathcal{B}_n^{0,1}(V_n)$ and $\mathbf{s} \in \mathbb{R}^{n+1}$. Then*

$$\sum_{n\geq 0} T^n \mathbf{s}^{-1-n} = -(T^2 - 2\mathrm{Re}[\mathbf{s}]T + |\mathbf{s}|^2 \mathcal{I})^{-1}(T - \overline{\mathbf{s}}\mathcal{I}), \tag{3.4}$$

for $\|T\| < |\mathbf{s}|$.

Proof. In Theorem 2.7.4 the components of $\mathbf{p}$ and $\mathbf{s}$ are real numbers and therefore such components obviously commute. When we formally replace $\mathbf{p}$ by the operator T we do not want to assume that necessarily $T_\mu T_\nu = T_\nu T_\mu$, and so we need to verify independently that (3.4) still holds. In what follows, we assume the convergence of the series to be in the norm of $\mathcal{B}_n(V_n)$. We need to prove that

$$-(T^2 - 2\mathrm{Re}[\mathbf{s}]T + |\mathbf{s}|^2\mathcal{I}) \sum_{n\geq 0} T^n \mathbf{s}^{-1-n} = (T - \bar{\mathbf{s}}\mathcal{I}),$$

i.e., that

$$(-T^2 + 2\mathrm{Re}[\mathbf{s}]T - |\mathbf{s}|^2\mathcal{I}) \sum_{n\geq 0} T^n \mathbf{s}^{-1-n} = T + (\mathbf{s} - 2\mathrm{Re}[\mathbf{s}])\mathcal{I}.$$

Observing that $-T^2 + 2\mathrm{Re}[\mathbf{s}]T - |\mathbf{s}|^2\mathcal{I}$ commutes with T^n we can rewrite the assertion as

$$\sum_{n\geq 0} T^n(-|\mathbf{s}|^2 - T^2 + 2\mathrm{Re}[\mathbf{s}]T)\mathbf{s}^{-1-n} = T + (\mathbf{s} - 2\mathrm{Re}[\mathbf{s}])\mathcal{I}.$$

We can rewrite the left-hand side of this equality by expanding the series

$$\begin{aligned}
&\sum_{n\geq 0} T^n(-|\mathbf{s}|^2\mathcal{I} - T^2 + 2\mathrm{Re}[\mathbf{s}]T)\mathbf{s}^{-1-n} \\
&= (-|\mathbf{s}|^2\mathcal{I} - T^2 + 2\mathrm{Re}[\mathbf{s}]T)\mathbf{s}^{-1} + T^1(-|\mathbf{s}|^2\mathcal{I} - T^2 + 2\mathrm{Re}[\mathbf{s}]T)\mathbf{s}^{-2} \\
&\quad + T^2(-|\mathbf{s}|^2\mathcal{I} - T^2 + 2\mathrm{Re}[\mathbf{s}]T)\mathbf{s}^{-3} + \dots \\
&= -\Big(|\mathbf{s}|^2\mathbf{s}^{-1} + T(-2\mathbf{s}\mathrm{Re}[\mathbf{s}] + |\mathbf{s}|^2)\mathbf{s}^{-2} + T^2(\mathbf{s}^2 - 2\mathbf{s}\mathrm{Re}[\mathbf{s}] + |\mathbf{s}|^2)\mathbf{s}^{-3} + \dots\Big),
\end{aligned}$$

and using the identity $\mathbf{s}^2 - 2\mathbf{s}\mathrm{Re}[\mathbf{s}] + |\mathbf{s}|^2 = 0$, we get

$$\begin{aligned}
&\sum_{n\geq 0} T^n(-|\mathbf{s}|^2 - T^2 + 2\mathrm{Re}[\mathbf{s}]T)\mathbf{s}^{-1-n} = -|\mathbf{s}|^2\mathbf{s}^{-1}\mathcal{I} + T\mathbf{s}^2\mathbf{s}^{-2} \\
&= -|\mathbf{s}|^2\mathbf{s}^{-1}\mathcal{I} + T = -\bar{\mathbf{s}}\mathbf{s}\mathbf{s}^{-1}\mathcal{I} + T = -\bar{\mathbf{s}}\mathcal{I} + T = (\mathbf{s} - 2\ \mathrm{Re}[\mathbf{s}])\mathcal{I} + T,
\end{aligned}$$

which concludes the proof. □

The S-resolvent operator series is the analog for operators of the noncommutative Cauchy series for s-monogenic functions. Theorem 3.1.3 shows that a functional calculus for n-tuples of noncommuting operators can be constructed using the Cauchy formula II, simply by replacing in the s-monogenic Cauchy kernel,

$$S^{-1}(\mathbf{s}, \mathbf{x}) = -(\mathbf{x}^2 - 2\mathbf{x}\mathrm{Re}[\mathbf{s}] + |\mathbf{s}|^2)^{-1}(\mathbf{x} - \bar{\mathbf{s}}),$$

the paravector $\mathbf{x}$ by an operator T of the form (3.2), even though the components of T do not commute.

Theorem 3.1.3 shows that the sum of the S-resolvent series exists in a set larger than the ball $\|T\| < |\mathbf{s}|$ and, given an operator of the form (3.2), it is possible to associate to it a new notion of spectrum.

Definition 3.1.4 (The S-spectrum and the S-resolvent set). *Let $T \in \mathcal{B}_n^{0,1}(V_n)$ and $\mathbf{s} \in \mathbb{R}^{n+1}$. We define the S-spectrum $\sigma_S(T)$ of T as*

$$\sigma_S(T) = \{\mathbf{s} \in \mathbb{R}^{n+1} \ : \ T^2 - 2\,\mathrm{Re}[\mathbf{s}]T + |\mathbf{s}|^2\mathcal{I} \quad \textit{is not invertible}\}.$$

The S-resolvent set $\rho_S(T)$ is defined by

$$\rho_S(T) = \mathbb{R}^{n+1} \setminus \sigma_S(T).$$

Definition 3.1.5 (The S-resolvent operator). *Let $T \in \mathcal{B}_n^{0,1}(V_n)$ and $\mathbf{s} \in \rho_S(T)$. We define the S-resolvent operator as*

$$S^{-1}(\mathbf{s}, T) := -(T^2 - 2\mathrm{Re}[\mathbf{s}]T + |\mathbf{s}|^2\mathcal{I})^{-1}(T - \bar{\mathbf{s}}\mathcal{I}). \tag{3.5}$$

Note that we are using the same symbol for both the S-resolvent operator series and the S-resolvent operator: no confusion can arise since it will always be clear which object we will be considering.

Proposition 3.1.6. *Let $T \in \mathcal{B}_n^{0,1}(V_n)$ and $\mathbf{s} \in \rho_S(T)$. Then if $T\mathbf{s} = \mathbf{s}T$, we have*

$$S^{-1}(\mathbf{s}, T) = (\mathbf{s}\mathcal{I} - T)^{-1}.$$

Proof. It follows by the commutativity since we have

$$(T^2 - 2\mathrm{Re}[\mathbf{s}]T + |\mathbf{s}|^2\mathcal{I})^{-1}(T - \bar{\mathbf{s}}\mathcal{I}) = (T - \mathbf{s}\mathcal{I})^{-1}(T - \bar{\mathbf{s}}\mathcal{I})^{-1}(T - \bar{\mathbf{s}}\mathcal{I}). \qquad \square$$

Example 3.1.7 (Pauli matrices). As an example, we compute the S-spectrum of two Pauli matrices σ_3, σ_1 (compare with example 4.10 in [62]):

$$\sigma_3 = \begin{bmatrix} 1 & 0 \\ 0 & -1 \end{bmatrix} \qquad \sigma_1 = \begin{bmatrix} 0 & 1 \\ 1 & 0 \end{bmatrix}.$$

Let us consider the matrix $T = \sigma_3 e_1 + \sigma_1 e_2$ and let us compute $T^2 - 2\mathrm{Re}[\mathbf{s}]T + |\mathbf{s}|^2\mathcal{I}$. We obtain the matrix

$$\begin{bmatrix} |\mathbf{s}|^2 - 2 - 2\mathrm{Re}[\mathbf{s}]e_1 & 2(e_1 - \mathrm{Re}[\mathbf{s}])e_2 \\ -2(e_1 + \mathrm{Re}[\mathbf{s}])e_2 & |\mathbf{s}|^2 - 2 + 2\mathrm{Re}[\mathbf{s}]e_1 \end{bmatrix}$$

whose S-spectrum is $\sigma_S(T) = \{0\} \cup \{\mathbf{s} \in \mathbb{R}^3 \ : \ \mathrm{Re}[\mathbf{s}] = 0, |\mathbf{s}| = 2\}$.

Remark 3.1.8. Note that if we embed the pair (σ_3, σ_1) as $\sigma_1 e_1 + \sigma_3 e_2$, the S-spectrum does not change. In general, when we embed an n-tuple of operators using a different order of the imaginary units, the S-spectrum will not be affected, as we will see in the next section.

Theorem 3.1.9. *Let $T \in \mathcal{B}_n^{0,1}(V_n)$ and $\mathbf{s} \in \rho_S(T)$. Let $S^{-1}(\mathbf{s}, T)$ be the S-resolvent operator defined in (3.5). Then $S^{-1}(\mathbf{s}, T)$ satisfies the (S-resolvent) equation*

$$S^{-1}(\mathbf{s}, T)\mathbf{s} - TS^{-1}(\mathbf{s}, T) = \mathcal{I}. \tag{3.6}$$

Proof. Replacing (3.5) in the above equation we have

$$\mathcal{I} = -(T^2 - 2\mathrm{Re}[\mathbf{s}]T + |\mathbf{s}|^2\mathcal{I})^{-1}(T - \bar{\mathbf{s}}\mathcal{I})\mathbf{s} + T(T^2 - 2\mathrm{Re}[\mathbf{s}]T + |\mathbf{s}|^2\mathcal{I})^{-1}(T - \bar{\mathbf{s}}\mathcal{I}) \tag{3.7}$$

and applying $(T^2 - 2\mathrm{Re}[\mathbf{s}]T + |\mathbf{s}|^2\mathcal{I})$ to both sides of (3.7), we get

$$T^2 - 2\mathrm{Re}[\mathbf{s}]T + |\mathbf{s}|^2\mathcal{I} = -(T - \bar{\mathbf{s}}\mathcal{I})\mathbf{s} + (T^2 - 2\mathrm{Re}[\mathbf{s}]T + |\mathbf{s}|^2\mathcal{I})T(T^2 - 2\mathrm{Re}[\mathbf{s}]T + |\mathbf{s}|^2\mathcal{I})^{-1}(T - \bar{\mathbf{s}}\mathcal{I}).$$

Since T and $T^2 - 2\mathrm{Re}[\mathbf{s}]T + |\mathbf{s}|^2\mathcal{I}$ commute, we obtain the identity

$$T^2 - 2\mathrm{Re}[\mathbf{s}]T + |\mathbf{s}|^2\mathcal{I} = -(T - \bar{\mathbf{s}}\mathcal{I})\mathbf{s} + T(T - \bar{\mathbf{s}}\mathcal{I})$$

which proves the statement. □

3.2 Properties of the S-spectrum

We state here some properties of the S-spectrum. In particular, we show that the S-spectrum consists of $(n-1)$-spheres (which, in particular, may reduce to points on the real axis) and therefore it has a structure that is compatible with the admissible domain for s-monogenic functions. We also show that the S-spectrum for n-tuples of bounded operators is compact.

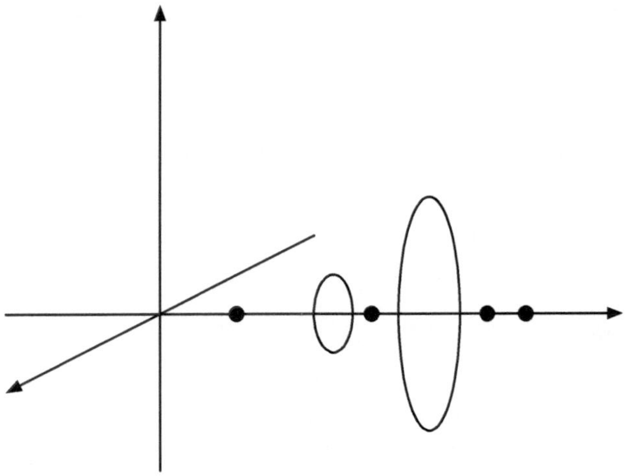

Figure 3.1: Structure of the S-spectrum in $\mathbb{R}^3$

Theorem 3.2.1 (Structure of the S-spectrum). *Let $T \in \mathcal{B}^{0,1}_n(V_n)$. Then its S-spectrum consists of the union of $(n-1)$-spheres of the form $[\mathbf{p}]$, $\mathbf{p} \in \mathbb{R}^{n+1}$.*

Proof. Let $\mathbf{p} = \mathrm{Re}[\mathbf{p}] + \underline{p} \in \sigma_S(T)$. If $\mathbf{p}$ is not real, then $\underline{p} \neq 0$ and all the elements of the sphere $\mathbf{s} = \mathrm{Re}[\mathbf{s}] + \underline{s}$ with $\mathrm{Re}[\mathbf{s}] = \mathrm{Re}[\mathbf{p}]$ and $|\underline{s}| = |\underline{p}|$ belong to the S-spectrum of T. □

Definition 3.2.2. *Let $T = T_0 + \sum_{j=1}^n e_jT_j \in \mathcal{B}_n^{0,1}(V_n)$. Let $U \subset \mathbb{R}^{n+1}$ be an axially symmetric s-domain that contains the S-spectrum $\sigma_S(T)$ of T and such that $\partial(U \cap \mathbb{C}_I)$ is the union of a finite number of continuously differentiable Jordan curves for every $I \in \mathbb{S}$. Then U is said to be a T-admissible open set.*

Definition 3.2.3. *Let $T = T_0 + \sum_{j=1}^n e_jT_j \in \mathcal{B}_n^{0,1}(V_n)$. Let $U \subset \mathbb{R}^{n+1}$ be a T-admissible open set. Suppose that $\overline{U}$ is contained in a domain of s-monogenicity of a function f. Then such a function f is said to be locally s-monogenic on $\sigma_S(T)$. We will denote by $\mathcal{M}_{\sigma_S(T)}$ the set of locally s-monogenic functions on $\sigma_S(T)$.*

Remark 3.2.4. Let W be an open set in $\mathbb{R}^{n+1}$ and let $f \in \mathcal{M}(W)$. In the Cauchy formula (2.37) the open set $U \subset W$ need not be necessarily connected. Indeed formula (2.37) obviously holds when $U = \cup_{i=1}^r U_i$, $\overline{U}_i \cap \overline{U}_j = \emptyset$ for $i \neq j$, where U_i are axially symmetric s-domains for all $i = 1, \dots, r$ and the boundaries of $U_i \cap \mathbb{C}_I$ consist of a finite number of continuously differentiable Jordan curves for $I \in \mathbb{S}$ for all $i = 1, \dots, r$. So when we choose $f \in \mathcal{M}_{\sigma_S(T)}$ the associated open set U need not be connected. With an abuse of language, we will call T-admissible such an open set U.

We now give a result that motivates the functional calculus.

Theorem 3.2.5. *Let $\mathbf{s} \in \mathbb{R}^{n+1}$, $a \in \mathbb{R}_n$, $m \in \mathbb{N} \cup \{0\}$ and consider the monomial $\mathbf{s}^m a$. Consider $T \in \mathcal{B}_n^{0,1}(V_n)$, let $U \subset \mathbb{R}^{n+1}$ be a T-admissible open set. Then, for every choice of $I \in \mathbb{S}$, we have*

$$T^m a = \frac{1}{2\pi} \int_{\partial(U\cap\mathbb{C}_I)} S^{-1}(\mathbf{s}, T)\; d\mathbf{s}_I\; \mathbf{s}^m a. \tag{3.8}$$

Proof. Let us consider the power series expansion of the S-resolvent operator $S^{-1}(\mathbf{s}, T)$ and a circle C_r centered in the origin and of radius $r > \|T\|$. We have

$$\begin{aligned} &\frac{1}{2\pi} \int_{\partial(U\cap\mathbb{C}_I)} S^{-1}(\mathbf{s}, T)\; d\mathbf{s}_I\; \mathbf{s}^m a \\ &= \frac{1}{2\pi} \sum_{n\geq 0} T^n \int_{\partial(U\cap\mathbb{C}_I)} \mathbf{s}^{-1-n+m}\; d\mathbf{s}_I a. \end{aligned} \tag{3.9}$$

Since

$$\int_{C_r} d\mathbf{s}_I \mathbf{s}^{-n-1+m} = 0 \text{ if } n \neq m, \quad \int_{C_r} d\mathbf{s}_I \mathbf{s}^{-n-1+m} = 2\pi \text{ if } n = m, \tag{3.10}$$

and since, by the Cauchy theorem, the above integrals are not affected if we replace C_r by $\partial(U \cap \mathbb{C}_I)$ for any $I \in \mathbb{S}$, we have

$$\frac{1}{2\pi} \sum_{n\geq 0} T^n \int_{C_r} \mathbf{s}^{-1-n+m}\; d\mathbf{s}_I a = \frac{1}{2\pi} \sum_{n\geq 0} T^n \int_{\partial(U\cap\mathbb{C}_I)} \mathbf{s}^{-1-n+m}\; d\mathbf{s}_I a = T^m a.$$ □

As it is well known, when we consider a single operator T acting on a Banach space (real or complex) we know that its spectrum is a compact set contained in the ball with center at the origin and radius $\|T\|$. In our case, we have that a similar property holds for the S-spectrum:

Theorem 3.2.6 (Compactness of S-spectrum). *Let $T \in \mathcal{B}_n^{0,1}(V_n)$. Then the S-spectrum $\sigma_S(T)$ is a compact nonempty set. Moreover $\sigma_S(T)$ is contained in $\{\mathbf{s} \in \mathbb{R}^{n+1} : |\mathbf{s}| \leq \|T\| \}$.*

Proof. Let $U \subset \mathbb{R}^{n+1}$ be a T-admissible open set and let $I \in \mathbb{S}$. Then

$$\frac{1}{2\pi} \int_{\partial(U\cap\mathbb{C}_I)} S^{-1}(\mathbf{s},T)\, d\mathbf{s}_I\, \mathbf{s}^m = T^m.$$

In particular, for $m = 0$, we have

$$\frac{1}{2\pi} \int_{\partial(U\cap\mathbb{C}_I)} S^{-1}(\mathbf{s},T)\, d\mathbf{s}_I = \mathcal{I},$$

where $\mathcal{I}$ denotes the identity operator. This shows that $\sigma_S(T)$ is a nonempty set, otherwise the integral would be zero by the vector-valued version of Cauchy's theorem. We now show that the S-spectrum is bounded. The series $\sum_{n\geq 0} T^n \mathbf{s}^{-1-n}$ converges if and only if $\|T\| < |\mathbf{s}|$ so the S-spectrum is contained in the set $\{\mathbf{s} \in \mathbb{R}^{n+1} : |\mathbf{s}| \leq \|T\|\}$, which is bounded and closed because the complement of $\sigma_S(T)$, i.e., $\rho_S(T)$, is open. Indeed, the function

$$g : \mathbf{s} \mapsto T^2 - 2\mathrm{Re}[\mathbf{s}]T + |\mathbf{s}|^2\mathcal{I}$$

is trivially continuous and, by Theorem 10.12 in [91], the set $\mathcal{U}(V_n)$ of all invertible elements of $\mathcal{B}_n(V_n)$ is an open set in $\mathcal{B}_n(V_n)$. Therefore $g^{-1}(\mathcal{U}(V_n)) = \rho_S(T)$ is an open set in $\mathbb{R}^{n+1}$. □

3.3 The functional calculus

The following result is an immediate consequence of the Hahn-Banach theorem for Banach modules over $\mathbb{R}_n$ (see [7], §2.10) and it will be used in the proof of the next result.

Corollary 3.3.1. *Let V_n be a right module over $\mathbb{R}_n$ and let $v \in V_n$. If $\langle \phi, v\rangle = 0$ for every linear and continuous functional ϕ in V_n', then $v = 0$.*

We now state and prove a crucial result that will allow us to define the functional calculus for n-tuples of not necessarily commuting operators. More precisely, when we replace in the Cauchy formula (2.37) the variable $\mathbf{x}$ with an operator $T \in \mathcal{B}_n^{0,1}(V_n)$, we have to verify that the integral remains independent of U and of $I \in \mathbb{S}$.

Theorem 3.3.2. *Let $T \in \mathcal{B}_n^{0,1}(V_n)$ and $f \in \mathcal{M}_{\sigma_S(T)}$. Let $U \subset \mathbb{R}^{n+1}$ be a T-admissible open set and set $d\mathbf{s}_I = -d\mathbf{s}I$ for $I \in \mathbb{S}$. Then the integral*

$$\frac{1}{2\pi}\int_{\partial(U\cap\mathbb{C}_I)} S^{-1}(\mathbf{s},T)\ d\mathbf{s}_I\ f(\mathbf{s}) \tag{3.11}$$

does not depend on the open set U and on the choice of the imaginary unit $I \in \mathbb{S}$.

Proof. We first observe that we can replace $\mathbf{x}$ by an operator $T \in \mathcal{B}_n^{0,1}(V_n)$ in the Cauchy formula (2.37); in fact Theorem 3.1.3 allows us to replace $\mathbf{x}$ by T in the function $S^{-1}(\mathbf{s},\mathbf{x})$, thus we can do the same substitution in the Cauchy formula (2.37) and write

$$\frac{1}{2\pi}\int_{\partial(U\cap\mathbb{C}_I)} S^{-1}(\mathbf{s},T)\ d\mathbf{s}_I\ f(\mathbf{s}).$$

For every linear and continuous functional $\phi \in V_n'$, consider the duality $\langle\phi, S^{-1}(\mathbf{s},T)v\rangle$, for $v \in V_n$ and define the function

$$g(\mathbf{s}) := \langle\phi, S^{-1}(\mathbf{s},T)v\rangle, \quad \text{for} \quad v \in V_n, \quad \phi \in V_n'. \tag{3.12}$$

We observe that the function $S^{-1}(\mathbf{s},\mathbf{x})$ is right s-monogenic in the variable $\mathbf{s}$ in its domain of definition, thanks to Proposition 2.7.9. The function g is right s-monogenic in the variable $\mathbf{s}$ on $\rho_S(T)$, and since $g(\mathbf{s}) \to 0$ as $\mathbf{s} \to \infty$ we have that g is s-monogenic also at infinity. The independence of the integral (3.11) from the choice of the T-admissible open set U, as long as $\partial(U \cap \mathbb{C}_I)$ does not cross the S-spectrum of T for every $I \in \mathbb{S}$, is a consequence of the Hahn-Banach theorem and of the Cauchy formula. Indeed, for any fixed $I \in \mathbb{S}$, the integral

$$\frac{1}{2\pi}\int_{\partial(U\cap\mathbb{C}_I)} g(\mathbf{s})d\mathbf{s}_I\ f(\mathbf{s}) \tag{3.13}$$

does not depend on U, by the Cauchy theorem. As a consequence, also the integral (3.11) does not depend on U, by Corollary 3.3.1. We now prove that the integral (3.13) does not depend on $I \in \mathbb{S}$. Since g is a right s-monogenic function on $\rho_S(T)$, we can consider a domain U' in $\rho_S(T)$ such that U' satisfies the hypothesis of Theorem 2.8.4 and such that $\partial U' \subset U$. We now choose $J \neq I$, $J \in \mathbb{S}$ and we write the function $g(\mathbf{s})$ using the Cauchy integral formula (2.39):

$$g(\mathbf{s}) = -\frac{1}{2\pi}\int_{\partial(U'\cap\mathbb{C}_J)^-} g(\mathbf{t})\ d\mathbf{t}_J\ S^{-1}(\mathbf{s},\mathbf{t}) \tag{3.14}$$

where the Jordan curve $\partial(U' \cap \mathbb{C}_J)^-$ is oriented clockwise. Observe that the orientation is chosen in order to include the singular points of $S^{-1}(\mathbf{s},\mathbf{t})$, i.e., $[\mathbf{s}] \cap \mathbb{C}_J \in \partial(U \cap \mathbb{C}_J)$ and to exclude the points belonging to the S-spectrum of T. Taking into account the orientation of $\partial(U' \cap \mathbb{C}_J)^-$ we can rewrite the integral (3.14) as

$$g(\mathbf{s}) = \frac{1}{2\pi}\int_{\partial(U'\cap\mathbb{C}_J)} g(\mathbf{t})\ d\mathbf{t}_J\ S^{-1}(\mathbf{s},\mathbf{t}). \tag{3.15}$$

Let us now substitute the expression of $g(\mathbf{s})$ in (3.15) into the integral (3.13) so that we obtain

$$\begin{aligned}
&\frac{1}{2\pi}\int_{\partial(U\cap\mathbb{C}_I)} g(\mathbf{s})\, d\mathbf{s}_I\, f(\mathbf{s}) \qquad (3.16)\\
&= \frac{1}{2\pi}\int_{\partial(U\cap\mathbb{C}_I)} \Big[\frac{1}{2\pi}\int_{\partial(U'\cap\mathbb{C}_J)} g(\mathbf{t})\, dt_J\, S^{-1}(\mathbf{s},\mathbf{t})\Big]\, d\mathbf{s}_I\, f(\mathbf{s})\\
&= \frac{1}{2\pi}\int_{\partial(U'\cap\mathbb{C}_J)} g(\mathbf{t})\, d\mathbf{t}_J\Big[\frac{1}{2\pi}\int_{\partial(U\cap\mathbb{C}_I)} S^{-1}(\mathbf{s},\mathbf{t})\, d\mathbf{s}_I\, f(\mathbf{s})\Big]\\
&= \frac{1}{2\pi}\int_{\partial(U'\cap\mathbb{C}_J)} g(\mathbf{t})\, d\mathbf{t}_J f(\mathbf{t}),
\end{aligned}$$

where we have used the Fubini theorem and Theorem 2.8.4. Since $\partial(U'\cap\mathbb{C}_J)$ is positively oriented and surrounds the S-spectrum of T, by the first part of the statement, we can substitute it by $\partial(U\cap\mathbb{C}_J)$, because of the independence of the integral on the open set U, and we get

$$\frac{1}{2\pi}\int_{\partial(U\cap\mathbb{C}_I)} g(\mathbf{s})\, d\mathbf{s}_I\, f(\mathbf{s}) = \frac{1}{2\pi}\int_{\partial(U\cap\mathbb{C}_J)} g(\mathbf{t})\, d\mathbf{t}_J\, f(\mathbf{t}). \qquad (3.17)$$

Since $g(\mathbf{t}) = \langle \phi, S^{-1}(\mathbf{s},T)v\rangle$ and the formula (3.17) holds for every $v\in V_n$, $\phi\in V_n'$, and for I, $J\in\mathbb{S}$, by Corollary 3.3.1 the integral (3.11) does not depend on $I\in\mathbb{S}$. □

We can now define our functional calculus.

Definition 3.3.3. *Let $T\in\mathcal{B}_n^{0,1}(V_n)$ and $f\in\mathcal{M}_{\sigma_S(T)}$. Let $U\subset\mathbb{R}^{n+1}$ be a T-admissible open set and set $d\mathbf{s}_I = -dsI$ for $I\in\mathbb{S}$. We define*

$$f(T) := \frac{1}{2\pi}\int_{\partial(U\cap\mathbb{C}_I)} S^{-1}(\mathbf{s},T)\, d\mathbf{s}_I\, f(\mathbf{s}). \qquad (3.18)$$

3.4 Algebraic rules

We recall that in general it is not true that the product of two s-monogenic functions is still s-monogenic. However this is true for a subset of s-monogenic functions, as we will show in the next proposition.

Let U be an open set in $\mathbb{R}^{n+1}$ and let $f\in\mathcal{M}(U)$. Choose $I=I_1\in\mathbb{S}$ and let $I_2,\dots,I_n$ be a completion to a basis of $\mathbb{R}_n$ such that $I_iI_j+I_jI_i=-2\delta_{ij}$. Denote by f_I the restriction of f to $\mathbb{C}_I$. By the Splitting Lemma we have

$$f_I(z) = \sum_{|A|=0}^{n-1} F_A(z)I_A, \quad I_A = I_{i_1}\dots I_{i_s}, \quad z=u+Iv$$

where $F_A : U\cap\mathbb{C}_I\to\mathbb{C}_I$ are holomorphic functions. The multi-index $A=i_1\dots i_s$ is such that $i_\ell\in\{2,\dots,n\}$, with $i_1<\dots<i_s$, or, when $|A|=0$, $I_\emptyset=1$.

Definition 3.4.1. *Let U be an open set in $\mathbb{R}^{n+1}$ and let $f \in \mathcal{M}(U)$. We denote by $\widetilde{\mathcal{M}}(U)$ the subclass of $\mathcal{M}(U)$ consisting of those functions f such that for all the possible choices of $I \in \mathbb{S}$ there is a completion $I_2, \dots, I_n$ to an orthonormal basis of $\mathbb{R}_n$ such that*

$$f_I(z) = \sum_{|A| \text{even}} F_A(z) I_A, \quad I_A = I_{i_1} \dots I_{i_s}, \quad z = u + Iv.$$

Proposition 3.4.2. *Let U be an open set in $\mathbb{R}^{n+1}$. Let $f \in \widetilde{\mathcal{M}}(U)$, $g \in \mathcal{M}(U)$, then $fg \in \mathcal{M}(U)$.*

Proof. Let $I \in \mathbb{S}$ and set $z = u + Iv$, then we have

$$\begin{aligned}
&\left(\frac{\partial}{\partial u} + I\frac{\partial}{\partial v}\right)(fg)_I(z) \\
&= \frac{\partial f_I}{\partial u}(z) g_I(z) + f_I(z)\frac{\partial g_I}{\partial u}(z) + I\frac{\partial f_I}{\partial v}(z) g_I(z) + I f_I(z)\frac{\partial g_I}{\partial v}(z).
\end{aligned}$$

Since $f \in \widetilde{\mathcal{M}}(U)$, the components F_A of $f_I(z)$ commute with I and with all the I_A since $|A|$ is even so f commutes with I, thus we obtain

$$\begin{aligned}
&\left(\frac{\partial}{\partial u} + I\frac{\partial}{\partial v}\right)(fg)_I(z) \\
&= \Big(\frac{\partial f_I}{\partial u}(z) + I\frac{\partial f_I}{\partial v}(z)\Big) g_I(z) + f_I(z)\Big(\frac{\partial g_I}{\partial u}(z) + I\frac{\partial g_I}{\partial v}(z)\Big) = 0.
\end{aligned}$$

□

Remark 3.4.3. The condition $f \in \widetilde{\mathcal{M}}(U)$ is not only sufficient but also necessary to have that $fg \in \mathcal{M}(U)$, when $f, g \in \mathcal{M}(U)$ and g is nonconstant. Indeed, for every choice of $I \in \mathbb{S}$ we have that

$$\left(\frac{\partial}{\partial u} + I\frac{\partial}{\partial v}\right)(fg)_I(z) = f_I(z)\frac{\partial g_I}{\partial u}(z) + I f_I(z)\frac{\partial g_I}{\partial v}(z) = 0.$$

Since g is s-monogenic we can write:

$$-f_I(z) I \frac{\partial g_I}{\partial v}(z) + I f_I(z)\frac{\partial g_I}{\partial v}(z) = (-f_I(z) I + I f_I(z))\frac{\partial g_I}{\partial v}(z) = 0.$$

Since $\frac{\partial g_I}{\partial v}(z) \neq 0$ in view of the fact that g is s-monogenic and nonconstant, the last equality implies that $f_I(z) I = I f_I(z)$ for all $I \in \mathbb{S}$. Using the Splitting Lemma on each plane $\mathbb{C}_I$ we obtain that the only possibility is that only the indices such that $|A|$ is even can appear.

Example 3.4.4. Let us consider a function f defined by the power series

$$f(\mathbf{x}) = \sum_{m \geq 0} \mathbf{x}^m a_m \qquad a_m \in [\mathbb{R}_n]_2,$$

converging on a suitable ball B. Assume, for simplicity, that the coefficients a_m are the product of two imaginary units e_i. Consider $I \in \mathbb{S}$ and a completion $I_2, \dots, I_n$ to an orthonormal basis of $\mathbb{R}_n$, $n \geq 2$. Then any a_m is the product of two linear combinations of the imaginary units I_j. Thus

$$\begin{aligned} f_I(u+Iv) &= \sum_{m\geq 0}(u+Iv)^m a_m = \sum_{m\geq 0}\sum_{|A|=0,2}(u+Iv)^m a'_{mA} I_A \\ &= \sum_{|A|=0,2}\Big(\sum_{m\geq 0}(u+Iv)^m a'_{mA}\Big) I_A = \sum_{|A|=0,2} F_A(u+Iv) I_A, \end{aligned}$$

where $a'_m \in \mathbb{R}$, belongs to $\widetilde{\mathcal{M}}(B)$.

Definition 3.4.5. *In Definition* 3.2.3 *consider instead of s-monogenic functions, the subset of functions in $\widetilde{\mathcal{M}}$. This subclass of $\mathcal{M}_{\sigma_S(T)}$ will be denoted by $\widetilde{\mathcal{M}}_{\sigma_S(T)}$.*

Theorem 3.4.6. *Let $T \in \mathcal{B}^{0,1}_n(V_n)$.*

(a) *Let f and $g \in \mathcal{M}_{\sigma_S(T)}$. Then we have*

$$(f+g)(T) = f(T) + g(T), \qquad (f\lambda)(T) = f(T)\lambda, \qquad \text{for all } \lambda \in \mathbb{R}_n.$$

(b) *Let $\phi \in \widetilde{\mathcal{M}}_{\sigma_S(T)}$ and $g \in \mathcal{M}_{\sigma_S(T)}$. Then we have*

$$(\phi g)(T) = \phi(T) g(T).$$

(c) *Let $f(\mathbf{s}) = \sum_{m\geq 0}\mathbf{s}^m p_m$ where $p_m \in \mathbb{R}_n$ be such that $f \in \mathcal{M}_{\sigma_S(T)}$. Then we have*

$$f(T) = \sum_{m\geq 0} T^m p_m.$$

Proof. Part (a) is a direct consequence of Definition 3.3.3.

Part (b): Denote by U a T-admissible open set on which g is s-monogenic and let $\phi \in \widetilde{\mathcal{M}}(U)$. By Proposition 3.4.2 the product ϕg belongs to $\mathcal{M}(U)$. Let G_1 and G_2 be two T-admissible open sets such that $G_1 \cup \partial G_1 \subset G_2$ and $G_2 \cup \partial G_2 \subset U$. Take $\mathbf{s} \in \partial G_1$ and $\mathbf{t} \in \partial G_2$ and observe that, for $I \in \mathbb{S}$, we have

$$g(\mathbf{s}) = \frac{1}{2\pi}\int_{\partial(G_2\cap\mathbb{C}_I)} S^{-1}(\mathbf{t},\mathbf{s})\, d\mathbf{t}_I\, g(\mathbf{t}).$$

Now consider

$$\begin{aligned} (\phi g)(T) &= \frac{1}{2\pi}\int_{\partial(G_1\cap\mathbb{C}_I)} S^{-1}(\mathbf{s},T)\, d\mathbf{s}_I\, \phi(\mathbf{s})\, g(\mathbf{s}) \\ &= \frac{1}{2\pi}\int_{\partial(G_1\cap\mathbb{C}_I)} S^{-1}(\mathbf{s},T)\, d\mathbf{s}_I\, \phi(\mathbf{s}) \Big[\frac{1}{2\pi}\int_{\partial(G_2\cap\mathbb{C}_I)} S^{-1}(\mathbf{t},\mathbf{s})\, d\mathbf{t}_I\, g(\mathbf{t})\Big]. \end{aligned}$$

By the vectorial version of the Fubini theorem we have

$$(\phi g)(T) = \frac{1}{(2\pi)^2} \int_{\partial(G_2\cap\mathbb{C}_I)} \Big[\int_{\partial(G_1\cap\mathbb{C}_I)} S^{-1}(\mathbf{s},T)\ d\mathbf{s}_I\ \phi(\mathbf{s})\ S^{-1}(\mathbf{t},\mathbf{s}) \Big]\ d\mathbf{t}_I\ g(\mathbf{t}).$$

Observe that $S^{-1}(\mathbf{t},\mathbf{s})$ is left s-monogenic in the variable $\mathbf{s} \in \partial G_1$ for $\mathbf{t} \in \partial G_2$ by Proposition 2.7.9, and since $\phi \in \widetilde{\mathcal{M}}_{\sigma_S(T)}$ by Proposition 3.4.2 it follows that $\phi(\mathbf{s})S^{-1}(\mathbf{t},\mathbf{s})$ is s-monogenic in the variable $\mathbf{s}$. So we obtain

$$\begin{aligned}
(\phi g)(T) &= \frac{1}{2\pi} \int_{\partial(G_2\cap\mathbb{C}_I)} \phi(T) S^{-1}(\mathbf{t},T)\ d\mathbf{t}_I\ g(\mathbf{t}) \\
&= \phi(T) \frac{1}{2\pi} \int_{\partial(G_2\cap\mathbb{C}_I)} S^{-1}(\mathbf{t},T)\ d\mathbf{t}_I\ g(\mathbf{t}) \\
&= \phi(T) g(T).
\end{aligned}$$

Part (c): For a suitable $R > 0$ the series $\sum_{m\geq 0} \mathbf{s}^m p_m$ converges in a ball $B(0,R)$ that contains $\sigma_S(T)$. So we can choose another ball $B_\varepsilon := \{\mathbf{s} : |\mathbf{s}| \leq \|T\| + \varepsilon\}$, for sufficiently small $\varepsilon > 0$, such that $B_\varepsilon \subset B(0,R)$. Since the series converges uniformly on ∂B_ε we have

$$\begin{aligned}
f(T) &= \frac{1}{2\pi} \int_{\partial(B_\varepsilon\cap\mathbb{C}_I)} S^{-1}(\mathbf{s},T)\ d\mathbf{s}_I \sum_{m\geq 0} \mathbf{s}^m p_m \\
&= \frac{1}{2\pi} \sum_{m\geq 0} \int_{\partial(B_\varepsilon\cap\mathbb{C}_I)} S^{-1}(\mathbf{s},T)\ d\mathbf{s}_I\ \mathbf{s}^m p_m \\
&= \frac{1}{2\pi} \sum_{m\geq 0} \int_{\partial(B_\varepsilon\cap\mathbb{C}_I)} \sum_{k\geq 0} T^k \mathbf{s}^{-1-k}\ d\mathbf{s}_I\ \mathbf{s}^m\ p_m = \sum_{m\geq 0} T^m\ p_m.
\end{aligned}$$

□

In the next section we study some further important properties of our functional calculus.

3.5 The spectral mapping and the S-spectral radius theorems

As we have already pointed out, the composition of two s-monogenic functions is not, in general, s-monogenic. When we deal with s-monogenic functions f, g which can be expanded into power series, the composition $f(g(\mathbf{x}))$ is defined when g has real coefficients and its range is contained in the ball on which f is defined. Here we prove a sufficient condition in order for the composition of two s-monogenic functions to be s-monogenic.

Definition 3.5.1. *Let $f : U \to \mathbb{R}_n$ be an s-monogenic function where U is an open set in $\mathbb{R}^{n+1}$. We define*

$$\mathcal{N}(U) = \{f \in \mathcal{M}(U)\ :\ f(U \cap \mathbb{C}_I) \subseteq \mathbb{C}_I,\ \ \forall I \in \mathbb{S}\}.$$

Lemma 3.5.2. *Let U be an open set in $\mathbb{R}^{n+1}$. We have $\mathcal{N}(U) \subset \widetilde{\mathcal{M}}(U)$.*

Proof. By definition, a function $f \in \mathcal{N}(U)$ has a splitting of the type $f(z) = F(z)$ with $F : \mathbb{C}_I \to \mathbb{C}_I$ holomorphic, thus $f \in \widetilde{\mathcal{M}}(U)$. □

Remark 3.5.3. Example 3.4.4 shows that, for $n \geq 3$, there are functions in $\widetilde{\mathcal{M}}(U)$ not belonging to $\mathcal{N}(U)$, thus the inclusion $\mathcal{N}(U) \subset \widetilde{\mathcal{M}}(U)$ is proper.

Let us first study the behavior of the product of functions in $\mathcal{N}(U)$.

Lemma 3.5.4. *Let U be an open set in $\mathbb{R}^{n+1}$.*

(a) *Let f and $g \in \mathcal{N}(U)$, then $f\,g$ and $g\,f$ belong to $\mathcal{N}(U)$.*

(b) *Let P, $Q \in \mathcal{N}(U)$ with $Q(\mathbf{x}) \neq 0$ in U. Then $Q^{-1}P$ and PQ^{-1} belong to $\mathcal{N}(U)$.*

Proof. It follows by restricting the functions in (a), (b) to the complex plane $\mathbb{C}_I$ and observing that the functions we are considering are holomorphic from $\mathbb{C}_I$ to $\mathbb{C}_I$ for every $I \in \mathbb{S}$. □

Remark 3.5.5. Using the same proof of Lemma 3.5.4 we have that if $f(\mathbf{x}) = \sum_{m \in \mathbb{Z}} (\mathbf{x} - p_0)^m a_m$, with $p_0, a_m \in \mathbb{R}$ is a series converging in a suitable set U, then $f \in \mathcal{N}(U)$.

Lemma 3.5.6. *Let U, U' be two open sets in $\mathbb{R}^{n+1}$ and let $f \in \mathcal{N}(U')$, $g \in \mathcal{N}(U)$ with $g(U) \subseteq U'$. Then $f(g(\mathbf{x}))$ is s-monogenic for $\mathbf{x} \in U$.*

Proof. Set $\mathbf{x} = u + Iv$. By hypothesis, $g(u + Iv) = \alpha(u, v) + I\beta(u, v)$, where α, β are real-valued functions and

$$f(g(u + Iv)) = f(\alpha(u, v) + I\beta(u, v)) \subseteq \mathbb{C}_I.$$

The function $f(g(u + Iv))$ is holomorphic on each plane $\mathbb{C}_I$ since it satisfies the condition

$$\bar{\partial}_I f(g(u + Iv)) = 0$$

for all $I \in \mathbb{S}$ and so $f(g(\mathbf{x}))$ is an s-monogenic function in $\mathbf{x}$. □

The following lemma will be used in the sequel.

Lemma 3.5.7. *Let U be an open set in $\mathbb{R}^{n+1}$ and assume $f \in \mathcal{N}(U)$. For $\nu \in \mathbb{R}^{n+1}$ define $U_{[\nu]} = \{\mathbf{x} \in U \;:\; f(\mathbf{x}) \notin [\nu]\}$. Define:*

(a) $h_0(\mathbf{x}) = f^2(\mathbf{x}) - 2\mathrm{Re}[f(\nu)]f(\mathbf{x}) + |f(\nu)|^2$,

(b) $h(\mathbf{x}) = (f^2(\mathbf{x}) - 2\mathrm{Re}[\nu]f(\mathbf{x}) + |\nu|^2)^{-1}$,

(c) $h_1(\mathbf{x}) = (f(\mathbf{x})^2 - 2\mathrm{Re}[\nu]f(\mathbf{x}) + |\nu|^2)^{-1}(f(\mathbf{x}) - \overline{\nu})$.

Then: $h_0 \in \mathcal{N}(U)$, $h \in \mathcal{N}(U_{[\nu]})$, $h_1 \in \mathcal{M}(U_{[\nu]})$.

Proof. To prove that $h_0 \in \mathcal{N}(U)$, observe that since $f \in \mathcal{N}(U)$, by Lemma 3.5.4 we have that f^2 belongs to $\mathcal{N}(U)$ so also h_0 belongs to $\mathcal{N}(U)$. The fact that $h \in \mathcal{N}(U_{[\nu]})$ follows from the previous result and Lemma 3.5.4. Finally, since $h_0 \in \mathcal{N}(U_{[\nu]}) \subset \widetilde{\mathcal{M}}(U_{[\nu]})$ thanks to Lemma 3.5.2, and $f - \overline{\nu} \in \mathcal{M}(U)$, the function $h_1(\mathbf{x}) = h_0(\mathbf{x})(f(\mathbf{x}) - \overline{\nu}) \in \mathcal{M}(U_{[\nu]})$ by Proposition 3.4.2. □

Definition 3.5.8. *In Definition* 3.2.3 *consider instead of the set of s-monogenic functions* $\mathcal{M}(U)$*, the subset of functions in* $\mathcal{N}(U)$*. This subclass of* $\mathcal{M}_{\sigma_S(T)}$ *will be denoted by* $\mathcal{N}_{\sigma_S(T)}$.

Theorem 3.5.9 (Spectral Mapping Theorem). *Let* $T \in \mathcal{B}_n^{0,1}(V_n)$, $f \in \mathcal{N}_{\sigma_S(T)}$, *and* $\lambda \in \sigma_S(T)$. *Then*

$$\sigma_S(f(T)) = f(\sigma_S(T)) = \{f(\mathbf{s}) : \mathbf{s} \in \sigma_S(T)\}.$$

Proof. Since $f \in \mathcal{N}_{\sigma_S(T)}$ there exists a T-admissible open set $U \subset \mathbb{R}^{n+1}$ such that $f \in \mathcal{N}(U)$. Let us fix $\lambda \in \sigma_S(T)$. For $\mathbf{x} \notin [\lambda]$, let us define the function $\tilde{g}(\mathbf{x})$ by

$$\tilde{g}(\mathbf{x}) = (\mathbf{x}^2 - 2\mathrm{Re}[\lambda]\mathbf{x} + |\lambda|^2)^{-1}(f^2(\mathbf{x}) - 2\mathrm{Re}[f(\lambda)]f(\mathbf{x}) + |f(\lambda)|^2).$$

Observe that $f \in \mathcal{N}(U)$ implies that $f^2(\mathbf{x}) - 2\mathrm{Re}[f(\lambda)]f(\mathbf{x}) + |f(\lambda)|^2 \in \mathcal{N}(U)$ by Lemma 3.5.7. The function $(\mathbf{x}^2 - 2\mathrm{Re}[\lambda]\mathbf{x} + |\lambda|^2)^{-1} \in \mathcal{N}(U \setminus \{[\lambda]\})$, by Lemma 3.5.4, thus $\tilde{g}(\mathbf{x}) \in \mathcal{N}(U \setminus \{[\lambda]\})$ by the same Lemma.

We can extend $\tilde{g}(\mathbf{x})$ to an s-monogenic function whose domain is U. We have to consider two cases. Suppose first that the $(n-1)$-sphere $[\lambda]$ does not reduce to a point on the real axis. Then we define

$$g(\mathbf{x}) = \begin{cases} \tilde{g}(\mathbf{x}) \quad \text{if } \mathbf{x} \notin [\lambda], \\ \dfrac{\partial f(\mu)}{\partial u} \dfrac{f(\mu) - \overline{f(\mu)}}{\mu - \overline{\mu}} \quad \text{if } \mathbf{x} = \mu = \lambda_0 + I\lambda_1 \in [\lambda], \quad I \in \mathbb{S}. \end{cases}$$

Given the $(n-1)$-sphere $[\lambda]$, on each plane $\mathbb{C}_I$, $I \in \mathbb{S}$, the function $\tilde{g}$ has two singularities $\lambda_0 \pm I\lambda_1 \in [\lambda]$. If we set $z = u + Iv$, we can compute the limit of $\tilde{g}$ on the plane $\mathbb{C}_I$ for $z \to \mu = \lambda_0 + I\lambda_1$ and for $z \to \overline{\mu} = \lambda_0 - I\lambda_1$. The restriction of f to the plane $\mathbb{C}_I$ is a holomorphic function from $U \cap \mathbb{C}_I$ with values in the complex plane $\mathbb{C}_I$, and, by Theorem 2.2.18 (ii),

$$f(\lambda_0 + I\lambda_1) = \eta_K(\lambda_0, \lambda_1) + I\theta_K(\lambda_0, \lambda_1).$$

However, $\eta_K, \theta_K : U \cap \mathbb{C}_K \to \mathbb{C}_K$ for all $K \in \mathbb{S}$, see Theorem 2.2.18 (ii), so $\eta_K = \eta, \theta_K = \theta$ are real-valued functions depending only on λ_0, λ_1. We can write

$$f(\lambda_0 + I\lambda_1) = \eta(\lambda_0, \lambda_1) + I\theta(\lambda_0, \lambda_1)$$

and we deduce that $\mathrm{Re}[f(\lambda)] = \mathrm{Re}[f(\mu)]$ and $|f(\lambda)|^2 = |f(\mu)|^2$ for any choice of μ and λ on the same $(n-1)$-sphere. We have:

$$\lim_{z\to\mu} g_I(z) = \lim_{z\to\mu}(z^2 - 2\mathrm{Re}[\mu]z + |\mu|^2)^{-1}(f^2(z) - 2\mathrm{Re}[f(\mu)]f(z) + |f(\mu)|^2)$$
$$= \lim_{z\to\mu} \frac{(f(z)-f(\mu))(f(z)-\overline{f(\mu)})}{(z-\mu)(z-\overline{\mu})} = f'(\mu)\frac{f(\mu)-\overline{f(\mu)}}{\mu-\overline{\mu}},$$

and similarly for the limit when $z \to \overline{\mu}$. Note that the derivative $f'(\mu)$ coincides with $\frac{\partial}{\partial u}f(\mu)$ since f is s-monogenic. In the second case, assume that $\lambda \in \mathbb{R}$. We define

$$g(\mathbf{x}) = \begin{cases} \tilde{g}(\mathbf{x}) & \text{if } \mathbf{x} \neq \lambda, \\ \left(\frac{\partial}{\partial u}f(\lambda)\right)^2 & \text{if } \mathbf{x} = \lambda \in \mathbb{R}. \end{cases}$$

Consider any $J \in \mathbb{S}$ and the restriction of f to the plane $\mathbb{C}_J$. Then $f : U \cap \mathbb{C}_J \to \mathbb{C}_J$ is a holomorphic function and $f(\lambda) \in \mathbb{R}$, indeed $f(\lambda) \in \mathbb{C}_J$ for all $J \in \mathbb{S}$. Let us set $z = u + Jv$. We have

$$\lim_{z\to\lambda} g_J(z) = \lim_{z\to\lambda}(z^2 - 2\mathrm{Re}[\lambda]z + |\lambda|^2)^{-1}(f^2(z) - 2\mathrm{Re}[f(\lambda)]f(z) + |f(\lambda)|^2)$$
$$= \lim_{z\to\lambda} \frac{(f(z)-f(\lambda))^2}{(z-\lambda)^2} = f'(\lambda)^2,$$

so the value of the limit is independent of the plane $\mathbb{C}_J$. The function $g_I : U \cap \mathbb{C}_I \to \mathbb{C}_I$ is extended by continuity to $U \cap \mathbb{C}_I$, so it is holomorphic on $U \cap \mathbb{C}_I$ for all $I \in \mathbb{S}$. We conclude that the function $g : U \to \mathbb{R}_n$ is an s-monogenic function.

Thanks to Theorem 3.4.6, we can write

$$f^2(T) - 2\mathrm{Re}[f(\lambda)]f(T) + |f(\lambda)|^2\mathcal{I} = (T^2 - 2\mathrm{Re}[\lambda]T + |\lambda|^2\mathcal{I})g(T).$$

If $f^2(T) - 2\mathrm{Re}[f(\lambda)]f(T) + |f(\lambda)|^2\mathcal{I}$ admits a bounded inverse

$$B := (f^2(T) - 2\mathrm{Re}[f(\lambda)]f(T) + |f(\lambda)|^2\mathcal{I})^{-1} \in \mathcal{B}_n(V_n),$$

then

$$(T^2 - 2\mathrm{Re}[\lambda]T + |\lambda|^2\mathcal{I})g(T)B = \mathcal{I},$$

i.e., $g(T)B$ is the inverse of $T^2 - 2\mathrm{Re}[\lambda]T + |\lambda|^2\mathcal{I}$. Thus $f(\sigma_S(T)) \subset \sigma_S(f(T))$. Now we take $\nu \in \sigma_S(f(T))$ such that $\nu \notin f(\sigma_S(T))$. The function

$$h(\mathbf{x}) := (f^2(\mathbf{x}) - 2\mathrm{Re}[\nu]f(\mathbf{x}) + |\nu|^2)^{-1}$$

is s-monogenic on $\sigma_S(T)$ by Lemma 3.5.7. By Theorem 3.4.6 we get

$$h(T)(f^2(T) - 2\mathrm{Re}[\nu]f(T) + |\nu|^2\mathcal{I}) = \mathcal{I}$$

this means that $\nu \notin \sigma_S(f(T))$, but this contradicts the assumption. So $\nu \in f(\sigma_S(T))$. □

Theorem 3.5.10. *Let $T \in \mathcal{B}_n^{0,1}(V_n)$, $f \in \mathcal{N}_{\sigma_S(T)}$, $\phi \in \mathcal{N}_{\sigma_S(f(T))}$ and let $F(\mathbf{s}) = \phi(f(\mathbf{s}))$. Then $F \in \mathcal{M}_{\sigma_S(T)}$ and $F(T) = \phi(f(T))$.*

Proof. The statement $F \in \mathcal{M}_{\sigma_S(T)}$ follows from Lemma 3.5.6 and the Spectral Mapping Theorem. Let $U \supset \sigma_S(f(T))$ be an $f(T)$-admissible open bounded set whose boundary is denoted by ∂U. Suppose that $U \cup \partial U$ is contained in the domain in which ϕ is s-monogenic. Let W be another T-admissible neighborhood of $\sigma_S(T)$ and let ∂W be its boundary. Suppose that $W \cup \partial W$ is contained in the domain where f is s-monogenic and that $f(W \cup \partial W) \subseteq U$. Let $I \in \mathbb{S}$ and define the operator

$$S^{-1}(\lambda, f(T)) = \frac{1}{2\pi} \int_{\partial(W \cap \mathbb{C}_I)} S^{-1}(\mathbf{s}, T)\, d\mathbf{s}_I\, S^{-1}(\lambda, f(\mathbf{s}))$$

where

$$S^{-1}(\lambda, f(\mathbf{s})) = -(f(\mathbf{s})^2 - 2\mathrm{Re}[\lambda] f(\mathbf{s}) + |\lambda|^2)^{-1}(f(\mathbf{s}) - \overline{\lambda}).$$

Observe that $S^{-1}(\lambda, f(\mathbf{s}))$, where it is defined, is left s-monogenic in the variable $\mathbf{s}$ by Lemma 3.5.7 and it is right s-monogenic in the variable λ by Proposition 2.7.9. In particular, if λ is a real number the function

$$S(\lambda, f(\mathbf{s})) = (f(\mathbf{s}) - \overline{\lambda})^{-1}(f(\mathbf{s})^2 - 2\mathrm{Re}[\lambda] f(\mathbf{s}) + |\lambda|^2)$$

is s-monogenic in the variable $\mathbf{s}$. Since

$$S^{-1}(\lambda, f(\mathbf{s})) S(\lambda, f(\mathbf{s})) = S(\lambda, f(\mathbf{s})) S^{-1}(\lambda, f(\mathbf{s}))$$

is the identity function, by Theorem 3.4.6, the operator $S^{-1}(\lambda, f(T))$ satisfies the equation:

$$\begin{aligned}
&[(f(T) - \overline{\lambda}\mathcal{I})^{-1}\, \lambda\, (f(T) - \overline{\lambda}\mathcal{I}) - f(T)] S^{-1}(\lambda, f(T)) \\
&= S^{-1}(\lambda, f(T))[(f(T) - \overline{\lambda}\mathcal{I})^{-1}\, \lambda\, (f(T) - \overline{\lambda}\mathcal{I}) - f(T)] = \mathcal{I}.
\end{aligned} \tag{3.19}$$

It is immediate to observe that if λ is not necessarily real, the relation (3.19) still holds. In fact, replacing the explicit expression for $S^{-1}(\lambda, f(T))$ we get:

$$S^{-1}(\lambda, f(T)) = -(f(T)^2 - 2\mathrm{Re}[\lambda] f(T) + |\lambda|^2)^{-1}(f(T) - \overline{\lambda});$$

in (3.19) we get an identity. As a consequence, we have

$$\begin{aligned}
\phi(f(T)) &= \frac{1}{2\pi} \int_{\partial(W \cap \mathbb{C}_I)} S^{-1}(\lambda, f(T))\, d\lambda_I\, \phi(\lambda) \\
&= \frac{1}{2\pi} \int_{\partial(W \cap \mathbb{C}_I)} \Big(\frac{1}{2\pi} \int_{\partial(U \cap \mathbb{C}_I)} S^{-1}(\mathbf{s}, T)\, d\mathbf{s}_I\, S^{-1}(\lambda, f(\mathbf{s})) \Big)\, d\lambda_I\, \phi(\lambda) \\
&= \frac{1}{2\pi} \int_{\partial(U \cap \mathbb{C}_I)} S^{-1}(\mathbf{s}, T)\, d\mathbf{s}_I \Big(\frac{1}{2\pi} \int_{\partial(W \cap \mathbb{C}_I)} S^{-1}(\lambda, f(\mathbf{s}))\, d\lambda_I\, \phi(\lambda) \Big) \\
&= \frac{1}{2\pi} \int_{\partial(U \cap \mathbb{C}_I)} S^{-1}(\mathbf{s}, T)\, d\mathbf{s}_I\, \phi(f(\mathbf{s})) \\
&= \frac{1}{2\pi} \int_{\partial(U \cap \mathbb{C}_I)} S^{-1}(\mathbf{s}, T)\, d\mathbf{s}_I\, F(\mathbf{s}) = F(T).
\end{aligned}$$

□

Theorem 3.5.11. *Let $T \in \mathcal{B}_n^{0,1}(V_n)$, $f_m \in \mathcal{M}_{\sigma_S(T)}$, $m \in \mathbb{N}$, and let $W \supset \sigma_S(T)$ be a T-admissible domain. Then if f_m converges uniformly to f on $W \cap \mathbb{C}_I$, for some $I \in \mathbb{S}$, then $f_m(T)$ converges to $f(T)$ in $\mathcal{B}_n(V_n)$.*

Proof. Let U be an axially symmetric s-domain such that $\overline{U} \subset W$ and assume that $\partial(U \cap \mathbb{C}_I)$ consists of a finite number of continuously differentiable Jordan arcs. Then $f_m \to f$ converges uniformly on $\partial(U \cap \mathbb{C}_I)$ and consequently

$$f_m(T) = \frac{1}{2\pi} \int_{\partial(U\cap\mathbb{C}_I)} S^{-1}(\mathbf{s}, T)\, d\mathbf{s}_I\, f_m(\mathbf{s})$$

converges, in the uniform topology of operators, to

$$f(T) = \frac{1}{2\pi} \int_{\partial(U\cap\mathbb{C}_I)} S^{-1}(\mathbf{s}, T)\, d\mathbf{s}_I\, f(\mathbf{s}). \qquad \square$$

Definition 3.5.12 (The S-spectral radius of T). *For any $T \in \mathcal{B}_n^{0,1}(V_n)$ we define the S-spectral radius of T to be the nonnegative real number*

$$r_S(T) := \sup \{ |\mathbf{s}| \ : \ \mathbf{s} \in \sigma_S(T) \}.$$

Theorem 3.5.13 (The S-spectral radius theorem). *Let $T \in \mathcal{B}_n^{0,1}(V_n)$ and let $r_S(T)$ be the S-spectral radius of T. Then*

$$r_S(T) = \lim_{m\to\infty} \|T^m\|^{1/m}.$$

Proof. For every $\mathbf{s} \in \mathbb{R}^{n+1}$ such that $|\mathbf{s}| > r_S(T)$ the series $\sum_{m\geq 0} T^m\, \mathbf{s}^{-1-m}$ converges in $\mathcal{B}_n(V_n)$ to the S-resolvent operator $S^{-1}(\mathbf{s}, T)$. So the sequence $T^m\, \mathbf{s}^{-1-m}$ is bounded in the norm of $\mathcal{B}_n(V_n)$ and

$$\limsup_{m\to\infty} \|T^m\|^{1/m} \leq r_S(T). \tag{3.20}$$

The Spectral Mapping Theorem implies that $\sigma_S(T^m) = (\sigma_S(T))^m$, so we have

$$(r_S(T))^m = r_S(T^m) \leq \|T^m\|,$$

from which we get

$$r_S(T) \leq \liminf \|T^m\|^{1/m}. \tag{3.21}$$

From (3.20), (3.21) we have

$$r_S(T) \leq \liminf_{m\to\infty} \|T^m\|^{1/m} \leq \limsup_{m\to\infty} \|T^m\|^{1/m} \leq r_S(T). \qquad \square$$

3.6 Projectors

We begin this section by proving a technical lemma that generalizes the S-resolvent equation (3.6).

Lemma 3.6.1. *Let $T \in \mathcal{B}_n^{0,1}(V_n)$. Set*

$$\mathcal{Q}_m(\mathbf{s},T) := \mathcal{I}\mathbf{s}^{[m-1]_+} + T\mathbf{s}^{[m-2]_+} + T^2\mathbf{s}^{[m-3]_+} + \ldots + T^{m-1}, \quad m \geq 1$$

where $\mathbf{s}^{[k]_+} = \mathbf{s}^k$ if $k \geq 0$, $\mathbf{s}^{[k]_+} = 0$ otherwise, and $\mathcal{Q}_0(\mathbf{s},T) := 0$. Then

$$T^m S^{-1}(\mathbf{s},T) = S^{-1}(\mathbf{s},T)s^m - \mathcal{Q}_m(\mathbf{s},T), \quad \text{for all } m = 0,1,2,\ldots. \tag{3.22}$$

Proof. Formula (3.22) holds trivially for $m = 0$ and holds for $m = 1$ because it follows from the S-resolvent equation (3.6).

We now suppose that (3.22) holds for the natural numbers less than or equal m and we show that it holds for $m+1$. By the induction step, we have

$$T^m S^{-1}(\mathbf{s},T) = S^{-1}(\mathbf{s},T)\mathbf{s}^m - \mathcal{Q}_m(\mathbf{s},T),$$

so we can write:

$$\begin{aligned} T^{m+1}S^{-1}(\mathbf{s},T) &= TS^{-1}(\mathbf{s},T)\mathbf{s}^m - T\mathcal{Q}_m(\mathbf{s},T) \\ &= TS^{-1}(\mathbf{s},T)\mathbf{s}^m - (T\mathbf{s}^{[m-1]_+} + \ldots + T^m) \\ &= TS^{-1}(\mathbf{s},T)\mathbf{s}^m + \mathcal{I}\mathbf{s}^m - (\mathcal{I}\mathbf{s}^m + T\mathbf{s}^{[m-1]_+} + \ldots + T^m) \end{aligned}$$

and, using the S-resolvent equation (3.6), we have

$$T^{m+1}S^{-1}(\mathbf{s},T) = S^{-1}(\mathbf{s},T)\mathbf{s}^{m+1} - \mathcal{Q}_{m+1}(\mathbf{s},T),$$

which is the formula we had to prove. □

Keeping in mind Remark 3.2.4, we can now prove the following theorem.

Theorem 3.6.2. *Let $T \in \mathcal{B}_n^{0,1}(V_n)$, $f \in \mathcal{M}_{\sigma_S(T)}$ and assume that $\sigma_S(T) = \sigma_{1S}(T) \cup \sigma_{2S}(T)$ with $dist(\sigma_{1S}(T), \sigma_{2S}(T)) > 0$. Let U be a T-admissible open set and let U_1, U_2 open sets such that $U = U_1 \cup U_2$ with $\overline{U}_1 \cap \overline{U}_2 = \emptyset$ and such that $\sigma_{1S}(T) \subset U_1$ and $\sigma_{2S}(T) \subset U_2$. For $I \in \mathbb{S}$ set $ds_I = -dsI$ and define*

$$P_j := \frac{1}{2\pi} \int_{\partial(U_j \cap \mathbb{C}_I)} S^{-1}(\mathbf{s},T)\, ds_I,$$

$$T_j^m := \frac{1}{2\pi} \int_{\partial(U_j \cap \mathbb{C}_I)} S^{-1}(\mathbf{s},T)\, ds_I\, \mathbf{s}^m, \quad m = 1,2,3,\ldots, \quad j = 1,2.$$

Then P_j are projectors and

(a) $P_1 + P_2 = \mathcal{I}$,

(b) $TP_j = T_j,\ \ j = 1, 2,$

(c) $T^m = T_1^m + T_2^m,\ \ m \geq 1.$

Proof. Observe that $P_j = T_j^0$ and note that equation (3.22) for $m = 0$ is trivially $T_j^0 S^{-1}(\mathbf{s}, T) = S^{-1}(\mathbf{s}, T)\mathbf{s}^0 = S^{-1}(\mathbf{s}, T)$. So we have

$$P_j^2 = P_j \frac{1}{2\pi} \int_{\partial(U_j \cap \mathbb{C}_I)} S^{-1}(\mathbf{s}, T)\, d\mathbf{s}_I = \frac{1}{2\pi} \int_{\partial(U_j \cap \mathbb{C}_I)} P_j S^{-1}(\mathbf{s}, T)\, d\mathbf{s}_I$$
$$= \frac{1}{2\pi} \int_{\partial(U_j \cap \mathbb{C}_I)} S^{-1}(\mathbf{s}, T)\, d\mathbf{s}_I = P_j.$$

To prove (a) we use the Cauchy integral theorem. Since $U_1 \cup U_2 = U$, we have

$$\frac{1}{2\pi} \int_{\partial(U_1 \cap \mathbb{C}_I)} S^{-1}(\mathbf{s}, T)\, d\mathbf{s}_I + \frac{1}{2\pi} \int_{\partial(U_2 \cap \mathbb{C}_I)} S^{-1}(\mathbf{s}, T)\, d\mathbf{s}_I$$
$$= \frac{1}{2\pi} \int_{\partial(U \cap \mathbb{C}_I)} S^{-1}(\mathbf{s}, T)\, d\mathbf{s}_I.$$

Since

$$\frac{1}{2\pi} \int_{\partial(U \cap \mathbb{C}_I)} S^{-1}(\mathbf{s}, T)\, d\mathbf{s}_I = \mathcal{I}$$

this gives $P_1 + P_2 = \mathcal{I}$.

To prove (b) we recall the resolvent relation, $TS^{-1}(\mathbf{s}, T) = S^{-1}(\mathbf{s}, T)\mathbf{s} - \mathcal{I}$, so

$$TP_j = \frac{1}{2\pi} \int_{\partial(U_j \cap \mathbb{C}_I)} TS^{-1}(\mathbf{s}, T)\, d\mathbf{s}_I = \frac{1}{2\pi} \int_{\partial(U_j \cap \mathbb{C}_I)} [S^{-1}(\mathbf{s}, T)\mathbf{s} - \mathcal{I}]\, d\mathbf{s}_I$$
$$= \frac{1}{2\pi} \int_{\partial(U_j \cap \mathbb{C}_I)} S^{-1}(\mathbf{s}, T)\, d\mathbf{s}_I\, \mathbf{s} = T_j.$$

Now adding the relations $T_j = TP_j$ we get, using (a),

$$T_1 + T_2 = TP_1 + TP_2 = T(P_1 + P_2) = T,$$

which is part (c) in the case $m = 1$.

To prove (c) for $m \geq 2$ observe that by Lemma 3.6.1 we get

$$S^{-1}(\mathbf{s}, T)\mathbf{s}^m - T^m S^{-1}(\mathbf{s}, T) = \mathcal{I}\mathbf{s}^{[m-1]_+} + T\mathbf{s}^{[m-2]_+} + \ldots + T^{m-1}.$$

Now, for $m \geq 2$, consider

$$T^m P_j = \frac{1}{2\pi} \int_{\partial(U_j \cap \mathbb{C}_I)} T^m S^{-1}(\mathbf{s}, T)\, d\mathbf{s}_I$$
$$= \frac{1}{2\pi} \int_{\partial(U_j \cap \mathbb{C}_I)} [S^{-1}(\mathbf{s}, T)\mathbf{s}^m - (\mathcal{I}\mathbf{s}^{[m-1]_+} + T\mathbf{s}^{[m-2]_+} + \ldots + T^{m-1})]\, d\mathbf{s}_I$$
$$= \frac{1}{2\pi} \int_{\partial(U_j \cap \mathbb{C}_I)} S^{-1}(\mathbf{s}, T)\, d\mathbf{s}_I\, \mathbf{s}^m = T_j^m.$$

So adding $T^m P_1 = T_1^m$ and $T^m P_2 = T_2^m$ and recalling (a) we get (c). □

3.7 Functional calculus for unbounded operators and algebraic properties

Let V be a real Banach space and $T = T_0 + \sum_{j=1}^{m} e_j T_j$ where $T_\mu : \mathcal{D}(T_\mu) \to V$ are linear operators for $\mu = 0, 1, \dots, n$ where at least one of the T_μ's is an unbounded operator. Then we will say that the operator T is unbounded.

Definition 3.7.1. *Let V be a Banach space and let V_n be the two-sided Banach module over $\mathbb{R}_n$ corresponding to $V \otimes \mathbb{R}_n$. Let $T_\mu : \mathcal{D}(T_\mu) \subset V \to V$ be linear closed densely defined operators for $\mu = 0, 1, \dots, n$. Let*

$$\mathcal{D}(T) = \{v \in V_n \ : \ Tv \in V_n \ \} \tag{3.23}$$

be the domain of the operator $T = T_0 + \sum_{j=1}^{n} e_j T_j$. We denote by $\mathcal{K}(V_n)$ the set of all operators T such that:

(1) *$\mathcal{D}(T) = \bigcap_{\mu=0}^{n} \mathcal{D}(T_\mu)$ is dense in V_n,*

(2) *$T - \overline{s}\mathcal{I}$ is densely defined in V_n,*

(3) *$\mathcal{D}(T^2) \subset \mathcal{D}(T)$ is dense in V_n.*

Observe that, when $T \in \mathcal{K}(V_n)$, the operator

$$-(T^2 - 2\mathrm{Re}[\mathbf{s}]T + |\mathbf{s}|^2\mathcal{I})^{-1}(T - \overline{\mathbf{s}}\mathcal{I})$$

is the restriction to the dense subspace $\mathcal{D}(T)$ of V_n of a bounded linear operator defined on V_n. This fact follows by the commutation relation

$$(T^2 - 2T\mathrm{Re}[\mathbf{s}] + |\mathbf{s}|^2\mathcal{I})^{-1}Tv = T(T^2 - 2T\mathrm{Re}[\mathbf{s}] + |\mathbf{s}|^2\mathcal{I})^{-1}v$$

which holds for all $v \in \mathcal{D}(T)$ and for all $\mathbf{s} \in \mathbb{R}^{n+1}$ such that

$$(T^2 - 2T\mathrm{Re}[\mathbf{s}] + |\mathbf{s}|^2\mathcal{I})^{-1} \in \mathcal{B}_n(V_n), \tag{3.24}$$

since the polynomial operator

$$T^2 - 2T\mathrm{Re}[\mathbf{s}] + |\mathbf{s}|^2\mathcal{I} : \mathcal{D}(T^2) \to V$$

has real coefficients. The operator

$$T(T^2 - 2T\mathrm{Re}[\mathbf{s}] + |\mathbf{s}|^2\mathcal{I})^{-1} : V_n \to \mathcal{D}(T)$$

is continuous for those $\mathbf{s} \in \mathbb{R}^{n+1}$ such that relation (3.24) holds. We introduce the following auxiliary operator.

Definition 3.7.2. *Let us set*

$$\mathcal{Q}_{\mathbf{s}}(T) := (T^2 - 2T\mathrm{Re}[\mathbf{s}] + |\mathbf{s}|^2\mathcal{I})^{-1}. \tag{3.25}$$

The S-resolvent set $\rho_S(T)$ of T is defined as

$$\rho_S(T) := \{\mathbf{s} \in \mathbb{R}^{n+1} \mid \mathcal{Q}_{\mathbf{s}}(T) \in \mathcal{B}_n(V_n)\}$$

and the S-spectrum $\sigma_S(T)$ of T is defined by

$$\sigma_S(T) = \mathbb{R}^{n+1} \setminus \rho_S(T).$$

Let us consider $\overline{\mathbb{R}}^{n+1} = \mathbb{R}^{n+1} \cup \{\infty\}$ endowed with the natural topology.

Definition 3.7.3. *We define the extended S-spectrum as*

$$\overline{\sigma}_S(T) := \sigma_S(T) \cup \{\infty\}.$$

Definition 3.7.4. *We say that f is an s-monogenic function at ∞ if $f(\mathbf{x})$ is an s-monogenic function in a set $D'(\infty, r) = \{\mathbf{x} \in \mathbb{R}^{n+1} : |\mathbf{x}| > r\}$, for some $r > 0$, and $\lim_{\mathbf{x}\to\infty} f(\mathbf{x})$ exists and it is finite. We set $f(\infty)$ to be the value of this limit.*

Remark 3.7.5. We know that if T is a linear and bounded operator, then $\sigma_S(T)$ is a compact nonempty set, but for unbounded operators, as in the classical case, the S-spectrum can be bounded (and even empty) or unbounded (and even all of $\mathbb{R}^{n+1}$). In the sequel we will assume that $\rho_S(T) \neq \emptyset$.

Observe that the operator

$$-(T^2 - 2T Re[\mathbf{s}] + |\mathbf{s}|^2\mathcal{I})^{-1}(T - \overline{\mathbf{s}}\mathcal{I}). \tag{3.26}$$

is bounded from $\mathcal{D}(T) \to \mathcal{D}(T^2)$ for all $\mathbf{s} \in \rho_S(T)$. We will consider the operator in (3.26) extended to all V_n as in the following definition.

Definition 3.7.6. *The S-resolvent operator for, $\mathbf{s} \in \rho_S(T)$, is defined by*

$$S^{-1}(\mathbf{s}, T) := \mathcal{Q}_{\mathbf{s}}(T)\overline{\mathbf{s}} - T\mathcal{Q}_{\mathbf{s}}(T) : \ V_n \to \mathcal{D}(T). \tag{3.27}$$

Remark 3.7.7. Observe that for the unbounded case the S-spectrum is not necessarily compact and nonempty, but the theorem on the structure still holds.

Theorem 3.7.8 (Structure of the spectrum). *Let $T \in \mathcal{K}(V_n)$ such that $\rho_S(T) \neq \emptyset$. If $\mathbf{p} \in \mathbb{R}^{n+1}$ belongs to $\sigma_S(T)$, then all the elements of the sphere $[\mathbf{p}]$ belong to $\sigma_S(T)$. The S-spectrum $\sigma_S(T)$ is a union of points on the real axis and/or $(n-1)$-spheres.*

Proof. The proof is analogous to the one of the bounded case. It immediately follows from the structure of the operator $\mathcal{Q}_{\mathbf{s}}(T)$. □

Theorem 3.7.9. *Let $T \in \mathcal{K}(V_n)$ such that $\rho_S(T) \neq \emptyset$. Then, for $\mathbf{s} \in \rho_S(T)$, the S-resolvent operator defined in (3.27) satisfies the equation*

$$S^{-1}(\mathbf{s}, T)\mathbf{s}v - TS^{-1}(\mathbf{s}, T)v = \mathcal{I}v, \quad \text{for all} \ \ v \in V_n. \tag{3.28}$$

Proof. It follows by direct computation. For $\mathbf{s} \in \rho_S(T)$ replace the S-resolvent operator in the equation (3.28) we get

$$[Q_{\mathbf{s}}(T)\overline{\mathbf{s}} - TQ_{\mathbf{s}}(T)]\mathbf{s}v - T[Q_{\mathbf{s}}(T)\overline{\mathbf{s}} - TQ_{\mathbf{s}}(T)]v = \mathcal{I}v, \quad \text{for all } v \in V_n.$$

Observe that $T[Q_{\mathbf{s}}(T)\overline{\mathbf{s}} - TQ_{\mathbf{s}}(T)]v \in V_n$ since $Q_{\mathbf{s}}(T) : V_n \to \mathcal{D}(T^2)$ and by trivial computations we get the identity

$$(T^2 - 2T\mathrm{Re}[\mathbf{s}] + |\mathbf{s}|^2\mathcal{I})Q_{\mathbf{s}}(T)v = v, \quad \text{for all } v \in V_n,$$

which proves the statement. □

Definition 3.7.10. *Let* $\mathbf{s} \in \rho_S(T) \neq \emptyset$. *The equation*

$$S^{-1}(\mathbf{s}, T)\mathbf{s} - TS^{-1}(\mathbf{s}, T) = \mathcal{I} \tag{3.29}$$

will be called the S-resolvent equation.

Recalling the notion of T-admissible open set given in Definition 3.2.2, we now give the following definition:

Definition 3.7.11. *Let $T \in \mathcal{K}(V_n)$. A function f is said to be locally s-monogenic on $\overline{\sigma}_S(T)$ if there exists a T-admissible open set U such that f is s-monogenic on U and at infinity. We will denote by $\mathcal{M}_{\overline{\sigma}_S(T)}$ the set of locally s-monogenic functions on $\overline{\sigma}_S(T)$.*

Remark 3.7.12. As we have pointed out in Remark 3.2.4, the open set U related to $f \in \mathcal{M}_{\overline{\sigma}_S(T)}$ need not be connected. Moreover, as in the classical functional calculus, U can depend on f and can be unbounded.

Definition 3.7.13. *Let $k \in \mathbb{R}$ and define the homeomorphism*

$$\Phi : \overline{\mathbb{R}}^{n+1} \to \overline{\mathbb{R}}^{n+1},$$

$$\mathbf{p} = \Phi(\mathbf{s}) = (\mathbf{s} - k)^{-1}, \quad \Phi(\infty) = 0, \quad \Phi(k) = \infty.$$

Definition 3.7.14. *Let $T : \mathcal{D}(T) \in \mathcal{K}(V_n)$ with $\rho_S(T) \cap \mathbb{R} \neq \emptyset$ and suppose that $f \in \mathcal{M}_{\overline{\sigma}_S(T)}$. Let us consider*

$$\phi(\mathbf{p}) := f(\Phi^{-1}(\mathbf{p}))$$

and the operator

$$A := (T - k\mathcal{I})^{-1}, \quad \textit{for some} \quad k \in \rho_S(T) \cap \mathbb{R}.$$

We define

$$f(T) := \phi(A). \tag{3.30}$$

Remark 3.7.15. Observe that, since $k \in \mathbb{R}$, we have that:

i) the function ϕ is s-monogenic because it is the composition of the function f which is s-monogenic and $\Phi^{-1}(\mathbf{p}) = \mathbf{p}^{-1} + k$ which is s-monogenic with real coefficients;

ii) if $k \in \rho_S(T) \cap \mathbb{R}$ we have that $(T - k\mathcal{I})^{-1} = -S^{-1}(k, T)$.

We now need a lemma.

Lemma 3.7.16. *Let* $\mathbf{s}$, $\mathbf{p} \in \mathbb{R}^{n+1}$ *and* $k \in \mathbb{R}$ *such that* $\mathbf{p} = (\mathbf{s} - k)^{-1}$. *Then the following identities hold*

$$s_0|\mathbf{p}|^2 = k|\mathbf{p}|^2 + p_0, \tag{3.31}$$

$$|\mathbf{p}|^2|\mathbf{s}|^2 = k^2|\mathbf{p}|^2 + 2p_0k + 1, \tag{3.32}$$

$$(2k\overline{\mathbf{p}} - 2s_0\overline{\mathbf{p}} + 1)\frac{1}{|\mathbf{p}|^2} = -\mathbf{p}^{-2}, \tag{3.33}$$

$$\frac{k^2\overline{\mathbf{p}} - |\mathbf{s}|^2\overline{\mathbf{p}} + k}{|\mathbf{p}|^2} = -\overline{\mathbf{s}}\mathbf{p}^{-2} \tag{3.34}$$

Proof. Identity (3.31) follows from

$$Re[\mathbf{s} - k] = Re[\mathbf{p}^{-1}] = Re[\overline{\mathbf{p}}|\mathbf{p}|^{-2}]$$

from which we have

$$s_0 - k = p_0|\mathbf{p}|^{-2}.$$

Identity (3.32) follows from the chain of identities

$$\begin{aligned} |\mathbf{s}|^2 = \mathbf{s}\overline{\mathbf{s}} &= (k + \mathbf{p}^{-1})\overline{(k + \mathbf{p}^{-1})} = (k + \mathbf{p}^{-1})(k + \overline{\mathbf{p}}^{-1}) \\ &= k^2 + k(\mathbf{p}^{-1} + \overline{\mathbf{p}}^{-1}) + \mathbf{p}^{-1}\overline{\mathbf{p}}^{-1} = k^2 + k\frac{2p_0}{|\mathbf{p}|^2} + \frac{1}{|\mathbf{p}|^2}. \end{aligned}$$

To prove (3.33) we consider the chain of identities

$$\begin{aligned} (2k - 2s_0 + \overline{\mathbf{p}}^{-1})\frac{\overline{\mathbf{p}}}{|\mathbf{p}|^2} &= (2k - 2s_0 + \overline{\mathbf{p}}^{-1})\mathbf{p}^{-1} \\ &= (2k - 2s_0 + \overline{\mathbf{s}} - k)(\mathbf{s} - k) = -(\mathbf{s} - k)^2 = -\mathbf{p}^{-2}. \end{aligned}$$

Finally we verify that (3.34) reduces to (3.32). Multiply (3.34) by $\mathbf{p}^2$ on the right to get

$$\frac{(k^2 - |\mathbf{s}|^2)|\mathbf{p}|^2\mathbf{p} + k\mathbf{p}^2}{|\mathbf{p}|^2} = -\overline{\mathbf{s}}$$

and now multiply by $\overline{\mathbf{p}}$ on the right to get

$$(k^2 - |\mathbf{s}|^2)\mathbf{p}\overline{\mathbf{p}} + \frac{k}{|\mathbf{p}|^2}\mathbf{p}^2\overline{\mathbf{p}} = -\overline{\mathbf{s}}\,\overline{\mathbf{p}} \tag{3.35}$$

and from $\mathbf{p} = (\mathbf{s} - k)^{-1}$ we get that

$$\overline{\mathbf{s}}\,\overline{\mathbf{p}} = 1 + k\overline{\mathbf{p}} \tag{3.36}$$

so from (3.35) and (3.36) we have

$$(k^2 - |\mathbf{s}|^2)|\mathbf{p}|^2 + k\mathbf{p} + 1 + k\overline{\mathbf{p}} = 0$$

which is identity (3.32). □

We can now prove a crucial result.

Theorem 3.7.17. *If $k \in \rho_S(T) \cap \mathbb{R} \neq \emptyset$ and Φ, ϕ are as above, then $\Phi(\overline{\sigma}_S(T)) = \sigma_S(A)$ and the relation $\phi(\mathbf{p}) := f(\Phi^{-1}(\mathbf{p}))$ determines a one-to-one correspondence between $f \in \mathcal{M}_{\overline{\sigma}_S(T)}$ and $\phi \in \mathcal{M}_{\overline{\sigma}_S(A)}$.*

Proof. From the definition of A we also have, for $k \in \rho_S(T) \cap \mathbb{R} \neq 0$,

$$A := (T - k\mathcal{I})^{-1} : V_n \to \mathcal{D}(T),$$

$$A^{-1} = T - k\mathcal{I} : \mathcal{D}(T) \to V_n$$

and

$$A^2 := (T^2 - 2kT + k^2\mathcal{I})^{-1} : V_n \to \mathcal{D}(T^2),$$
$$A^{-2} = T^2 - 2kT + k^2\mathcal{I} : \mathcal{D}(T^2) \to V_n.$$

Observe that for $\mathbf{p} \in \rho_S(A)$

$$Q_{\mathbf{p}}(A) := (A^2 - 2p_0A + |\mathbf{p}|^2\mathcal{I})^{-1} \in \mathcal{B}_n(V_n)$$

and

$$S^{-1}(\mathbf{p}, A) = Q_{\mathbf{p}}(A)\overline{\mathbf{p}} - AQ_{\mathbf{p}}(A).$$

Let us consider the relation

$$\begin{aligned}
Q_{\mathbf{p}}(A) &= \Big[(T - k\mathcal{I})^{-2} - 2p_0(T - k\mathcal{I})^{-1} + |\mathbf{p}|^2\mathcal{I}\Big]^{-1} \\
&= \Big[[\mathcal{I} - 2p_0(T - k\mathcal{I}) + |\mathbf{p}|^2(T - k\mathcal{I})^2](T - k\mathcal{I})^{-2}\Big]^{-1} \\
&= (T - k\mathcal{I})^2[\mathcal{I} - 2p_0(T - k\mathcal{I}) + |\mathbf{p}|^2(T - k\mathcal{I})^2]^{-1} \\
&= |\mathbf{p}|^{-2}(T - k\mathcal{I})^2[T^2 - 2(k + p_0/|\mathbf{p}|^2)T + (k^2|\mathbf{p}|^2 + 2p_0k + 1)/|\mathbf{p}|^2\mathcal{I}]^{-1}.
\end{aligned}$$

Using (3.31) and (3.32) we get

$$Q_{\mathbf{p}}(A) = |\mathbf{p}|^{-2}(T - k\mathcal{I})^2[T^2 - 2s_0T + |\mathbf{s}|^2\mathcal{I}]^{-1} : \ V_n \to V_n,$$

that is

$$Q_{\mathbf{p}}(A) = |\mathbf{p}|^{-2}(T - k\mathcal{I})^2 Q_{\mathbf{s}}(T). \tag{3.37}$$

Since A is a bounded operator $S^{-1}(\mathbf{p}, A) = Q_{\mathbf{p}}(A)\overline{\mathbf{p}} - AQ_{\mathbf{s}}(A) : V_n \to V_n$, we have

$$\begin{aligned}
S^{-1}(\mathbf{p}, A) &= |\mathbf{p}|^{-2}(T - k\mathcal{I})^2 Q_{\mathbf{s}}(T)\overline{\mathbf{p}} - |\mathbf{p}|^{-2}(T - k\mathcal{I})Q_{\mathbf{s}}(T) \\
&= |\mathbf{p}|^{-2}\Big[(T^2 - 2kT + k^2\mathcal{I})Q_{\mathbf{s}}(T)\overline{\mathbf{p}} - (T - k)Q_{\mathbf{s}}(T)\Big] \\
&= |\mathbf{p}|^{-2}\Big[(T^2 - 2s_0T + |\mathbf{s}|^2\mathcal{I})Q_{\mathbf{s}}(T)\overline{\mathbf{p}} \\
&\quad + (-2kT + k^2 + 2s_0T - |\mathbf{s}|^2)Q_{\mathbf{s}}(T)\overline{\mathbf{p}} - (T - k\mathcal{I})Q_{\mathbf{s}}(T)\Big] \\
&= |\mathbf{p}|^{-2}\Big[\mathcal{I}\overline{\mathbf{p}} + Q_{\mathbf{s}}(T)[k^2\overline{\mathbf{p}} - |\mathbf{s}|^2\overline{\mathbf{p}} + k] \\
&\quad - TQ_{\mathbf{s}}(T)[2k\overline{\mathbf{p}} - 2s_0\overline{\mathbf{p}} + 1]\Big] \\
&= \Big[\mathcal{I}\mathbf{p}^{-1} + Q_{\mathbf{s}}(T)\frac{k^2\overline{\mathbf{p}} - |\mathbf{s}|^2\overline{\mathbf{p}} + k}{|\mathbf{p}|^2} - TQ_{\mathbf{s}}(T)\frac{2k\overline{\mathbf{p}} - 2s_0\overline{\mathbf{p}} + 1}{|\mathbf{p}|^2}.\Big]
\end{aligned}$$

Now we use the identities (3.33) and (3.34) to get

$$S^{-1}(\mathbf{p}, A) = \mathcal{I}\mathbf{p}^{-1} - Q_{\mathbf{s}}(T)\overline{\mathbf{s}}\mathbf{p}^{-2} + TQ_{\mathbf{s}}(T)\mathbf{p}^{-2}$$

and finally

$$S^{-1}(\mathbf{p}, A) = \mathcal{I}\mathbf{p}^{-1} - S^{-1}(\mathbf{s}, T)\mathbf{p}^{-2}. \tag{3.38}$$

So $\mathbf{p} \in \rho_S(A)$, $\mathbf{p} \neq 0$, then $\mathbf{s} \in \rho_S(T)$.

Now take $\mathbf{s} \in \rho_S(T)$. We verify that

$$S^{-1}(\mathbf{s}, T) = -AS^{-1}(\mathbf{p}, A)\mathbf{p}$$

holds. Indeed, by (3.37) we get the qualities:

$$\begin{aligned}
-AS^{-1}(\mathbf{p}, A)\mathbf{p} &= -A[Q_{\mathbf{p}}(A)\overline{\mathbf{p}} - AQ_{\mathbf{p}}(A)]\mathbf{p} \\
&= -(T - k\mathcal{I})^{-1}\Big[[|\mathbf{p}|^{-2}(T - k\mathcal{I})^2 Q_{\mathbf{s}}(T)]\overline{\mathbf{p}} \\
&\quad - (T - k\mathcal{I})^{-1}[|\mathbf{p}|^{-2}(T - k\mathcal{I})^2 Q_{\mathbf{s}}(T)]\Big]\mathbf{p} \\
&= -TQ_{\mathbf{s}}(T) + Q_{\mathbf{s}}(T)(\frac{\mathbf{p}}{|\mathbf{p}|^2} + k) = S^{-1}(\mathbf{s}, T).
\end{aligned}$$

So if $\mathbf{s} \in \rho_S(T)$, then $\mathbf{p} \in \rho_S(A)$, $\mathbf{p} \neq 0$.

The point $\mathbf{p} = 0$ belongs to $\sigma_S(A)$ since $S^{-1}(0, A) = A^{-1} = T - k\mathcal{I}$ is unbounded. The last part of the statement is evident from the definition of Φ. □

Bearing in mind Definition 3.7.14, we can state the following result:

Theorem 3.7.18. *Let $T \in \mathcal{K}(V_n)$ with $\rho_S(T) \cap \mathbb{R} \neq \emptyset$ and suppose that $f \in \mathcal{M}_{\overline{\sigma}_S(T)}$. Then the operator $f(T) = \phi(A)$ is independent of $k \in \rho_S(T) \cap \mathbb{R}$. Let W be a T-admissible open set and let f be an s-monogenic function such that its domain of* *s-monogenicity contains $\overline{W}$. Set $ds_I = -dsI$ for $I \in \mathbb{S}$, then we have*

$$f(T) = f(\infty)\mathcal{I} + \frac{1}{2\pi}\int_{\partial(W\cap\mathbb{C}_I)} S^{-1}(\mathbf{s},T)d\mathbf{s}_I f(\mathbf{s}). \tag{3.39}$$

Proof. The first part of the statement follows from the validity of formula (3.39) since the integral is independent of k.

Given $k \in \rho_S(T)\cap\mathbb{R}$ and the set W we can assume that $k \notin \overline{W}\cap\mathbb{C}_I$, $\forall I \in \mathbb{S}$ since otherwise, by the Cauchy theorem, we can replace W by W', on which f is s-monogenic, such that $k \notin \overline{W}'\cap\mathbb{C}_I$, without altering the value of the integral (3.39). Moreover, the integral (3.39) is independent of the choice of $I \in \mathbb{S}$, thanks to the structure of the spectrum (see Theorem 3.2.1) and an argument similar to the one used to prove Theorem 3.3.2.

We have that $\mathcal{V}\cap\mathbb{C}_I := \Phi^{-1}(W\cap\mathbb{C}_I)$ is an open set that contains $\sigma_S(T)$ and its boundary $\partial(\mathcal{V}\cap\mathbb{C}_I) = \Phi^{-1}(\partial(W\cap\mathbb{C}_I))$ is positively oriented and consists of a finite number of continuously differentiable Jordan curves. Using the relation (3.38) we have

$$\begin{aligned}
&\frac{1}{2\pi}\int_{\partial(W\cap\mathbb{C}_I)} S^{-1}(\mathbf{s},T)d\mathbf{s}_I f(\mathbf{s})\\
&= -\frac{1}{2\pi}\int_{\partial(\mathcal{V}\cap\mathbb{C}_I)} \left(\mathbf{p}\mathcal{I} - S^{-1}(\mathbf{p},A)\mathbf{p}^2\right)\mathbf{p}^{-2}d\mathbf{p}_I\phi(\mathbf{p})\\
&= -\frac{1}{2\pi}\int_{\partial(\mathcal{V}\cap\mathbb{C}_I)} \mathbf{p}^{-1}d\mathbf{p}_I\phi(\mathbf{p}) + \frac{1}{2\pi}\int_{\partial(\mathcal{V}\cap\mathbb{C}_I)} S^{-1}(\mathbf{p},A)d\mathbf{p}_I\phi(\mathbf{p})\\
&= -\mathcal{I}\phi(0) + \phi(A).
\end{aligned}$$

Now by definition $\phi(A) = f(T)$ and $\phi(0) = f(\infty)$ we obtain

$$\frac{1}{2\pi}\int_{\partial(W\cap\mathbb{C}_I)} S^{-1}(\mathbf{s},T)d\mathbf{s}_I f(\mathbf{s}) = -\mathcal{I}f(\infty) + f(T). \qquad \square$$

Theorem 3.7.19. *Let f and $g \in \mathcal{M}_{\overline{\sigma}_S(T)}$. Then*

$$(f+g)(T) = f(T) + g(T).$$

Let $g \in \mathcal{M}_{\overline{\sigma}_S(T)}$ and let $f \in \widetilde{\mathcal{M}}_{\overline{\sigma}_S(T)}$. Then

$$(fg)(T) = f(T)g(T).$$

Proof. Observe that $fg \in \mathcal{M}_{\overline{\sigma}_S(T)}$ thanks to Proposition 3.4.2. Let $\phi(\mu) = f(\Phi^{-1}(\mu))$ and $\psi(\mu) = g(\Phi^{-1}(\mu))$. Thanks to Proposition 3.4.2 and Lemma 3.5.6 the product $\phi\psi$ is s-monogenic. By definition we have

$$f(T) = \phi(A), \qquad g(T) = \psi(A)$$

By Theorem 3.4.6 we have

$$(\phi + \psi)(A) = \phi(A) + \psi(A), \qquad (\phi\psi)(A) = \phi(A)\psi(A)$$

so we get the statement. □

Theorem 3.7.20. *Let $T \in \mathcal{K}(V_n)$ with $\rho_S(T) \cap \mathbb{R} \neq \emptyset$ and let $f \in \mathcal{N}_{\overline{\sigma}_S(T)}$. Then*

$$\sigma_S(f(T)) = f(\overline{\sigma}_S(T)).$$

Proof. Let $\phi(\mu) = f(\Phi^{-1}(\mu))$. By the Spectral Mapping Theorem we have $\phi(\sigma_S(A)) = \sigma_S(\phi(A))$ and by Theorem 3.7.17 we also have $\Phi(\sigma_S(T) \cup \{\infty\}) = \sigma_S(A)$. So we obtain

$$\phi(\Phi(\sigma_S(T) \cup \{\infty\})) = \phi(\sigma_S(A)) = \sigma_S(\phi(A)) = \sigma_S(f(T)).$$

On the other hand

$$\phi(\Phi(\sigma_S(T) \cup \{\infty\})) = f(\Phi^{-1}(\Phi(\sigma_S(T) \cup \{\infty\}))) = f(\overline{\sigma}_S(T)).$$ □

We conclude the section with an example.

Example 3.7.21. Let $T \in \mathcal{K}(V_n)$ such that T^{-1} is a bounded operator. From the definition of S-resolvent operator we get $S^{-1}(0,T) = -T^{-1}$ so $\rho_S(T)$ contains 0. Moreover the function $f(\mathbf{x}) = \mathbf{x}^{-1}$ is s-monogenic in a T-admissible open set U such that $0 \notin \overline{U}$. Since $f(\mathbf{x}) = \mathbf{x}^{-1} \to 0$ as $\mathbf{x} \to \infty$, by Theorem 3.7.18, we have

$$T^{-1} = \frac{1}{2\pi} \int_{\partial(U \cap \mathbb{C}_I)} S^{-1}(\mathbf{s}, T) d\mathbf{s}_I \mathbf{s}^{-1}$$

and thanks to Theorem 3.7.20 we obtain:

$$\sigma_S(T^{-1}) = \{\lambda^{-1} \ : \ \lambda \in \overline{\sigma}_S(T)\ \}.$$

3.8 Notes

Note 3.8.1. Further readings. The material in this chapter covers the contents of [25], where the authors started the study of this functional calculus, and its further developments due to Colombo and Sabadini, [15] and [18]. For further readings see also [19], [22] and [20].

Note 3.8.2. Monogenic functions. A functional calculus based on the classical notion of monogenic functions was extensively studied by Jefferies, McIntosh and their coworkers. We mention here, with no claim of completeness, the works [60], [61], [65], [66], [77], the book [62] and the references therein. In this note we mention some of their ideas. To start with, we will quickly recall the basic notions on monogenic functions.

The well-known notion of monogenic functions with values in a Clifford algebra (see [7]) is based on the so-called Dirac operator

$$\partial_{\underline{x}} = \sum_{j=1}^{n} e_j \partial_{x_j}. \tag{3.40}$$

Remark 3.8.3. A variation of the Dirac operator is the Cauchy–Riemann operator:

$$\partial_{\mathbf{x}} = \partial_{x_0} + \partial_{\underline{x}},$$

whose nullsolutions $f : U \subseteq \mathbb{R}^{n+1} \to \mathbb{R}_n$ on an open set U are still called (left) monogenic. Moreover, in the literature, the Cauchy–Riemann operator is often called a Dirac operator since it is possible to obtain it from the Dirac operator in (3.40) by grouping the imaginary units and making some identifications in a suitable way.

In the sequel we will consider functions which are monogenic according to the following definition:

Definition 3.8.4. *A real differentiable function $f : U \subseteq \mathbb{R}^{n+1} \to \mathbb{R}_n$ on an open set U is called (left) monogenic in U if it satisfies $\partial_{\mathbf{x}} f(\mathbf{x}) = 0$ on U.*

Monogenic functions can be expanded into power series in terms of the building blocks $z_j = e_j x_0 - e_0 x_j$, $1 \leq j \leq n$: one has to consider these symmetric polynomials and the sum of all their possible permutations for any given degree k according to the following definition:

Definition 3.8.5. *Homogeneous monogenic polynomials of degree k are defined as*

$$V^{\ell_1,\dots,\ell_k}(\mathbf{x}) = \frac{1}{k!} \sum_{\ell_1,\dots,\ell_k} z_{\ell_1} \dots z_{\ell_k}, \tag{3.41}$$

where $z_j = x_j e_0 - x_0 e_j$ and the sum is taken over all different permutation of $\ell_1, \dots, \ell_k$.

Definition 3.8.6. *Denote by Σ_n the surface area of the unit sphere in $\mathbb{R}^{n+1}$ and by $\bar{\mathbf{x}} = x_0 - \underline{x}$ the conjugate of $\mathbf{x} = x_0 + \underline{x}$. For each $\mathbf{x} \in \mathbb{R}^{n+1}$, define the function $G(\cdot, \mathbf{x})$ as*

$$G(\omega, \mathbf{x}) = \frac{1}{\Sigma_n} \frac{\bar{\omega} - \bar{\mathbf{x}}}{|\omega - \mathbf{x}|^{n+1}}. \tag{3.42}$$

Note that $G(\omega, \mathbf{x})$, for $\omega \neq \mathbf{x}$, is both left and right monogenic as a function of ω. It plays the role of the Cauchy kernel as shown in the following result (see [7]).

Theorem 3.8.7. *Let $\Omega \subset \mathbb{R}^{n+1}$ be a bounded open set with smooth boundary $\partial\Omega$ and exterior unit normal $n(\omega)$ defined for all $\omega \in \partial\Omega$. For any left monogenic function f defined in a neighborhood of U of $\overline{\Omega}$, we have the Cauchy formula*

$$\int_{\partial\Omega} G(\omega, \mathbf{x}) n(\omega) f(\omega) d\mu(\omega) = \begin{cases} f(\mathbf{x}), & if \ \ \mathbf{x} \in \Omega, \\ 0, & if \ \ \mathbf{x} \notin \Omega, \end{cases} \tag{3.43}$$

where μ is the surface measure of $\partial\Omega$.

Note 3.8.8. The monogenic functional calculus. The Cauchy formula (3.43) is the starting point for the monogenic functional calculus. To this purpose, it is useful to consider a suitable series expansion of the kernel $G(\omega, x) = G_\omega(\mathbf{x})$:

$$G_\omega(\mathbf{x}) = \sum_{k\geq 0} \left(\sum_{(\ell_1,\dots,\ell_k)} V^{\ell_1,\dots,\ell_k}(\mathbf{x}) W_{\ell_1,\dots,\ell_k}(\omega) \right)$$

in the region $|\mathbf{x}| > |\omega|$ (see [7]) where, for each $\omega \in \mathbb{R}^{n+1}$, $\omega \neq 0$,

$$W_{\ell_1,\dots,\ell_k}(\omega) = (-1)^k \partial_{\omega_{\ell_1}} \dots \partial_{\omega_{\ell_k}} G_\omega(0)$$

and $V^{\ell_1,\dots,\ell_k}(\mathbf{x})$ are defined in (3.41).

Keeping in mind the definition of Banach modules, see the beginning of this chapter, consider now an n-tuple $T = (T_1, \dots, T_n)$ of bounded linear operator acting on a Banach space X and let

$$R > (1 + \sqrt{2}) \| \sum_{j=1}^{n} T_j e_j \|. \tag{3.44}$$

Let us formally replace z_j by T_j and 1 by the identity operator $\mathcal{I}$ in the Cauchy kernel series. It can be shown (see [65] Lemma 3.12, [62], Lemma 4.7) that

$$\sum_{k\geq 0} \left(\sum_{(\ell_1,\dots,\ell_k)} V^{\ell_1,\dots,\ell_k}(T) W_{\ell_1,\dots,\ell_k}(\omega) \right) \tag{3.45}$$

where

$$V^{\ell_1,\dots,\ell_k}(T) = \frac{1}{k!} \sum_{\ell_1,\dots,\ell_k} T_{\ell_1} \dots T_{\ell_k},$$

converges uniformly for all $\omega \in \mathbb{R}^{n+1}$ such that $|\omega| \geq R$, where R is given in (3.44). We set the sum of the series (3.45) equal to $G_\omega(T)$ which turns out to be a bounded operator.

Remark 3.8.9. In [65] the so-called resolvent set is the set of $\omega \in \mathbb{R}^n$ such that the series (3.45) converges. The spectral set $\sigma_C(T)$ of T is defined as the set complement of the resolvent set.

An important result is the following (see [65]):

Theorem 3.8.10. *Let $(T_1, \dots, T_n)$ be an n-tuple of bounded self-adjoint operators. Let Ω be a domain with piecewise smooth boundary whose complement is connected, and suppose that $\sigma_C(T) \subseteq \Omega$. Then, for every $f \in \mathcal{M}(\Omega)$ the mapping*

$$f(\mathbf{x}) \mapsto f(T) = \int_{\partial\Omega} G_\omega(T) n(\omega) f(\omega) d\mu(\omega)$$

defines a functional calculus.

Remark 3.8.11. Assume that T is an n-tuple of bounded self adjoint operators, and Ω is an open set with piecewise smooth boundary with connected complement containing the spectral set $\sigma_C(T)$. Then, see [65], the map in Theorem 3.8.10 defines a functional calculus for functions which are monogenic on Ω. This fact is guaranteed by a Runge type approximation theorem.

In [62] the definition of spectrum is different:

Definition 3.8.12. *The monogenic spectrum $\gamma(T)$ of the n-tuple T is the complement of the largest connected open set U in $\mathbb{R}^{n+1}$ in which the function $G_\omega(T)$ defined by the series above is the restriction of a monogenic function with domain U.*

Let $\langle T, \underline{\xi}\rangle := \sum_{j=1}^n T_j\xi_j$ and suppose that $\sigma(\langle T, \underline{\xi}\rangle)$ is real for all $\underline{\xi} \in \mathbb{R}^n$ (here σ denotes the spectrum in the classical sense, i.e., the set of singularities of $(\lambda\mathcal{I} - \langle T, \underline{\xi}\rangle)^{-1}$). Then we have the following result (see [62]):

Theorem 3.8.13. *Let $T = (T_1, \dots, T_n)$ be an n-tuple of noncommuting bounded linear operator acting on a Banach space X and suppose that $\sigma(\langle T, \underline{\xi}\rangle) \subseteq \mathbb{R}$ for all $\underline{\xi} \in \mathbb{R}^n$. Then the $\mathcal{B}_n(X_n)$-valued function $G_\omega(T)$ defined in* (3.45) *is the restriction to the region*

$$\Gamma = \{\omega \in \mathbb{R}^{n+1} \ : \ |\omega| > (1+\sqrt{2})\|\sum_{j=1}^n T_j e_j\|\}$$

of a left and right monogenic function on $\mathbb{R}^{n+1}\setminus\mathbb{R}^n$.

This result guarantees that $G_\omega(T)$ is monogenic outside a ball. However, $G_\omega(T)$ can be monogenic in a larger set containing Γ. Denote with the same symbol $G_\omega(T)$ its maximal monogenic extension and let Ω be the union of all open sets containing the open set Γ on which is defined a two-sided monogenic function whose restriction on Γ equals the series $G_\omega(T)$. Then the extension is unique because the domain Ω is connected, contains Γ and the spectrum is a subset of $\mathbb{R}^n$ and hence it cannot disconnect a set in $\mathbb{R}^{n+1}$.

Let $T = (T_1, \dots, T_n)$ be an n-tuple of noncommuting bounded linear operators acting on a Banach space X and $\underline{\xi} \in \mathbb{R}^n$. Suppose that $\sigma(\langle T, \underline{\xi}\rangle)$ is real for all $\underline{\xi} \in \mathbb{R}^n$. Let $\Omega \subseteq \mathbb{R}^{n+1}$ be a bounded open neighborhood of $\gamma(T)$ with smooth boundary and exterior normal $n(\omega)$, for all $\omega \in \partial\Omega$. Let f be a monogenic function defined in an open neighborhood of $\overline{\Omega}$. Define the operator $f(T)$ by

$$f(T) = \int_{\partial\Omega} G_\omega(T) n(\omega) f(\omega) d\mu(\omega). \tag{3.46}$$

Denote by $\mathcal{M}(\gamma(T), \mathbb{R}_n)$ the right module of monogenic functions defined in a neighborhood of $\gamma(T)$ in $\mathbb{R}^{n+1}$.

Definition 3.8.14. *The map $f \mapsto f(T)$, $f \in \mathcal{M}(\gamma(T), \mathbb{R}_n)$ is called monogenic functional calculus.*

The map defined above is a right-module homomorphism.

Remark 3.8.15. Let $p_{\underline{\xi}}$ be a complex-valued polynomial, $\underline{\xi} \in \mathbb{R}^n$ and consider

$$p_{\underline{\xi}}(x_1\xi_1 + \ldots + x_n\xi_n).$$

Then we define

$$p_{\underline{\xi}}(T_1, \ldots, T_n) := p_{\underline{\xi}}(T_1\xi_1 + \ldots + T_n\xi_n).$$

Note that the function $G_\omega(T)$ admits a plane wave expansion as follows

Proposition 3.8.16. *Let $\omega \in \mathbb{R}^{n+1}$, $\omega = \omega_0 + \underline{\omega}$, $\omega_0 \neq 0$. Then*

$$G_\omega(T) = \frac{(n-1)!}{2}\left(\frac{i}{2\pi}\right)^n \operatorname{sgn}(\omega_0)^{n-1}\int_{S^{n-1}} (1+is)(\langle \omega\mathcal{I} - T, s\rangle - \omega_0 s)^{-n}\, ds. \tag{3.47}$$

When the condition $\sigma(\langle T, \underline{\xi}\rangle) \in \mathbb{R}$ is satisfied for all $\underline{\xi} \in \mathbb{R}^n$, then $\gamma(T) \subset \mathbb{R}^n$ is the complement in $\mathbb{R}^{n+1}$ of the points ω at which the function defined by the integral above is continuous. In the case of commuting bounded linear operators the spectrum can be determined directly as shown in the next result (see Theorem 3.3 and Corollary 3.4 in [77]).

Theorem 3.8.17. *Let $T = (T_1, \ldots, T_n)$ be a n-tuple of commuting bounded linear operator acting on a Banach space X and suppose that $\sigma(T_j) \subseteq \mathbb{R}$ for all $j = 1, \ldots, n$. Then $\gamma(T)$ is the complement in $\mathbb{R}^n$ of the set of all $\underline{\lambda} \in \mathbb{R}^n$ for which the operator $\sum_{j=1}^n (\lambda_j\mathcal{I} - A_j)^2$ is invertible in $\mathcal{B}(X)$ (equivalently: $(\underline{\lambda}\mathcal{I} - T)$ is invertible in $End(X)$).*

Chapter 4

Quaternionic Functional Calculus

The first section of this chapter collects the main results on the theory of slice regular functions. Similarly to what happens in the theory of regular functions in the sense of Cauchy–Fueter, whose results sometimes resemble the analogous results for monogenic functions, also for slice regular functions we have that some statements and their proofs mimic those we proved in Chapter 2. They are repeated here for the reader's convenience, especially because the notation in the quaternionic case might be simpler. Note that the richer structure of quaternions allows results which are not necessarily true for s-monogenic functions. The results that are specific to the quaternionic case or those for which the proofs are significantly different or simpler will be followed by their proofs.

4.1 Notation and definition of slice regular functions

The Clifford algebra over two units $\mathbb{R}_2$ is the algebra of quaternions. It is usually denoted by $\mathbb{H}$ in honor of Hamilton who introduced them in 1843. Instead of the imaginary units e_1, e_2 and e_1e_2, the imaginary units in $\mathbb{H}$ are denoted by i, j and k respectively and an element in $\mathbb{H}$ is of the form $q = x_0 + ix_1 + jx_2 + kx_3$, for $x_\ell \in \mathbb{R}$. The real algebra $\mathbb{H}$ is a skew field and there are no higher-dimensional Clifford algebras which are division algebras. The real part, imaginary part and modulus of a quaternion are defined as

$$\text{Re } q = x_0, \qquad \text{Im } q = ix_1 + jx_2 + kx_3, \qquad |q| = \sqrt{x_0^2 + x_1^2 + x_2^2 + x_3^2}.$$

The conjugate of the quaternion $q = x_0 + ix_1 + jx_2 + kx_3$ is defined by

$$\bar{q} = \text{Re } q - \text{Im } q = x_0 - ix_1 - jx_2 - kx_3$$

(compare with Definition 2.1.11) and it satisfies

$$|q| = \sqrt{q\bar{q}} = \sqrt{\bar{q}q}.$$

The inverse of any nonzero element q is given by

$$q^{-1} = \frac{\bar{q}}{|q|^2}.$$

Notice that a generic element q of $\mathbb{H}$ can be written as the linear combination of two complex numbers:

$$q = (x_0 + x_1 i) + (x_2 + x_3 i)j. \tag{4.1}$$

By identifying an element in $\mathbb{H}$ with pairs of complex numbers, each in the complex plane $\mathbb{R}+\mathbb{R}i$, it is possible to define the algebra of quaternions using the well-known Cayley–Dickson process, see [70]. From this point of view, a quaternion q is a pair of complex numbers (a, b) endowed with an operation of addition componentwise and with the multiplication defined by

$$qp = (a, b)(c, d) = (ac - d\bar{b}, ad + b\bar{c})$$

where $\bar{a}$ denotes the complex conjugate of a.

Let us denote by $\mathbb{S}$ the unit sphere of purely imaginary quaternions, i.e.,

$$\mathbb{S} = \{q = ix_1 + jx_2 + kx_3 \text{ such that } x_1^2 + x_2^2 + x_3^2 = 1\}.$$

Notice that if $I \in \mathbb{S}$, then $I^2 = -1$; for this reason the elements of $\mathbb{S}$ are also called imaginary units. Note that $\mathbb{S}$ is a 2-dimensional sphere in $\mathbb{R}^4$. Given a nonreal quaternion $q = x_0 + \mathrm{Im}q = x_0 + I|\mathrm{Im}q|$, $I = \mathrm{Im}q/|\mathrm{Im}q| \in \mathbb{S}$, we can associate to it the 2-dimensional sphere defined by

$$[q] = \{x_0 + I|\mathrm{Im}q| \mid I \in \mathbb{S}\}.$$

This sphere has center at the real point x_0 and radius $|\mathrm{Im}q|$. In this chapter we will denote an element in the complex plane $\mathbb{R} + I\mathbb{R}$ by $x + Iy$.

Definition 4.1.1. *Let U be an open set in $\mathbb{H}$. A real differentiable function $f : U \to \mathbb{H}$ is said to be slice left regular (or* s-regular *for short) if, for every $I \in \mathbb{S}$, its restriction f_I to the complex plane $\mathbb{C}_I = \mathbb{R} + I\mathbb{R}$ passing through the origin and containing* 1 *and I satisfies*

$$\overline{\partial}_I f(x + Iy) := \frac{1}{2}\left(\frac{\partial}{\partial x} + I\frac{\partial}{\partial y}\right) f_I(x + Iy) = 0,$$

on $U \cap \mathbb{C}_I$. We will denote by $\mathcal{R}(U)$, or by $\mathcal{R}^L(U)$ when confusion may arise, the set of left s-regular functions on the open set U.

Analogously, a function is said to be right slice regular (or right s-regular *for short) if*

$$(f_I\overline{\partial}_I)(x+Iy) := \frac{1}{2}\left(\frac{\partial}{\partial x}f_I(x+Iy) + \frac{\partial}{\partial y}f_I(x+Iy)I\right) = 0,$$

on $U \cap \mathbb{C}_I$. We will denote by $\mathcal{R}^R(U)$ the set of right s-regular functions on the open set U.

Remark 4.1.2. It is easy to verify that the (left) s-regular functions on $U \subseteq \mathbb{H}$ form a right $\mathbb{H}$-vector space. The right s-regular functions on $U \subseteq \mathbb{H}$ form a left $\mathbb{H}$-vector space.

An immediate consequence of the definition of s-regularity is that the monomial $q^n a$, with $a \in \mathbb{H}$, is s-regular and so all polynomials with quaternionic coefficients on the right are s-regular. Obviously, polynomials with quaternionic coefficients on the left are right s-regular.

We define the I-derivative of f at q by

$$\partial_I f_I(x+Iy) := \frac{1}{2}\left(\frac{\partial}{\partial x}f_I(x+Iy) - I\frac{\partial}{\partial y}f_I(x+Iy)\right),$$

and the right I-derivative by

$$(f_I\partial_I)(x+Iy) := \frac{1}{2}\left(\frac{\partial}{\partial x}f_I(x+Iy) - \frac{\partial}{\partial y}f_I(x+Iy)I\right).$$

We are now ready to give the following definition:

Definition 4.1.3. *Let U be an open set in $\mathbb{H}$, and let $f : U \to \mathbb{H}$ be an s-regular function. The slice derivative (in short s-derivative) of f, $\partial_s f$, is defined as follows:*

$$\partial_s(f)(q) = \begin{cases} \partial_I(f)(q) & \text{if } q = x+Iy,\ y \neq 0, \\ \dfrac{\partial f}{\partial x}(x) & \text{if } q = x \in \mathbb{R}. \end{cases}$$

Notice that the definition of s-derivative is well posed because it is applied only to s-regular functions for which

$$\frac{\partial}{\partial x}f(x+Iy) = -I\frac{\partial}{\partial y}f(x+Iy) \qquad \forall I \in \mathbb{S},$$

and therefore, analogously to what happens in the complex case,

$$\partial_s(f)(x+Iy) = \partial_I(f)(x+Iy) = \partial_x(f)(x+Iy).$$

Note that if f is an s-regular function, then its s-derivative is still regular because

$$\overline{\partial}_I(\partial_s f(x+Iy)) = \partial_s(\overline{\partial}_I f(x+Iy)) = 0, \tag{4.2}$$

and therefore

$$\partial_s^n f(x+Iy) = \frac{\partial^n f}{\partial x^n}(x+Iy).$$

We will now identify a class of domains that naturally qualify as domains of definition of regular functions. Other reasons for this definition will appear in the sequel.

Definition 4.1.4. *Let $U \subseteq \mathbb{H}$ be a domain in $\mathbb{H}$. We say that U is a* slice domain (s-domain for short) *if $U \cap \mathbb{R}$ is nonempty and if $U \cap \mathbb{C}_I$ is a domain in $\mathbb{C}_I$ for all $I \in \mathbb{S}$.*

In order to study s-regular functions, we will need a representation of the restriction of an s-regular function as a pair of holomorphic functions. To do so, we need a preliminary, simple result which is the quaternionic version of formula (2.2):

Proposition 4.1.5. *Let I and J be two elements in $\mathbb{S}$. Then their quaternionic product IJ can be computed through the following formula:*

$$IJ = -\langle I, J\rangle + I \wedge J.$$

Two elements I and J in $\mathbb{S}$ are orthogonal if and only if $\langle I, J\rangle = 0$. The previous proposition shows, in particular, that the product of two orthogonal elements of $\mathbb{S}$ lies in $\mathbb{S}$ as well. We will use this simple fact to build orthogonal bases in $\mathbb{S}$.

Proposition 4.1.6. *Let I and J be two orthogonal elements in $\mathbb{S}$, and let $K = IJ$. Then:*

(1) *$K = IJ = -JI$ is an element of $\mathbb{S}$,*

(2) *K is orthogonal to both I and J,*

(3) *$JK = I = -KJ$ and $KI = J = -IK$.*

Proof. (1) Since I and J are orthogonal, and $I \wedge J = -J \wedge I$, the result is an immediate consequence of the previous proposition.

(2) Using the arguments above, we obtain

$$\langle K, I\rangle = \langle IJ, I\rangle = \langle I \wedge J, I\rangle = 0.$$

(3) By a repeated application of Proposition 4.1.5 we obtain

$$\begin{aligned} JK &= J(IJ) = J(-\langle I, J\rangle + I \wedge J) \\ &= -\langle J, -\langle I, J\rangle + I \wedge J\rangle + J \wedge (-\langle I, J\rangle + I \wedge J). \end{aligned}$$

The orthogonality of I and J implies

$$JK = -\langle J, I \wedge J\rangle + J \wedge (I \wedge J).$$

Now note that $\langle J, I \wedge J\rangle = 0$ because $I \wedge J$ is orthogonal to J, and that, by the same reason, $IJ = I \wedge J$. Thus to conclude the proof we only need to show that $J \wedge K = I$. This is an immediate consequence of the orthogonality of I and J. □

This result shows that we can use I, J, and K as a basis for $\mathbb{S}$, and that, given any element I in $\mathbb{S}$, we can always construct such a basis, though not in a unique way. We now state the Splitting Lemma:

Lemma 4.1.7 (Splitting Lemma). *If f is an s-regular function on an open set U, then for every $I \in \mathbb{S}$, and every $J \in \mathbb{S}$, perpendicular to I, there are two holomorphic functions $F, G : U \cap \mathbb{C}_I \to \mathbb{C}_I$ such that for any $z = x + Iy$, it is*

$$f_I(z) = F(z) + G(z)J.$$

Proof. Let $K \in \mathbb{S}$ be such that I, J, K is an orthogonal basis and write $f_I(x+Iy) = f(x+Iy)$ as $f = f_0 + If_1 + Jf_2 + Kf_3$. Since f is s-regular, we know that

$$(\frac{\partial}{\partial x} + I\frac{\partial}{\partial y})f_I(x+Iy) = 0.$$

This expression becomes

$$\frac{\partial f_0}{\partial x} - \frac{\partial f_1}{\partial y} + I(\frac{\partial f_0}{\partial y} + \frac{\partial f_1}{\partial x}) + J(\frac{\partial f_2}{\partial x} - \frac{\partial f_3}{\partial y}) + K(\frac{\partial f_3}{\partial x} + \frac{\partial f_2}{\partial y}) = 0.$$

Therefore the functions $f_0 + If_1$ and $f_2 + If_3$ satisfy the Cauchy–Riemann system and thus they are both holomorphic. In particular, if we set $f_0 + If_1 = F$, and $f_2 + If_3 = G$, we obtain that

$$f_I(x+Iy) = F(x+Iy) + G(x+Iy)J,$$

and the lemma follows with $z = x + Iy$. □

4.2 Properties of slice regular functions

As we saw previously, polynomials in q are s-regular. In this section we will show that in fact power series are s-regular as well and that every s-regular function defined on an s-domain can be expanded into power series into a small open ball centered at a real point of the domain.

The Splitting Lemma 4.1.7 shows that every s-regular function f on an s-domain U can be written on $U \cap \mathbb{C}_I$ as $f = F + GJ$, with J orthogonal to I and F, G holomorphic on the plane $\mathbb{C}_I$; it is therefore obvious that f admits, on that plane, a series expansion in powers of z. Such an expansion can be used to provide a series expansion for f in powers of q. The next few results are stated for s-regular functions on balls

$$B = B(0, R) = \{q \in \mathbb{H} \mid |q| < R\},$$

centered at the origin with positive radius R. It is easy to see how the proofs can be modified to account for the case of balls with center on any point on the real axis. Unlike the complex case, however, it is not possible in general to extend the theory to a ball centered in any point of the quaternionic space. Indeed, $f(q) = (q-p)^n$ is not regular unless p is real.

Theorem 4.2.1. *A function $f : B \to \mathbb{H}$ is s-regular if, and only if, it has a series expansion of the form*

$$f(q) = \sum_{n \geq 0} q^n \frac{1}{n!} \frac{\partial^n f}{\partial x^n}(0)$$

converging on B. In particular if f is s-regular, then it is infinitely differentiable on B.

Corollary 4.2.2. *Let $f : B \to \mathbb{H}$ be s-regular. If there exists $I \in \mathbb{S}$ such that $f(\mathbb{C}_I) \subseteq \mathbb{C}_I$, then the series expansion of f*

$$f(q) = \sum_{n \geq 0} q^n \frac{1}{n!} \frac{\partial^n f}{\partial x^n}(0)$$

has all its coefficients in $\mathbb{C}_I$. If, in particular, there are two different units $I, J \in \mathbb{S}$ such that $f(\mathbb{C}_I) \subseteq \mathbb{C}_I$ and $f(\mathbb{C}_J) \subseteq \mathbb{C}_J$, then the coefficients are real.

The previous results extend to the case of an s-domain as follows:

Corollary 4.2.3. *Let f be s-regular on an s-domain U. Then for any real point p_0 in U, the function f can be represented by power series*

$$f(q) = \sum_{n \geq 0} (q - p_0)^n \frac{1}{n!} \frac{\partial^n f}{\partial x^n}(p_0)$$

on the ball $B(p_0, R)$ where $R = R_{p_0}$ is the largest positive real number such that $B(p_0, R)$ is contained in U.

The power series expansion is the key ingredient in proving the analogs, for s-regular functions, of many well-known results from the theory of holomorphic functions in one variable.

Theorem 4.2.4 (Identity Principle). *Let $f : U \to \mathbb{H}$ be an s-regular function on an s-domain U. Denote by $Z_f = \{q \in U : f(q) = 0\}$ the zero set of f. If there exists $I \in \mathbb{S}$ such that $\mathbb{C}_I \cap Z_f$ has an accumulation point, then $f \equiv 0$ on U.*

This result immediately implies the following corollary.

Corollary 4.2.5. *Let f and g be s-regular functions on an s-domain U. If there exists $I \in \mathbb{S}$ such that $f \equiv g$ on a subset of $U \cap \mathbb{C}_I$ having an accumulation point in $U \cap \mathbb{C}_I$, then $f \equiv g$ everywhere on U.*

We will now prove a natural analog, for s-regular functions, of the Cauchy representation formula. In order to state it appropriately, we will adopt the following notation. If $q \in \mathbb{H}$, we set

$$I_q = \begin{cases} \dfrac{\operatorname{Im}(q)}{|\operatorname{Im}(q)|} \in \mathbb{S} & \text{if } \operatorname{Im}(q) \neq 0, \\ \text{any element of } \mathbb{S} & \text{otherwise.} \end{cases}$$

It is immediate that for any nonreal quaternion $q \in \mathbb{H}\backslash\mathbb{R}$, there exist, and are unique, $x, y \in \mathbb{R}$ with $y > 0$, and such that $q = x + I_q y$.

We can now prove an integral representation formula, which is of limited validity, but it is enough to prove several results.

Theorem 4.2.6 (Cauchy formula, I). *Let $U \subset \mathbb{H}$ be an axially symmetric s-domain and $q \in U$. Let $f : U \to \mathbb{H}$ be an s-regular function, and suppose $\gamma \subset \mathbb{C}_{I_q}$ be a continuously differentiable Jordan curve surrounding q. Then*

$$f(q) = \frac{1}{2\pi} \int_\gamma (\zeta - q)^{-1} d\zeta_{I_q} f(\zeta),$$

where $d\zeta_{I_q} = -d\zeta I_q$.

Proof. Notice that for any ζ belonging to the plane $\mathbb{C}_{I_q}$ containing q, $\zeta \neq q$ we have the equality

$$(\zeta - q)^{-1} d\zeta = d\zeta(\zeta - q)^{-1}.$$

The result now follows immediately from the Splitting Lemma and the classical Cauchy formula, as indicated by the following equalities:

$$\begin{aligned}
&\frac{1}{2\pi} \int_\gamma (\zeta - q)^{-1} d\zeta_{I_q} f(\zeta) = \frac{1}{2\pi} \int_\gamma (\zeta - q)^{-1} d\zeta_{I_q} f_{I_q}(\zeta) \\
&= \frac{1}{2\pi} \int_\gamma (\zeta - q)^{-1} d\zeta_{I_q} (F(\zeta) + G(\zeta)J) \qquad (4.3) \\
&= \frac{1}{2\pi} \int_\gamma (\zeta - q)^{-1} d\zeta_{I_q} F(\zeta) + \left(\frac{1}{2\pi} \int_\gamma (\zeta - q)^{-1} d\zeta_{I_q} G(\zeta)\right) J \\
&= F(q) + G(q)J = f(q).
\end{aligned}$$

□

As an immediate consequence we obtain:

Theorem 4.2.7 (Cauchy Estimates). *Let $U \subset \mathbb{H}$ be an axially symmetric s-domain and let $f : U \to \mathbb{H}$ be an s-regular function. Let $p_0 \in U \cap \mathbb{R}$, $I \in \mathbb{S}$, and $r > 0$ be such that*

$$\overline{\Delta_I}(p_0, r) = \{(x + Iy) : (x - p_0)^2 + y^2 \leq r^2\}$$

is contained in $U \cap \mathbb{C}_I$. If $M_I = \max\{|f(q)| : q \in \partial\Delta_I(p_0, r)\}$ and if $M = \inf\{M_I : I \in \mathbb{S}\}$, then

$$\frac{1}{n!} \left| \frac{\partial^n f}{\partial x^n}(p_0) \right| \leq \frac{M}{r^n}, \qquad n \geq 0.$$

We now have all the tools needed to prove the analog of the Liouville theorem.

Theorem 4.2.8 (Liouville). *Let $f : \mathbb{H} \to \mathbb{H}$ be an entire regular function (i.e., an s-regular function defined and s-regular everywhere on $\mathbb{H}$). If f is bounded by a positive constant M, then f is constant.*

We now show how to generalize the theory of Laurent series to the quaternionic case, and we show how the domain of convergence of such series $\sum_{n\in\mathbb{Z}} q^n a_n$ is a four-dimensional spherical shell $A(0, R_1, R_2) = \{q \in \mathbb{H} : R_1 < |q| < R_2\}$. More precisely one can prove, just as in the complex case, the following result.

Lemma 4.2.9. *Let $\{a_n\}_{n\in\mathbb{Z}} \subset \mathbb{H}$. There exist R_1, R_2 with $0 \leq R_1 < R_2 \leq \infty$ such that*

(1) *the series $\sum_{n\in\mathbb{N}} q^n a_n$ and $\sum_{n\in\mathbb{N}} q^{-n} a_{-n}$ both converge absolutely and uniformly on compact subsets of $A = A(0, R_1, R_2)$;*

(2) *for all $q \in \mathbb{H} \setminus \bar{A}$, either $\sum_{n\in\mathbb{N}} q^n a_n$ or $\sum_{n\in\mathbb{N}} q^{-n} a_{-n}$ diverge.*

As a consequence we have

Theorem 4.2.10. *Let $\sum_{n\in\mathbb{Z}} q^n a_n$ be a series having domain of convergence $A = A(0, R_1, R_2)$ with $R_1 < R_2$. Then $f : A \to \mathbb{H}$ $q \mapsto \sum_{n\in\mathbb{Z}} q^n a_n$ is an s-regular function.*

Proof. The proof follows by direct computation from the definition of s-regularity. □

We will now prove that all s-regular functions $f : A(0, R_1, R_2) \to \mathbb{H}$ admit Laurent series expansions.

Theorem 4.2.11 (Laurent Series Expansion). *Let $A = A(0, R_1, R_2)$ with $0 \leq R_1 < R_2 \leq +\infty$ and let $f : A \to \mathbb{H}$ be an s-regular function. There exists $\{a_n\}_{n\in\mathbb{Z}} \subset \mathbb{H}$ such that*

$$f(q) = \sum_{n\in\mathbb{Z}} q^n a_n \tag{4.4}$$

for all $q \in A$.

Proof. Choose a complex plane $\mathbb{C}_I$ and consider the annulus we get by intersecting $\mathbb{C}_I$ with the shell A:

$$A_I = A_I(0, R_1, R_2) = \{z \in \mathbb{C}_I : R_1 < |z| < R_2\}.$$

Consider the restriction $f_I = f_{|A_I}$ and choose $J \in \mathbb{S}, J \perp I$. By the Splitting Lemma there exist two holomorphic functions $F, G : A_I \to \mathbb{C}_I$ such that $f_I = F + GJ$. Let $F(z) = \sum_{n\in\mathbb{Z}} z^n \alpha_n$ and $G(z) = \sum_{n\in\mathbb{Z}} z^n \beta_n$ be the Laurent series expansions of the functions F and G (which have coefficients $\alpha_n, \beta_n \in \mathbb{C}_I$). If we let $a_n = \alpha_n + \beta_n J$ for all $n \in \mathbb{Z}$, then

$$f_I(z) = \sum_{n\in\mathbb{Z}} z^n a_n$$

for all $z \in A_I$. Now consider the quaternionic Laurent series $\sum_{n\in\mathbb{Z}} q^n a_n$, which converges in A by Lemma 4.2.9 and which defines an s-regular function on A by theorem 4.2.10. The statement now follows by the Identity Principle. □

4.3 Representation Formula for slice regular functions

We begin by stating the analog of the representation formula in the case of quaternions.

Definition 4.3.1. *Let $U \subseteq \mathbb{H}$. We say that U is* axially symmetric *if, for all $x+Iy \in U$, the whole 2-sphere $[x+Iy]$ is contained in U.*

Theorem 4.3.2 (Representation Formula). *Let f be an s-regular function on an axially symmetric s-domain $U \subseteq \mathbb{H}$. Choose any $J \in \mathbb{S}$. Then the following equality holds for all $q = x+Iy \in U$:*

$$f(x+Iy) = \frac{1}{2}\Big[f(x+Jy)+f(x-Jy)\Big] + I\frac{1}{2}\Big[J[f(x-Jy)-f(x+Jy)]\Big]. \tag{4.5}$$

Moreover, for all $x, y \in \mathbb{R}$ such that $x+\mathbb{S}y \subseteq U$, there exist $\alpha, \beta \in \mathbb{H}$ such that for all $K \in \mathbb{S}$ we have

$$\alpha := \frac{1}{2}\Big[f(x+yK)+f(x-Ky)\Big] \quad and \quad \beta := \frac{1}{2}\Big[K[f(x-Ky)-f(x+Ky)]\Big]. \tag{4.6}$$

Remark 4.3.3. The proof of this result follows the same lines of the corresponding result for s-monogenic functions, see Theorem 2.2.18; analogously, and with the same technique one can prove the version of the representation theorem with two different imaginary units.

Some immediate consequences are the following:

Corollary 4.3.4. *Let $U \subseteq \mathbb{H}$ be an axially symmetric s-domain, $D \subseteq \mathbb{R}^2$ such that $x+Iy \in U$ whenever $(x,y) \in D$ and let $f : U \to \mathbb{H}$. The function f is an s-regular function if and only if there exist two differentiable functions $\alpha, \beta : D \subseteq \mathbb{R}^2 \to \mathbb{R}_n$ satisfying $\alpha(x,y) = \alpha(x,-y)$, $\beta(x,y) = -\beta(x,-y)$ and the Cauchy–Riemann system*

$$\begin{cases} \partial_x\alpha - \partial_y\beta = 0, \\ \partial_x\beta + \partial_y\alpha = 0, \end{cases} \tag{4.7}$$

such that

$$f(x+Iy) = \alpha(x,y) + I\beta(x,y). \tag{4.8}$$

Corollary 4.3.5. *An s-regular function $f : U \to \mathbb{H}$ on an axially symmetric s-domain is infinitely differentiable on U.*

Corollary 4.3.6. *Let $U \subseteq \mathbb{H}$ be an axially symmetric s-domain and let $f : U \to \mathbb{H}$ be an s-regular function. For all $x, y \in \mathbb{R}$ such that $x+Iy \in U$ there exist $a, b \in \mathbb{H}$ such that*

$$f(x+Iy) = a + Ib \tag{4.9}$$

for all $I \in \mathbb{S}$. In particular, f is affine in $I \in \mathbb{S}$ on each 2-sphere $[x+Iy]$ and the image of the 2-sphere $[x+Iy]$ is the set $[a+Ib]$.

Proof. This is a direct application of Theorem 4.3.2. □

Corollary 4.3.7. *Let $U \subseteq \mathbb{H}$ be an axially symmetric s-domain and let $f : U \to \mathbb{H}$ be an s-regular function. If $f(x + Jy) = f(x + Ky)$ for $I \neq K$ in $\mathbb{S}$, then f is constant on $[x + Iy]$. In particular, if $f(x + Jy) = f(x + Ky) = 0$ for $I \neq K$ in $\mathbb{S}$, then f vanishes on the entire 2-sphere $[x + Iy]$.*

Lemma 4.3.8 (Extension Lemma). *Let $J \in \mathbb{S}$ and let D_J be a domain in $\mathbb{C}_J$, symmetric with respect to the real axis and such that $D_J \cap \mathbb{R} \neq \emptyset$. Let U_{D_J} be the axially symmetric s-domain defined by*

$$U_{D_J} = \bigcup_{x+Jy \in D_J, I \in \mathbb{S}} (x + Iy).$$

If $f : D_J \to \mathbb{H}$ satisfies $\overline{\partial}_J f = 0$, then the function $\tilde{f} : U_{D_J} \to \mathbb{H}$ defined by

$$\tilde{f}(x + Iy) = \frac{1}{2}\Big[f(x + Jy) + f(x - Jy)\Big] + I\frac{1}{2}\Big[J[f(x - Jy) - f(x + Jy)]\Big] \quad (4.10)$$

is the unique s-regular, infinitely differentiable extension of f to U_{D_J}. In particular any holomorphic function $f : D_J \to \mathbb{C}_J$ has a unique s-regular, infinitely differentiable extension to U_{D_J}. The function $\tilde{f}$ will be often denoted by ext(f).

If an s-regular function f is the extension of a holomorphic function of $\mathbb{C}_I$, for some $I \in \mathbb{S}$, then the following result holds:

Proposition 4.3.9. *Let $U \subseteq \mathbb{H}$ be an axially symmetric s-domain and let $f : U \to \mathbb{H}$ be an s-regular function. Suppose that there exists an imaginary unit $J \in \mathbb{S}$ such that $f(\mathbb{C}_J) \subset \mathbb{C}_J$. If there exists an imaginary unit $I \in \mathbb{S}$ such that $I \notin \mathbb{C}_J$ and that $f(x_0 + y_0 I) = 0$, then $f(x_0 + Ly_0) = 0$ for all $L \in \mathbb{S}$.*

Proof. We only have to consider the case $y_0 \neq 0$. By formula (4.3.2) we have, on $U \cap \mathbb{C}_I$,

$$f(x + Iy) = \frac{1}{2}\Big[f(x + Jy) + f(x - Jy)\Big] + I\frac{1}{2}\Big[J[f(x - Jy) - f(x + Jy)]\Big];$$

in particular $f(x_0 + Iy_0) = 0$ implies that

$$0 = \frac{1}{2}\Big[f(x_0 + y_0 J) + f(x_0 - Jy_0)\Big] + I\frac{1}{2}\Big[J[f(x_0 - Jy_0) - f(x_0 + Jy_0)]\Big],$$

and since $f(\mathbb{C}_J) \subset \mathbb{C}_J$ and 1 and I are linearly independent on $\mathbb{C}_J$ we get

$$f(x_0 + Jy_0) = -f(x_0 - Jy_0), \qquad f(x_0 - Jy_0) = f(x_0 + Jy_0)$$

whence $f(x_0 - Jy_0) = f(x_0 + Jy_0) = 0$. Again, the conclusion follows from Corollary 4.3.7. □

This result has an important consequence concerning the zeros of holomorphic functions defined on domains intersecting the real axis and symmetric with respect to it. Since any such holomorphic function f can be uniquely extended (see the Extension Lemma 4.3.8) to an s-regular function over quaternions, the question of distinguishing which zeros of f will remain isolated after the extension, and which will become "spherical", naturally arises. The answer is in the following proposition.

Proposition 4.3.10. *Let $J \in \mathbb{S}$ and let D_J be a domain in $\mathbb{C}_J$, symmetric with respect to the real axis and such that $D_J \cap \mathbb{R} \neq \emptyset$. Let U_{D_J} be the axially symmetric s-domain defined by*

$$U_{D_J} = \bigcup_{x+Jy \in D_J} (x + \mathbb{S}y).$$

Let $f : D_J \to \mathbb{C}_J$ be a holomorphic function and $\tilde{f} : U_{D_J} \to \mathbb{H}$ be its s-regular extension. If $\tilde{f}(x_0 + Jy_0) = 0$, for $y_0 \neq 0$, then the zero $(x_0 + Jy_0)$ of $\tilde{f}$ is not isolated, or equivalently $\tilde{f}(x_0 + Ly_0) = 0$ for all $L \in \mathbb{S}$, if, and only if,

$$f(x_0 + Jy_0) = f(x_0 - Jy_0) = 0.$$

Proof. The proof easily follows from Corollary 4.3.7. □

The first part of the following result can be proved as in Theorem 2.5.14, or it can be given a much simpler and more intuitive proof which exploits the specific nature of $\mathbb{H}$ and of s-regular functions.

Theorem 4.3.11 (Structure of the Zero Set). *Let $U \subseteq \mathbb{H}$ be an axially symmetric s-domain and let $f : U \to \mathbb{H}$ be an s-regular function. Suppose that f does not vanish identically. Then if the zero set of f is nonempty, it consists of the union of isolated 2-spheres and/or isolated points.*

Proof. Let $q_0 = x_0 + Jy_0$ be a zero of f. By Corollary 4.3.4 we know that $f(x + Iy) = \alpha(x, y) + I\beta(x, y)$, thus

$$f(q_0) = f(x_0 + Jy_0) = \alpha(x_0, y_0) + J\beta(x_0, y_0) = 0.$$

If $\beta(x_0, y_0) = 0$, then also $\alpha(x_0, y_0) = 0$ so $f(x_0 + Iy_0) = 0$ for every choice of an imaginary unit $I \in \mathbb{S}$ thus the whole sphere defined by q_0 is a solution of $f(q) = 0$. If $\beta(x_0, y_0) \neq 0$, then it is an invertible element in $\mathbb{H}$. In this case, $\alpha(x_0, y_0) \neq 0$ otherwise we get $J\beta(x_0, y_0) = 0$ and so $J = 0$ which is absurd. Since the inverse of $\beta(x_0, y_0)$ is unique, the element $J = -\alpha(x_0, y_0)\beta(x_0, y_0)^{-1}$ is also unique. If $J \in \mathbb{S}$, then q_0 is the only solution of $f(q) = 0$ on the sphere defined by q_0. The fact that the spheres are isolated can be proved as in Theorem 2.5.14, using the notion of symmetrization of an s-regular function (see Definition 4.3.18). □

Let $U \subseteq \mathbb{H}$ be an axially symmetric s-domain and let $f : U \to \mathbb{H}$ be an s-regular function. For any $I, J \in \mathbb{S}$, with $I \perp J$, the Splitting Lemma guarantees the existence of two holomorphic functions $F, G : U \cap \mathbb{C}_I \to \mathbb{C}_I$ such that, for all $z = x + Iy \in U \cap \mathbb{C}_I$,

$$f_I(z) = F(z) + G(z)J.$$

Let us define the function $f_I^c : U \cap \mathbb{C}_I \to \mathbb{H}$ as

$$f_I^c(z) = \overline{F(\bar{z})} - G(z)J. \tag{4.11}$$

Then $f_I^c(z)$ is obviously a holomorphic map and hence its unique s-regular extension to U defined, according to the Extension Lemma 4.3.8, by

$$f^c(q) = \mathrm{ext}(f_I^c)(q),$$

is s-regular on U.

Definition 4.3.12. *Let $U \subseteq \mathbb{H}$ be an axially symmetric s-domain and let $f : U \to \mathbb{H}$ be s-regular. The function*

$$f^c(q) = \mathrm{ext}(f_I^c)(q)$$

defined by the extension (4.11) *is called the s-regular conjugate of f.*

The s-regular conjugate of a function f, which in general does not coincide with the conjugate $\bar{f}$, has peculiar properties that will play a key role in the sequel.

Proposition 4.3.13. *Let $f : B(p_0, R) \to \mathbb{H}$ be an s-regular function on an open ball in $\mathbb{H}$ centered at a real point p_0. If*

$$f(q) = \sum_{n \geq 0} (q - p_0)^n a_n,$$

then

$$f^c(q) = \sum_{n \geq 0} (q - p_0)^n \bar{a}_n.$$

Proof. We will suppose without loss of generality that $p_0 = 0$. By Corollary 4.2.3, given any $I \in \mathbb{S}$, the coefficients of the power series expansion of f are obtainable as the coefficients of the power series of f_I. For all $z = x + Iy \in \mathbb{C}_I \cap B(0, R)$ and $J \in \mathbb{S}$ with $J \perp I$ we get

$$f_I(z) = F(z) + G(z)J = \sum_{n \geq 0} \frac{z^n}{n!} \Big(\frac{\partial F}{\partial x}(0) + \frac{\partial G}{\partial x}(0)J \Big) = \sum_{n \geq 0} \frac{z^n}{n!} \partial_s f(0)$$

and hence

$$f_I^c(z) = \overline{F(\bar{z})} - G(z)J = \sum_{n \geq 0} \frac{z^n}{n!} \Big(\overline{\frac{\partial F}{\partial x}}(0) - \frac{\partial G}{\partial x}(0)J \Big) = \sum_{n \geq 0} \frac{z^n}{n!} \overline{\partial_s f(0)}$$

which proves the assertion. □

Remark 4.3.14. As it can be expected, the product of two s-regular functions is not, in general, s-regular. In the case of s-regular polynomials, i.e., polynomials with quaternionic coefficients on the right, one can exploit the standard multiplication of polynomials in a skew field (see, e.g., [71]) to obtain an s-regular product. This product extends naturally to s-regular power series as

$$\Big(\sum_{n\geq 0} q^n a_n\Big) * \Big(\sum_{n\geq 0} q^n b_n\Big) = \sum_{i,n\geq 0} q^n a_i b_{n-i}, \qquad (a_n, b_n \in \mathbb{H}) \tag{4.12}$$

and is extensively used when studying the properties of their zero sets.

In the case of s-regular functions defined on axially symmetric s-domains, inspired by the case of power series, we will define an s-regular product as follows. Let $U \subseteq \mathbb{H}$ be an axially symmetric s-domain and let $f, g : U \to \mathbb{H}$ be s-regular functions. For any $I, J \in \mathbb{S}$, with $I \perp J$, the Splitting Lemma guarantees now the existence of four holomorphic functions $F, G, H, K : U \cap \mathbb{C}_I \to \mathbb{C}_I$ such that for all $z = x + Iy \in U \cap \mathbb{C}_I$,

$$f_I(z) = F(z) + G(z)J \qquad g_I(z) = H(z) + K(z)J.$$

We define the function $f_I * g_I : U \cap \mathbb{C}_I \to \mathbb{H}$ as

$$f_I * g_I(z) = [F(z)H(z) - G(z)\overline{K(\bar{z})}] + [F(z)K(z) + G(z)\overline{H(\bar{z})}]J. \tag{4.13}$$

Then $f_I * g_I(z)$ is obviously a holomorphic map and hence its unique s-regular extension to U defined, according to the Extension Lemma 4.3.8, by

$$f * g(q) = \text{ext}(f_I * g_I)(q),$$

is s-regular on U.

Definition 4.3.15. *Let $U \subseteq \mathbb{H}$ be an axially symmetric s-domain and let $f, g : U \to \mathbb{H}$ be s-regular. The function*

$$f * g(q) = \text{ext}(f_I * g_I)(q)$$

defined as the extension of (4.13) *is called the s-regular product of f and g. This product is called the $*$-product or the s-regular product.*

Remark 4.3.16. It is immediate to verify that the $*$-product is associative, distributive but, in general, not commutative.

Remark 4.3.17. Let $H(z)$ be a holomorphic function in the variable $z \in \mathbb{C}_I$ and let $J \in \mathbb{S}$ be orthogonal to I. Then by the definition of $*$-product we obtain $J * H(z) = \overline{H(\bar{z})}J$.

Using the notion of $*$-multiplication of s-regular functions, it is possible to associate to any s-regular function f its "symmetrization" also called "normal form", denoted by f^s. We will show that all the zeros of f^s are spheres of type $[x+Iy]$ (real points, in particular) and that, if $x+Iy$ is a zero of f (isolated or not), then $[x+Iy]$ is a zero of f^s.

Let $U \subseteq \mathbb{H}$ be an axially symmetric s-domain and let $f : U \to \mathbb{H}$ be an s-regular function. For any $I, J \in \mathbb{S}$, with $I \perp J$, the Splitting Lemma guarantees the existence of two holomorphic functions $F, G : U \cap \mathbb{C}_I \to \mathbb{C}_I$ such that, for all $z = x+Iy \in U \cap \mathbb{C}_I$,

$$f_I(z) = F(z) + G(z)J.$$

We define the function $f_I^s : U \cap \mathbb{C}_I \to \mathbb{C}_I$ as

$$\begin{aligned} f_I^s &= f_I * f_I^c = (F(z)+G(z)J) * (\overline{F(\bar{z})} - G(z)J) \\ &= [F(z)\overline{F(\bar{z})} + G(z)\overline{G(\bar{z})}] + [-F(z)G(z) + G(z)F(z)]J \\ &= F(z)\overline{F(\bar{z})} + G(z)\overline{G(\bar{z})} = f_I^c * f_I. \end{aligned} \tag{4.14}$$

Then f_I^s is obviously holomorphic and hence its unique s-regular extension to U defined by

$$f^s(q) = \operatorname{ext}(f_I^s)(q)$$

is s-regular.

Definition 4.3.18. *Let $U \subseteq \mathbb{H}$ be an axially symmetric s-domain and let $f : U \to \mathbb{H}$ be s-regular. The function*

$$f^s(q) = \operatorname{ext}(f_I^s)(q)$$

defined by the extension of (4.14) *is called the symmetrization (or normal form) of f.*

Remark 4.3.19. Notice that formula (4.14) yields that, for all $I \in \mathbb{S}$, $f^s(U \cap \mathbb{C}_I) \subseteq \mathbb{C}_I$.

The symmetrization process is well behaved with respect to the $*$-product, conjugation and reciprocal:

Proposition 4.3.20. *Let $U \subseteq \mathbb{H}$ be an axially symmetric s-domain and let $f, g : U \to \mathbb{H}$ be s-regular functions. Then*

$$(f*g)^c = g^c * f^c$$

and

$$(f*g)^s = f^s g^s = g^s f^s. \tag{4.15}$$

Proof. It is sufficient to show that $(f*g)^c = g^c * f^c$. As customary, we can use the Splitting Lemma to write on $U \cap \mathbb{C}_I$ that $f_I(z) = F(z) + G(z)J$ and $g_I(z) = H(z) + K(z)J$. We have

$$f_I * g_I(z) = [F(z)H(z) - G(z)\overline{K(\bar{z})}] + [F(z)K(z) + G(z)\overline{H(\bar{z})}]J$$

and hence

$$(f_I * g_I)^c(z) = [\overline{F(\bar{z})}\ \overline{H(\bar{z})} - \overline{G(\bar{z})}K(z)] - [F(z)K(z) + G(z)\overline{H(\bar{z})}]J.$$

We now compute

$$\begin{aligned} g_I^c(z) * f_I^c(z) &= (\overline{H(\bar{z})} - K(z)J) * (\overline{F(\bar{z})} - G(z)J) \\ &= \overline{H(\bar{z})} * \overline{F(\bar{z})} - \overline{H(\bar{z})} * G(z)J - K(z)J * \overline{F(\bar{z})} + K(z)J * G(z)J \end{aligned}$$

and conclude by Remark 4.3.17. □

Proposition 4.3.21. *Let $U \subseteq \mathbb{H}$ be an axially symmetric s-domain and let $f : U \to \mathbb{H}$ be an s-regular function. The function $(f^s(q))^{-1}$ is s-regular on $U \setminus \{q \in \mathbb{H} \mid f^s(q) = 0\}$.*

Proof. The function f^s is such that $f^s(U \cap \mathbb{C}_I) \subseteq \mathbb{C}_I$ for all $I \in \mathbb{S}$ by Remark 4.3.19. Thus, for any given $I \in \mathbb{S}$ the Splitting Lemma implies the existence of a holomorphic function $F : U \cap \mathbb{C}_I \to \mathbb{C}_I$ such that $f_I^s(z) = F(z)$ for all $z \in U \cap \mathbb{C}_I$. The inverse of the function F is holomorphic on $U \cap \mathbb{C}_I$ outside the zero set of F. The conclusion follows by the equality $(f_I^s)^{-1} = F^{-1}$. □

Proposition 4.3.22. *Let $U \subseteq \mathbb{H}$ be an axially symmetric s-domain and let $f, g : U \to \mathbb{H}$ be s-regular functions. Then*

$$f * g(q) = f(q)\ g(f(q)^{-1}qf(q)), \tag{4.16}$$

for all $q \in U$.

Proof. Let I be any element in $\mathbb{S}$ and let $q = x + Iy$. If $f(x + Iy) \neq 0$, simple computations show that

$$f(x + Iy)^{-1}(x + Iy)f(x + Iy) = x + yf(x + Iy)^{-1}If(x + Iy)$$

with $f(x+Iy)^{-1}If(x+Iy) \in \mathbb{S}$. Using now the representation formula (4.3.2) for the function g, we get

$$\begin{aligned} &g(f(q)^{-1}qf(q)) = g(x + yf(x + Iy)^{-1}If(x + Iy)) \\ &= \frac{1}{2}\{g(x + Iy) + g(x - Iy) - f(x + Iy)^{-1}If(x + Iy)[Ig(x + Iy) - Ig(x - Iy)]\} \end{aligned}$$

and

$$\begin{aligned} &\psi(q) := f(q)g(f(q)^{-1}qf(q)) \\ &= \frac{1}{2}\{f(x + Iy)[g(x + Iy) + g(x - Iy)] - If(x + Iy)[Ig(x + Iy) - Ig(x - Iy)]\}. \end{aligned}$$

If we prove that the function $f(q)g(f(q)^{-1}qf(q))$ is s-regular, then our assertion will follow by the Identity Principle since formula (4.16) holds on a small open ball of U centered at a point on the real axis (see Proposition 4.2.3). Let us compute, with obvious notation for the derivatives:

$$\begin{aligned}&\frac{\partial}{\partial x}\psi(x+Iy)\\&=\frac{1}{2}\{f_x(x+Iy)[g(x+Iy)+g(x-Iy)]-If_x(x+Iy)[Ig(x+Iy)-Ig(x-Iy)]\}\\&\quad+\frac{1}{2}\{f(x+Iy)[g_x(x+Iy)+g_x(x-Iy)]-If(x+Iy)[Ig_x(x+Iy)-Ig_x(x-Iy)]\}\end{aligned}$$

and

$$\begin{aligned}&I\frac{\partial}{\partial y}\psi(x+Iy)\\&=\frac{1}{2}\{If_y(x+Iy)[g(x+Iy)+g(x-Iy)]+f_y(x+Iy)[Ig(x+Iy)-Ig(x-!Iy)]\}\\&\quad+\frac{1}{2}\{If(x+Iy)[g_y(x+Iy)+g_y(x-Iy)]+f(x+Iy)[Ig_y(x+Iy)-Ig_y(x-Iy)]\}.\end{aligned}$$

By using the three relations

$$\begin{aligned}f_x(x+Iy)+If_y(x+Iy)&=g_x(x+Iy)+Ig_y(x+Iy)\\&=g_x(x-Iy)-Ig_y(x-Iy)=0,\end{aligned}$$

we obtain that

$$\left(\frac{\partial}{\partial x}+I\frac{\partial}{\partial y}\right)\psi(x+Iy)=0.$$

The fact that I is arbitrary proves the assertion. □

Theorem 4.3.23. *Let $U\subseteq\mathbb{H}$ be an axially symmetric s-domain and let $f,g:U\to\mathbb{H}$ be s-regular functions. Then $f*g(q)=0$ if and only if $f(q)=0$ or $f(q)\neq 0$ and $g(f(q)^{-1}qf(q))=0$.*

Proof. Theorem 4.16 implies that

$$f*g(q)=f(q)\ g(f(q)^{-1}qf(q)).$$

Therefore $f*g(q)=0$ if and only if $f(q)g(f(q)^{-1}qf(q))=0$ if and only if either $f(q)=0$ or $f(q)\neq 0$ but then $g(f(q)^{-1}qf(q))=0$. □

In particular, if $f*g$ has a zero in the sphere $S:=[x+Iy]$, then either f or g have a zero in S. However, the zeros of g in S need not be in one-to-one correspondence with the zeros of $f*g$ in S which are not zeros of f.

This theorem allows us to recover a well-known result for polynomials, see [71]:

Theorem 4.3.24. *Let p,r be polynomials in the quaternionic variable q with quaternionic coefficients. Assume that $p*r(q)=0$ and $s=p(q)\neq 0$. Then $r(s^{-1}qs)=0$.*

Remark 4.3.25. In the previous result, we implicitly assumed that the polynomials are s-regular and thus their coefficients are written on the right. In the case of polynomials with coefficients on the left (and so right s-regular) the statement guarantees that if $p * r(q) = 0$ and $s = r(q) \neq 0$, then $p(sqs^{-1}) = 0$.

The definitions we have introduced are useful to define an s-regular inverse of a function. First of all, observe that, by Proposition 4.3.21, $(f_I^s)^{-1} * f^c$ is s-regular for any s-regular function f. Since, again by Remark 4.3.19, the function $(f_I^s)^{-1} : U \cap \mathbb{C}_I \to \mathbb{C}_I$ is holomorphic for all $I \in \mathbb{S}$, where $f_I^s \neq 0$, then we can write $(f^s)^{-1} * f^c = (f^s)^{-1} f^c$. An easy computation shows that

$$(f^{-*} * f)(q) = (f^s(q))^{-1}(f^c * f)(q) = 1$$

and justifies the following definition:

Definition 4.3.26. *Let $U \subseteq \mathbb{H}$ be an axially symmetric s-domain and let $f : U \to \mathbb{H}$ be an s-regular function. We define the function f^{-*} as*

$$f^{-*}(q) := (f^s(q))^{-1} f^c(q).$$

It is now immediate to verify the validity of the following:

Proposition 4.3.27. *Let $U \subseteq \mathbb{H}$ be an axially symmetric s-domain and let $f : U \to \mathbb{H}$ be an s-regular function. The function f^{-*} is the inverse of f with respect to the s-regular product.*

4.4 The slice regular Cauchy kernel

The Cauchy kernel which we will define and study in this section was inspired by the need to have a suitable Cauchy formula to develop a functional calculus for quaternionic operators. This kernel, obtained in this section using algebraic techniques which rely on the structure of quaternions (in particular on the fact that the components of a quaternion are real numbers and therefore commute), will later on be shown to remain valid even when the quaternions are replaced by operators, whose components do not commute. Since in the complex case, the kernel $(\zeta - z)^{-1}$ is the sum of the series $\sum_{n\geq 0} z^n \zeta^{-1-n}$, for $|z| < |\zeta|$, we introduce the following definition:

Definition 4.4.1. *Let q, $s \in \mathbb{H}$ be such that $sq \neq qs$. We will call the series expansion*

$$S^{-1}(s, q) := \sum_{n\geq 0} q^n s^{-1-n},$$

for $|q| < |s|$ a noncommutative Cauchy kernel series (shortly Cauchy kernel series).

Just as in the complex case, we will be looking for the sum of the Cauchy kernel series which will be an s-regular function in a domain larger than the ball in which the series converges.

Theorem 4.4.2. *Let q and s be two quaternions such that $qs \neq sq$ and consider*

$$S^{-1}(s,q) := \sum_{n\geq 0} q^n s^{-1-n}.$$

Then the inverse $S(s,q)$ of $S^{-1}(s,q)$ is the nontrivial solution to the equation

$$S^2 + Sq - sS = 0. \tag{4.17}$$

Proof. Observe that

$$S^{-1}(s,q)s = \sum_{n\geq 0} q^n s^{-1-n} s = \sum_{n\geq 0} q^n s^{-n} = 1 + qs^{-1} + q^2 s^{-2} + \dots$$

and

$$qS^{-1}(s,q) = q\sum_{n\geq 0} q^n s^{-1-n} = \sum_{n\geq 0} q^{1+n} s^{-1-n} = qs^{-1} + q^2 s^{-2} + \dots$$

so that

$$S^{-1}(s,q)s - qS^{-1}(s,q) = 1.$$

Keeping in mind that $S^{-1}S = SS^{-1} = 1$ we get

$$S(S^{-1}s - qS^{-1})S = S^2$$

from which we obtain the proof. □

Remark 4.4.3. The polynomial $R(s,q) := s - q$ is a solution of equation (4.17) if and only if $sq = qs$ (and in particular, for example, if $s = s_0 + s_1 I$, $q = q_0 + q_1 I$ for some $I \in \mathbb{S}$). Indeed the result follows immediately from the chain of equalities

$$(s-q)^2 + (s-q)q - s(s-q) = s^2 - sq - qs + q^2 - s^2 + sq + sq - q^2 = -qs + sq$$

whose last term vanishes if and only if $sq = qs$.

We now compute the nontrivial solution to the equation (4.17). To this purpose, and unlike what we have done throughout this book, we will transform this equation into a polynomial equation with coefficients on the left. This technical detail is necessary in order to get a solution written in a form which will be suitable for the functional calculus. We will show in Note 4.18.3 how this solution was originally computed using Niven's algorithm.

Theorem 4.4.4. *Let q, $s \in \mathbb{H}$ be such that $qs \neq sq$. Then the nontrivial solution of*

$$S^2 + Sq - sS = 0 \tag{4.18}$$

is given by

$$S(s,q) = -(q - \overline{s})^{-1}(q^2 - 2\,\mathrm{Re}[s]\,q + |s|^2). \tag{4.19}$$

Proof. We transform the equation $S^2 + Sq - sS = 0$ into another one having coefficients on the left. Set

$$S := W - q$$

and replace it in the equation to get

$$(W - q)(W - q) + (W - q)q - s(W - q) = 0,$$

so the equation becomes

$$W^2 - (s + q)W + sq = (W - s) * (W - q) = 0$$

where $*$ denotes the s-regular product (on the left). One root is $W = q$, while the second root is $W = (q - \bar{s})^{-1}s(q - \bar{s})$, thus

$$S = (q - \bar{s})^{-1}s(q - \bar{s}) - q.$$

By grouping $(q - \bar{s})^{-1}$ on the left we obtain (4.19). □

Definition 4.4.5. *The function defined by*

$$-(q^2 - 2q\mathrm{Re}[s] + |s|^2)^{-1}(q - \bar{s}). \tag{4.20}$$

will be called the Cauchy kernel function and will be denoted again, with an abuse of notation, by $S^{-1}(s, q)$.

Note that we are using the same symbol S^{-1} to denote both the Cauchy kernel series and the Cauchy kernel function. In fact they coincide where they are both defined by virtue of their s-regularity (see Proposition 4.4.9 below) and in view of the identity principle, see Theorem 4.2.4. We have the following result which shows that the Cauchy kernel function $S^{-1}(s, q)$ is s-regular in the two variables s, q.

The proof can be obtained by direct computations, similarly to what we did to prove Proposition 2.7.9, but here we prove such a result by solving a second-degree equation.

Proposition 4.4.6. *For any $q, s \in \mathbb{H}$ such that $q \neq \bar{s}$ the following identity holds:*

$$(q - \bar{s})^{-1}s(q - \bar{s}) - q = -(s - \bar{q})q(s - \bar{q})^{-1} + s, \tag{4.21}$$

or, equivalently,

$$-(q - \bar{s})^{-1}(q^2 - 2q\mathrm{Re}[s] + |s|^2) = (s^2 - 2\mathrm{Re}[q]s + |q|^2)(s - \bar{q})^{-1}. \tag{4.22}$$

Proof. Let us solve equation $S^2 + Sq - sS = 0$ by transforming it into an equation with right coefficients by setting

$$S := W + s$$

and replacing it in the equation. We get

$$(W+s)(W+s)+(W+s)q-s(W+s)=W^2+W(s+q)+sq=0.$$

This equation can be split as $(W+s)*(W+q)=0$, where $*$ denotes the (right) s-regular product. It is immediate that one root is $W=-s$ while the second is $W=(-\bar{q}+s)(-q)q(-\bar{q}+s)^{-1}$. These two roots correspond to $S=0$ and $S=-(s-\bar{q})q(s-\bar{q})^{-1}+s$ which coincides with (4.19) when written in the form $S=(s^2-\mathrm{Re}[q]s+|q|^2)(s-\bar{q})^{-1}$. □

Remark 4.4.7. By writing $S(s,q)$ as $S(s,q)=(q-\overline{s})^{-1}s(q-\overline{s})-q$ we immediately see that $S(s,q)=S(s-u,q-u)$ for any $u\in\mathbb{R}$.

The Cauchy kernel function can also be obtained by taking, in a suitable way, the s-regular inverse of the function $R(s,q)=R_s(q)=s-q$. In principle we have four possibilities to construct an s-regular inverse: on the left (resp. on the right) with respect to q and on the left (resp. on the right) with respect to s. To obtain the desired function $S^{-1}(s,q)$ we can proceed as in the following result:

Proposition 4.4.8. *The right s-regular inverse with respect to q of the function $R(s,q)=R_s(q)=s-q$ is*

$$S^{-1}(s,q)=-(q^2-2q\mathrm{Re}[s]+|s|^2)^{-1}(q-\bar{s}).$$

Proof. We have that $R_s^s(q)=(q^2-2\mathrm{Re}[s]q+|s|^2)$ and $R_s^c(q)=\bar{s}-q$. □

We also have:

Proposition 4.4.9. *The function $S^{-1}(s,q)$ is left s-regular in the variable q and right s-regular in the variable s in its domain of definition.*

Proof. The s-regularity in q follows by construction. The right s-regularity in s follows by direct computation using the identity (4.22). □

Remark 4.4.10. Since the function $S^{-1}(s,q)$ turns out to be right s-regular with respect to s, it is also possible to construct it by taking the right s-regular inverse of $R(s,q)=R_q(s)=s-q$. The right s-regular inverse of $R(s,q)=R_s(q)=s-q$ in the variable q turns out to be the function

$$S_R^{-1}(s,q):=-(q-\bar{s})(q^2-2\mathrm{Re}[s]q+|s|^2)^{-1} \tag{4.23}$$

for $q^2-2\mathrm{Re}[s]q+|s|^2\neq 0$. Note that when $|q|<|s|$ we have

$$S_R^{-1}(s,q)=\sum_{n\geq 0}s^{-n-1}q^n.$$

We are now in a position to study the distribution of the singularities of the s-regular Cauchy kernel:

Proposition 4.4.11. *Let $q \in \mathbb{H}\setminus\mathbb{R}$. The singularities of the function $S^{-1}(s,q) = S_q^{-1}(s) = -(q^2 - \mathrm{Re}[s]q + |s|^2)^{-1}(q - \bar{s})$ lie on the 2-sphere S_q. More precisely: on the plane $\mathbb{C}_I$, $I \neq I_q$, $S^{-1}(s,q)$ has the two singularities $\mathrm{Re}[q] \pm I|\mathrm{Im}[q]|$ while on the plane $\mathbb{C}_{I_q}$ the only singularity is q. When $q \in \mathbb{R}$, then $S_q^{-1}(s) = -(q - s)^{-1}$ and the only singularity is q.*

Proof. Suppose $q \in \mathbb{H}\setminus\mathbb{R}$. The singularities of $S_q^{-1}(s)$ corresponds to the roots of $|s|^2 - 2\mathrm{Re}[s]q + q^2 = 0$. This equation can be written by splitting real and imaginary parts as $|s|^2 - 2\mathrm{Re}[s]\mathrm{Re}[q] + \mathrm{Re}[q]^2 - |\mathrm{Im}[q]|^2 = 0$, $(\mathrm{Re}[s] - \mathrm{Re}[q])\mathrm{Im}[q] = 0$. The assumption implies $\mathrm{Re}[s] = \mathrm{Re}[q]$ and so $|s| = |q|$, i.e., the sphere $[q]$. Consider the plane $\mathbb{C}_I$, $I \neq I_q$: it intersects the 2-sphere $[q]$ in $\mathrm{Re}[q] \pm I|\mathrm{Im}[q]|$. When $I = I_q$, then q and s commute, so

$$S_q^{-1}(s) = -(q-s)^{-1}(q-\bar{s})^{-1}(q-\bar{s}) = -(q-s)^{-1}$$

and the statement follows. When q is real the conclusion follows by the same commutation argument. □

We conclude the section by proving two different explicit series expansions of the regular Cauchy kernel.

Theorem 4.4.12. *Let q and $s = u + vI$ ($I \in \mathbb{S}$, $v > 0$) be two quaternions such that*

$$|q - u| < v. \tag{4.24}$$

Then the noncommutative Cauchy kernel admits the series expansion

$$S^{-1}(s,q) = \sum_{n\geq 0}(q-u)^n(vI)^{-n-1}. \tag{4.25}$$

Proof. Remark 4.4.7 shows that the inverse of the Cauchy kernel $S(s,q)$ is such that $S(s,q) = S(s-u, q-u)$ for any $u \in \mathbb{R}$. As a consequence we have that $S^{-1}(s,q) = S^{-1}(s-u, q-u)$ for any $u \in \mathbb{R}$. By setting $s = u + vI$ and considering the series expansion

$$S^{-1}(s-u, q-u) = \sum_{n\geq 0}(q-u)^n(vI)^{-n-1}$$

we get the statement. □

To conclude, we now examine what happens on the complement of the closure of the domain in which the series above converges. We will adopt a Laurent type approach:

Theorem 4.4.13. *Let q and $s = u + vI$ ($I \in \mathbb{S}$, $v > 0$) be two quaternions such that*

$$|q - u| > v. \tag{4.26}$$

Then the noncommutative Cauchy kernel can be represented by the series

$$S^{-1}(s,q) = -\sum_{n\geq 0}(q-u)^{-n-1}(vI)^n \tag{4.27}$$

Proof. Consider the equalities

$$\begin{aligned}(q^2 - 2qu + u^2 + v^2)^{-1} &= \left((q-u)^2 + v^2\right)^{-1}\\ &= \left((q-u)^2(1 + v^2(q-u)^{-2})\right)^{-1}\\ &= \left(1 + v^2(q-u)^{-2}\right)^{-1}(q-u)^{-2}\\ &= \sum_{n\geq 0}(-1)^n v^{2n}(q-u)^{-2n-2}.\end{aligned}$$

We now multiply the last expression by $-(q - u + vI)$ on the right-hand side and obtain

$$\begin{aligned}S^{-1}(s,q) &= -\sum_{n\geq 0}(-1)^n v^{2n}(q-u)^{-2n-2}(q-u+vI)\\ &= -\Big(\sum_{n\geq 0}(-1)^n v^{2n}(q-u)^{-2n-1} + \sum_{n\geq 0}(-1)^n v^{2n}(q-u)^{-2n-2}vI\Big)\end{aligned}$$

Since $(-1)^n v^{2n} = (vI)^{2n}$ is a real number we obtain

$$= -\Big(\sum_{n\geq 0}(q-u)^{-2n-1}(vI)^{2n} + \sum_{n\geq 0}(q-u)^{-2n-2}(vI)^{2n+1}\Big)$$

from which the statement follows. □

Remark 4.4.14. Theorems 4.4.12 and 4.4.13 provide the analogue of the complex series expansions $\frac{1}{z-w} = \sum z^n w^{-n-1}$ which holds for $z, w \in \mathbb{C}$ such that $|z| < |w|$ and $\frac{1}{z-w} = -\sum z^{-n-1}w^n$ which holds for $|z| > |w|$. The function $\frac{1}{z-w}$ is obviously defined on the larger set consisting of complex numbers z such that $z \neq w$ while $S^{-1}(s,q)$ is defined outside its singularities.

4.5 The Cauchy integral formula II

In this section we will present a new version of the Cauchy formula (4.2.6), in which the integral expressing $f(q)$ does not depend on the plane containing q. These results are similar to the corresponding results in the case of slice s-monogenic functions. We will repeat the statements that will be useful in the sequel in order to adapt the notation to the quaternionic setting, but we will omit the proofs.

Lemma 4.5.1. *Let f, g be quaternion-valued, continuously (real) differentiable functions on an open set U_I of the plane $\mathbb{C}_I$. Then for every open $W_I \subset U_I$ whose boundary is a finite union of continuously differentiable Jordan curves, we have*

$$\int_{\partial W_I} g\, ds_I f = 2\int_{W_I}((g\overline{\partial}_I)f + g(\overline{\partial}_I f))d\sigma$$

where $s = x + Iy$ is the variable on $\mathbb{C}_I$, $ds_I = -Ids$ and $d\sigma = dx \wedge dy$.

An immediate consequence of the lemma is the following:

Corollary 4.5.2. *Let f and g be a left s-regular and a right s-regular function, respectively, on an open set $U \in \mathbb{H}$. For any $I \in \mathbb{S}$ and every open $W \subset U_I$ whose boundary is a finite union of continuously differentiable Jordan curves, we have*

$$\int_{\partial W} g ds_I f = 0.$$

We are now ready to prove the Cauchy formula II:

Theorem 4.5.3. *Let $U \subseteq \mathbb{H}$ be an axially symmetric s-domain such that $\partial(U \cap \mathbb{C}_I)$ is the union of a finite number of continuously differentiable Jordan curves, for every $I \in \mathbb{S}$. Let f be an s-regular function on an open set containing $\bar{U}$ and, for any $I \in \mathbb{S}$, set $ds_I = -Ids$. Then for every $q = x + Iy_q \in U$ we have*

$$f(q) = \frac{1}{2\pi} \int_{\partial(U \cap \mathbb{C}_I)} -(q^2 - 2\mathrm{Re}[s]q + |s|^2)^{-1}(q - \bar{s}) ds_I f(s). \tag{4.28}$$

Moreover the value of the integral depends neither on U nor on the imaginary unit $I \in \mathbb{S}$.

If f is a right s-regular function on a set that contains $\overline{U}$, then

$$f(q) = \frac{1}{2\pi} \int_{\partial(U \cap \mathbb{C}_I)} f(s) ds_I S_R^{-1}(s,q) \tag{4.29}$$

$$= -\frac{1}{2\pi} \int_{\partial(U \cap \mathbb{C}_I)} f(s) ds_I S^{-1}(q,s) \tag{4.30}$$

and the integral (4.29) does not depend on the choice of the imaginary unit $I \in \mathbb{S}$ and on U.

An immediate consequence of the Cauchy formula is the following result:

Theorem 4.5.4 (Derivatives using the s-regular Cauchy kernel). *Let $U \subset \mathbb{H}$ be an axially symmetric s-domain. Suppose $\partial(U \cap \mathbb{C}_I)$ is a finite union of continuously differentiable Jordan curves for every $I \in \mathbb{S}$. Let f be an s-regular function on U and set $ds_I = ds/I$. Let q, s. Then*

$$\partial_x^n f(x) = \frac{n!}{2\pi} \int_{\partial(U \cap \mathbb{C}_I)} (q^2 - 2s_0 q + |s|^2)^{-n-1}(q - \bar{s})^{(n+1)*} ds_I f(s)$$

$$= \frac{n!}{2\pi} \int_{\partial(U \cap \mathbb{C}_I)} [S^{-1}(s,q)(q - \bar{s})^{-1}]^{n+1}(q - \bar{s})^{(n+1)*} ds_I f(s) \tag{4.31}$$

where

$$(q - \bar{s})^{n*} = \sum_{k=0}^{n} \frac{n!}{(n-k)!k!} q^{n-k} \bar{s}^k, \tag{4.32}$$

is the n-th power with respect to the $*$-product. Moreover, the integral does not depend on U and on the imaginary unit $I \in \mathbb{S}$.

4.6 Linear bounded quaternionic operators

In this section, we collect the main properties of the quaternionic functional calculus. Let V be a right vector space on $\mathbb{H}$. A map $T : V \to V$ is said to be a right linear operator if

$$T(u+v) = T(u) + T(v), \quad T(us) = T(u)s, \quad \text{for all } s \in \mathbb{H}, \ u, v \in V.$$

The multiplication of operators, and in particular the powers T^n of a quaternionic operator, are defined inductively by the relations $T^0 = \mathcal{I}$, where $\mathcal{I}$ denotes the identity operator, and $T^n = TT^{n-1}$. By $\mathrm{End}^R(V)$ we denote the set of right linear operators acting on V. In the sequel, we will consider only two-sided vector spaces V, otherwise the set $\mathrm{End}^R(V)$ is neither a left nor a right vector space over $\mathbb{H}$. With this assumption, $\mathrm{End}^R(V)$ becomes both a left and a right vector space on $\mathbb{H}$ with respect to the operations

$$(sT)(v) := sT(v), \quad (Ts)(v) := T(sv), \quad \text{for all } s \in \mathbb{H}, \ v \in V. \tag{4.33}$$

In particular (4.33) gives $(s\mathcal{I})(v) = (\mathcal{I}s)(v) = sv$. Similarly, we can consider V as a left vector space on $\mathbb{H}$ and a map $T : V \to V$ is said to be a left linear operator if

$$T(u+v) = T(u) + T(v), \quad T(su) = sT(u), \quad \text{for all } s \in \mathbb{H}, \ u, v \in V.$$

We denote by $\mathrm{End}^L(V)$ the set of left linear operators on V. $\mathrm{End}^L(V)$ is both a left and a right vector space on $\mathbb{H}$ with respect to the operations:

$$(Ts)(v) := T(v)s, \quad (sT)(v) := T(vs), \quad \text{for all } s \in \mathbb{H}, \quad \text{and for all } v \in V. \tag{4.34}$$

In particular (4.34) gives $(\mathcal{I}s)(v) = (s\mathcal{I})(v) = vs$.

Definition 4.6.1. *Given a ring $(R, +, *)$ where $+, *$ denote the addition and the multiplication operations, respectively, the opposite ring $(R^{op}, +^{op}, *^{op})$ has the same underlying set as R, i.e, $R^{op} = R$ and the same additive structure while the multiplication $*^{op}$ is defined by $r *^{op} s := s * r$.*

The following result can be found for example in [3], section 4:

Proposition 4.6.2. *The two rings $\mathrm{End}^R(V)$ and $\mathrm{End}^L(V)$ with respect to the addition and composition of operators are opposite rings of each other.*

Remark 4.6.3. The fact that $\mathrm{End}^R(V)$ and $\mathrm{End}^L(V)$ are opposite rings can be efficiently illustrated by considering the multiplication by a scalar. Let us denote by N_s the multiplication on the left by a scalar $s \in \mathbb{H}$. Obviously we have that $N_s \in \mathrm{End}^R(V)$. We can rewrite (4.33) as

$$(N_sT)(v) = N_s(T(v)), \quad (TN_s)(v) = T(N_s(v)), \quad \text{for all } s \in \mathbb{H}, \ v \in V.$$

Denoting by M_s the operator that multiplies on the right a vector v by a scalar $s \in \mathbb{H}$, i.e., $M_s(v) = vs$, we have $M_s \in \mathrm{End}^L(V)$ and the operations defined in (4.34) for left linear operators can be written as

$$(TM_s)(v) = M_s(T(v)), \quad (M_sT)(v) = T(M_s(v)), \quad \text{for all } s \in \mathbb{H}, \ v \in V.$$

It appears clearly that, according to (4.34), the composition of T and M_s has to be taken in the reverse order.

Remark 4.6.4. We have that V is a module on the left on the ring $\mathrm{End}^R(V)$ and it is a module on the right on the ring $\mathrm{End}^L(V)$ (see [3], section 4). For this reason, the action of a right linear operator T on a vector $v \in V$ is often denoted in the literature by Tv while if T is a left linear operator, its action on v is denoted by vT. In light of this notation, the properties (4.34) can be written as

$$v(Ts) = (vT)s, \quad v(sT) = (vs)T, \quad \text{for all } s \in \mathbb{H}, \ \text{ and for all } v \in V.$$

In light of Remark 4.6.3 it is evident that this notation for left linear operators is useful especially when dealing with the multiplication. In general, when we will write $T(v)$ we will mean, unless otherwise specified, $T(v) = Tv$ and T right linear operator.

Proposition 4.6.5. *Let $T \in \mathrm{End}^R(V)$. Then we have*

(1) $(s\mathcal{I})T(v) = (sT)(v)$, *for all* $v \in V$, $s \in \mathbb{H}$;

(2) $T(s\mathcal{I})(v) = (Ts)(v)$, *for all* $v \in V$, $s \in \mathbb{H}$.

Let $T \in \mathrm{End}^L(V)$. Then we have

(3) $v((s\mathcal{I})T) = v(sT)$, *for all* $v \in V$, $s \in \mathbb{H}$;

(4) $v(T(s\mathcal{I})) = v(Ts)$, *for all* $v \in V$, $s \in \mathbb{H}$.

Proof. All the properties can be easily shown by using (4.33) and (4.34). In particular, if $T \in \mathrm{End}^L(V)$ we have $v((s\mathcal{I})T) = vsT = v(sT)$ and $v(T(s\mathcal{I})) = vT(s\mathcal{I}) = vTs = v(Ts)$. □

Remark 4.6.6. Let T be a right linear operator and let $a \in \mathbb{R}$. Then $aT = Ta$, in fact: $(aT)(v) = aT(v) = T(v)a = T(va) = T(av) = (Ta)(v)$. A similar property holds when T is left linear.

To deal with bounded operators we need an additional hypothesis on the vector space V and some more notations. Thus, in the sequel:

(i) V is a two-sided quaternionic Banach space with norm $\|\cdot\|$,

(ii) $\mathcal{B}^R(V)$ is the two-sided vector space of all right linear bounded operators on V,

(iii) $\mathcal{B}^L(V)$ is the two-sided vector space of all left linear bounded operators on V,

(iv) when we do not differentiate between left or right linear bounded operators on V we use the symbol $\mathcal{B}(V)$ and we call an element in $\mathcal{B}(V)$ a "linear operator".

It is easy to verify that $\mathcal{B}^R(V)$ and $\mathcal{B}^L(V)$ are Banach spaces if they are endowed with their natural norms:

$$\|T\| := \sup_{v\in V} \frac{\|T(v)\|}{\|v\|}.$$

4.7 The S-resolvent operator series

In this section we prove that in the Cauchy formula (4.28) we can formally replace the quaternion q by a quaternionic linear operator. In fact, for example, the kernel $-(q^2-2q\mathrm{Re}[s]+|s|^2)^{-1}(q-\overline{s})$ has been obtained by summing the Cauchy kernel series $\sum_{n\geq 0} q^n s^{-1-n}$ when the series converges. Here we prove that the sum of the series $\sum_{n\geq 0} T^n s^{-1-n}$ equals $-(T^2-2Re[s]T+|s|^2\mathcal{I})^{-1}(T-\overline{s}\mathcal{I})$, that is, it is formally obtained by replacing the quaternion q by T, also when the components of T do not commute. This is the reason why our functional calculus can be developed in a natural way starting from the Cauchy formula (4.28).

Definition 4.7.1. *Let $T\in\mathcal{B}(V)$. We define the left Cauchy kernel operator series or S-resolvent operator series as*

$$S_L^{-1}(s,T)=\sum_{n\geq 0} T^n s^{-1-n}, \tag{4.35}$$

and the right Cauchy kernel operator series as

$$S_R^{-1}(s,T)=\sum_{n\geq 0} s^{-1-n}T^n, \tag{4.36}$$

for $\|T\|<|s|$.

Remark 4.7.2. It is important to note, one more time, that the action of the S-resolvent operators series $S_L^{-1}(s,T)$ and $S_R^{-1}(s,T)$ in the case of left linear operators T is on the right, i.e., for every $v\in V$ we have $v\mapsto vS_L^{-1}(s,T)$ and $v\mapsto vS_R^{-1}(s,T)$. In particular, for the left Cauchy kernel operator series we have $v\mapsto \sum_n vT^n s^{-n-1}=\sum_n T^n(v)s^{-n-1}$. Thus, even though $S_L^{-1}(s,T)$ is formally the same operator used for right linear operators, $S_L^{-1}(s,T)$ acts in a different way.

Proposition 4.7.3. *If $\|T\|<|s|$ the operator $T-\overline{s}\mathcal{I}$ is invertible.*

Proof. When $\|T\|<|s|$ the series $\sum_{n\geq 0}(s^{-1}T)^n s^{-1}\mathcal{I}$ is convergent in the operator norm. As we shall see in Theorem 4.14.6, the series provides the inverse of the operator $T-\overline{s}\mathcal{I}$. □

Theorem 4.7.4. *Let $T\in\mathcal{B}(V)$ and let $s\in\mathbb{H}$. Then, for $\|T\|<|s|$:*

(1) *the operator*

$$S_L(s,T) = (T - \overline{s}\mathcal{I})^{-1}\, s(T - \overline{s}\mathcal{I}) - T \tag{4.37}$$

is the inverse of $\sum_{n\geq 0} T^n s^{-1-n}$ *and*

$$\sum_{n\geq 0} T^n s^{-1-n} = -(T^2 - 2Re[s]T + |s|^2\mathcal{I})^{-1}(T - \overline{s}\mathcal{I}), \tag{4.38}$$

(2) *the operator*

$$S_R(s,T) = (T - \overline{s}\mathcal{I})\, s(T - \overline{s}\mathcal{I})^{-1} - T \tag{4.39}$$

is the inverse of $\sum_{n\geq 0} s^{-1-n}T^n$ *and*

$$\sum_{n\geq 0} s^{-1-n}T^n = -(T - \overline{s}\mathcal{I})(T^2 - 2Re[s]T + |s|^2\mathcal{I})^{-1}. \tag{4.40}$$

Proof. The proof that

$$S_L(s,T)S_L^{-1}(s,T) = \mathcal{I} \tag{4.41}$$

can be done similarly to the proof of Theorem 3.1.3, so we show that

$$S_R(s,T)^{-1}S_R(s,T) = \mathcal{I} \tag{4.42}$$

where $S_R(s,T)$ is given by (4.36) and the interpretation of the symbols is as in Remark 4.7.2. We rewrite (4.42) by multiplying both hand side by $T - \overline{s}\mathcal{I}$ on the right:

$$\Big(\sum_{n\geq 0} s^{-1-n}T^n\Big)(T - \overline{s}\mathcal{I})s - \Big(\sum_{n\geq 0} s^{-1-n}T^n\Big)T(T - \overline{s}\mathcal{I}) = T - \overline{s}\mathcal{I}$$

which, by Proposition 4.6.5, can be written as

$$\Big(\sum_{n\geq 0} s^{-1-n}T^n\Big)(Ts - |s|^2\mathcal{I} - T^2 + T\overline{s}) = T - \overline{s}\mathcal{I}$$

and then we get

$$\Big(\sum_{n\geq 0} s^{-1-n}T^n\Big)(-|s|^2\mathcal{I} - T^2 + 2Re[s]T) = T - \overline{s}\mathcal{I}.$$

Observe that $-|s|^2\mathcal{I} - T^2 + 2Re[s]T$ has real coefficients and so it commutes with T^n. By writing explicitly the terms of the series

$$\sum_{n\geq 0} s^{-1-n}T^n(-|s|^2 - T^2 + 2Re[s]T)$$

and using the identity

$$s^2 - 2sRe[s] + |s|^2 = 0 \quad \text{for all} \quad s \in \mathbb{H},$$

we get

$$\sum_{n\geq 0} s^{-1-n} T^n (-|s|^2 - T^2 + 2Re[s]T) = T - \overline{s}\mathcal{I}.$$

So we finally have the identity

$$T - \overline{s}\mathcal{I} = T - \overline{s}\mathcal{I}.$$

The equality (4.40) follows directly by taking the inverse of

$$\begin{aligned} S_R(s,T) &= [(T - \overline{s}\mathcal{I})\, s - T(T - \overline{s}\mathcal{I})](T - \overline{s}\mathcal{I})^{-1} \\ &= -[T^2 - 2Re[s]T + |s|^2\mathcal{I}](T - \overline{s}\mathcal{I})^{-1}, \end{aligned}$$

thus the operator $S_R(s,T)$ is the inverse of (4.36) for $\|T\| < |s|$. □

The S-resolvent operator admits another power series expansion which is described in the next result:

Theorem 4.7.5. *Let $T \in \mathcal{B}(V)$ and $s \notin \sigma_S(T)$ be such that*

$$\|T - Re[s]\mathcal{I}\| < |s - Re[s]|. \tag{4.43}$$

Then the S-resolvent operator admits the series expansion

$$S_L^{-1}(s,T) = -\sum_{n\geq 0} (T - Re[s]\mathcal{I})^n (s - Re[s])^{-n-1}. \tag{4.44}$$

Proof. Observe that

$$\begin{aligned} (T^2 - 2TRe[s] + |s|^2\mathcal{I})^{-1} &= [(T - Re[s]\mathcal{I})^2 + |s|^2\mathcal{I} - (Re[s])^2\mathcal{I}]^{-1} \\ &= (|s|^2 - (Re[s])^2)^{-1}\Big[\mathcal{I} + \frac{(T - Re[s]\mathcal{I})^2}{(|s|^2 - (Re[s])^2)}\Big]^{-1} \\ &= \sum_{n\geq 0} (-1)^n \frac{(T - Re[s]\mathcal{I})^{2n}}{(|s|^2 - (Re[s])^2)^{n+1}}. \end{aligned} \tag{4.45}$$

Since $|Im[s]|^2 = -(Im[s])^2$, then by replacing (4.45) in (4.38) we get

$$\begin{aligned} S_L^{-1}(s,T) = -\Big(&\sum_{n\geq 0} (T - Re[s]\mathcal{I})^{2n+1} (s - Re[s])^{-2n-2} \\ &+ \sum_{n\geq 0} (T - Re[s]\mathcal{I})^{2n} (s - Re[s])^{-2n-1}\Big). \end{aligned}$$

Adding the two terms we get (4.44) which converges when (4.43) holds. □

An analogous result, with obvious variations, can be proved for $S_R^{-1}(s,T)$.

4.8 The S-spectrum and the S-resolvent operators

Observe that equalities (4.38) and (4.40) hold only for $\|T\| < |s|$, but the right-hand sides are defined on a larger set. Thus, we give the following definitions which are the main tools to define the quaternionic functional calculus.

Definition 4.8.1 (The S-spectrum and the S-resolvent set). *Let $T \in \mathcal{B}(V)$. We define the S-spectrum $\sigma_S(T)$ of T as*

$$\sigma_S(T) = \{s \in \mathbb{H} : \ T^2 - 2\,Re[s]T + |s|^2\mathcal{I} \quad \textit{is not invertible}\}.$$

The S-resolvent set $\rho_S(T)$ is defined by

$$\rho_S(T) = \mathbb{H} \setminus \sigma_S(T).$$

Remark 4.8.2. Observe that in the definition of $\sigma_S(T)$ we mean that the operator $T^2 - 2\,Re[s]T + |s|^2\mathcal{I}$ is not invertible in $\mathcal{B}^R(V)$ if T is right linear. On the other hand, if T is left linear we mean that $T^2 - 2\,Re[s]T + |s|^2\mathcal{I}$ is not invertible in $\mathcal{B}^L(V)$.

The notion of the S-spectrum of a linear quaternionic operator T is suggested by the definition of the S-resolvent operator that is the kernel useful for the quaternionic functional calculus.

Definition 4.8.3 (The S-resolvent operator). *Let V be a two-sided quaternionic Banach space, $T \in \mathcal{B}(V)$ and $s \in \rho_S(T)$. We define the left S-resolvent operator as*

$$S_L^{-1}(s,T) := -(T^2 - 2Re[s]T + |s|^2\mathcal{I})^{-1}(T - \overline{s}\mathcal{I}), \tag{4.46}$$

and the right S-resolvent operator as

$$S_R^{-1}(s,T) := -(T - \overline{s}\mathcal{I})(T^2 - 2Re[s]T + |s|^2\mathcal{I})^{-1}. \tag{4.47}$$

Theorem 4.8.4. *Let $T \in \mathcal{B}(V)$ and let $s \in \rho_S(T)$. Then, the left S-resolvent operator satisfies the equation*

$$S_L^{-1}(s,T)s - TS_L^{-1}(s,T) = \mathcal{I}, \tag{4.48}$$

and the right S-resolvent operator satisfies the equation

$$sS_R^{-1}(s,T) - S_R^{-1}(s,T)T = \mathcal{I}. \tag{4.49}$$

Proof. It follows by direct computation. Indeed, replacing the operator (4.46) in the left-hand side of the equality (4.48) we have

$$\begin{aligned} &-(T^2 - 2Re[s]T + |s|^2\mathcal{I})^{-1}(T - \overline{s}\mathcal{I})s \\ &+ T(T^2 - 2Re[s]T + |s|^2\mathcal{I})^{-1}(T - \overline{s}\mathcal{I}). \end{aligned} \tag{4.50}$$

Since T and $T^2 - 2Re[s]T + |s|^2\mathcal{I}$ commute, we can group on the right $(T^2 - 2Re[s]T + |s|^2\mathcal{I})^{-1}$ in the previous expression thus obtaining

$$(T^2 - 2Re[s]T + |s|^2\mathcal{I})^{-1}[-(T - \overline{s}\mathcal{I})s + T(T - \overline{s}\mathcal{I})$$

which becomes

$$(T^2 - 2Re[s]T + |s|^2\mathcal{I})^{-1}(T^2 - 2Re[s]T + |s|^2\mathcal{I}) = \mathcal{I}$$

which proves the statement. With analogous computations we can prove (4.49). □

Definition 4.8.5. *Let $T \in \mathcal{B}(V)$ and let $s \in \rho_S(T)$. We call*

$$S_L^{-1}(s,T)s - TS_L^{-1}(s,T) = \mathcal{I}$$

left S-resolvent equation and

$$sS_R^{-1}(s,T) - S_R^{-1}(s,T)T = \mathcal{I}$$

right S-resolvent equation.

Theorem 4.8.6 (Structure of the S-spectrum). *Let $T \in \mathcal{B}(V)$ and let $p = p_0 + p_1 I \in \sigma_S(T)$. Then all the elements of the sphere $[p_0 + Ip_1]$ belong to $\sigma_S(T)$.*

Proof. Consider the equation $(T^2 - 2Re[p]T + |p|^2\mathcal{I})v = 0$ for $v \neq 0$ and for $p = p_0 + Ip_1$. The coefficients depend only on the real numbers p_0, p_1 and not on $I \in \mathbb{S}$. Therefore all $s = p_0 + Jp_1$ such that $J \in \mathbb{S}$ are in the S-spectrum of T. □

Definition 4.8.7. *Let V be a two-sided quaternionic Banach space, $T \in \mathcal{B}(V)$, and let $U \subset \mathbb{H}$ be an axially symmetric s-domain that contains the S-spectrum $\sigma_S(T)$, such that $\partial(U \cap \mathbb{C}_I)$ is the union of a finite number of continuously differentiable Jordan curves for every $I \in \mathbb{S}$. We say that U is a T-admissible open set.*

We can now introduce the class of functions for which we can define the two versions of the quaternionic functional calculus.

Definition 4.8.8. *Let V be a two-sided quaternionic Banach space, $T \in \mathcal{B}(V)$ and let W be an open set in $\mathbb{H}$.*

(i) *A function $f \in \mathcal{R}^L(W)$ is said to be locally left regular on $\sigma_S(T)$ if there exists a T-admissible domain $U \subset \mathbb{H}$ such that $\overline{U} \subset W$. We will denote by $\mathcal{R}^L_{\sigma_S(T)}$ the set of locally left regular functions on $\sigma_S(T)$.*

(ii) *A function $f \in \mathcal{R}^R(W)$ is said to be locally right regular on $\sigma_S(T)$ if there exists a T-admissible domain $U \subset \mathbb{H}$ such that $\overline{U} \subset W$. We will denote by $\mathcal{R}^R_{\sigma_S(T)}$ the set of locally right regular functions on $\sigma_S(T)$.*

Remark 4.8.9. Let W be an open set in $\mathbb{H}$ and let $f \in \mathcal{R}^L(W)$. In the Cauchy formula (4.28) the open set $U \subset W$ need not be necessarily connected. In fact, formula (4.28) obviously holds when $U = \cup_{i=1}^r U_i$, $\overline{U}_i \cap \overline{U}_j = \emptyset$ when $i \neq j$ where U_i are T-admissible for all $i = 1, \dots, r$ and the boundaries of $U_i \cap \mathbb{C}_I$ consists of a finite number of continuously differentiable Jordan curves for $I \in \mathbb{S}$ for all $i = 1, \dots, r$. So when we choose $f \in \mathcal{R}^L_{\sigma_S(T)}$, the related open set U need not be connected. In the sequel we will state our results relating them to a domain U but our results obviously hold for open sets $U = \cup_{i=1}^r U_i$ as above. We will call such an open set U a T-admissible open set.

Using the left S-resolvent operator S_L^{-1}, we now give a result that motivates the functional calculus; analogous considerations can apply using S_R^{-1} with obvious modifications.

Theorem 4.8.10. *Let s, $a \in \mathbb{H}$, $m \in \mathbb{N}$, $T \in \mathcal{B}(V)$ and let $U \subset \mathbb{H}$ be a T-admissible open set. Set $ds_I = -dsI$ for $I \in \mathbb{S}$. Then*

$$T^m a = \frac{1}{2\pi} \int_{\partial(U \cap \mathbb{C}_I)} S_L^{-1}(s,T)\, ds_I\, s^m\, a. \tag{4.51}$$

Proof. Consider the power series expansion for the operator $S_L^{-1}(s,T)$ and a circle C_r on $\mathbb{C}_I$ centered in the origin and of radius $r > \|T\|$. We have

$$\frac{1}{2\pi} \int_{\partial(U \cap \mathbb{C}_I)} S_L^{-1}(s,T)\, ds_I\, s^m\, a = \frac{1}{2\pi} \sum_{n \geq 0} T^n \int_{C_r} s^{-1-n+m}\, ds_I a = T^m\, a,$$

since

$$\int_{C_r} ds_I s^{-n-1+m} = 0 \quad if \quad n \neq m,$$

$$\int_{C_r} ds_I s^{-n-1+m} = 2\pi \quad if \quad n = m.$$

The Cauchy theorem on $\mathbb{C}_I$ shows that the above integrals are not affected if we replace C_r by $\partial(U \cap \mathbb{C}_I)$ independently of $I \in \mathbb{S}$. □

Theorem 4.8.11 (Compactness of S-spectrum). *Let $T \in \mathcal{B}(V)$. Then the S-spectrum $\sigma_S(T)$ is a compact nonempty set.*

Proof. Let $U \subset \mathbb{H}$ be a T-admissible open set. Set $ds_I = -dsI$ for $I \in \mathbb{S}$. Then

$$\frac{1}{2\pi} \int_{\partial(U \cap \mathbb{C}_I)} S^{-1}(s,T)\, ds_I\, s^m = T^m.$$

In particular, for $m = 0$, we have

$$\frac{1}{2\pi} \int_{\partial(U \cap \mathbb{C}_I)} S^{-1}(s,T)\, ds_I = \mathcal{I},$$

which shows that $\sigma_S(T)$ is a nonempty set (otherwise the integral would be zero by the vector-valued version of Cauchy's theorem). The S-spectrum is closed because the complement of $\sigma_S(T)$ is open. Indeed, the function

$$g : s \mapsto T^2 - 2Re[s]T + |s|^2\mathcal{I}$$

is trivially continuous and, by Theorem 10.12 in [91], the set $\mathcal{U}(V)$ of all invertible elements of $\mathcal{B}(V)$ is an open set in $\mathcal{B}(V)$. Therefore

$$g^{-1}(\mathcal{U}(V)) = \rho_S(T)$$

is an open set in $\mathbb{H}$. The S-spectrum is a bounded set because for $\|T\| < |s|$ the series $\sum_{n\geq 0} T^n s^{-1-n}$ and $\sum_{n\geq 0} s^{-1-n}T^n$ converge. So we conclude that it is compact. □

4.9 Examples of S-spectra

We now give some examples of the computation of the S-spectrum of some matrices. In section 4.14 we will compare the S-spectrum with a more standard spectrum, the so-called left spectrum, showing that, in general, there is no relation between them. In what follows s will always denote a quaternion.

Example 4.9.1. We determine the S-spectrum of the matrix

$$T_1 = \begin{bmatrix} 1 & 0 \\ 0 & j \end{bmatrix}.$$

We consider

$$\left(\begin{bmatrix} 1 & 0 \\ 0 & j \end{bmatrix}^2 - 2s_0 \begin{bmatrix} 1 & 0 \\ 0 & j \end{bmatrix} + |s|^2 \begin{bmatrix} 1 & 0 \\ 0 & 1 \end{bmatrix} \right) \begin{bmatrix} q_1 \\ q_2 \end{bmatrix} = 0$$

which gives

$$\begin{bmatrix} 1 - 2s_0 + |s|^2 & 0 \\ 0 & -1 - 2js_0 + |s|^2 \end{bmatrix} \begin{bmatrix} q_1 \\ q_2 \end{bmatrix} = 0.$$

Since q_1 and $q_2 \in \mathbb{H}$ cannot be simultaneously zero, we must have either

$$1 - 2s_0 + |s|^2 = 0$$

or

$$-1 - 2js_0 + |s|^2 = 0.$$

The first equation gives

$$\begin{aligned} & 1 - 2s_0 + |s|^2 = 0 \\ \Leftrightarrow \quad & 1 - 2s_0 + s_0^2 + s_1^2 + s_2^2 + s_3^2 = 0 \\ \Leftrightarrow \quad & (s_0 - 1)^2 + s_1^2 + s_2^2 + s_3^2 = 0 \\ \Leftrightarrow \quad & s_0 = 1, \quad s_j = 0, \quad j = 1, 2, 3; \end{aligned}$$

while the second equation gives

$$\begin{aligned} -1 - 2js_0 + |s|^2 = 0 \ &\Leftrightarrow \ s_0 = 0 \ \text{ and } \ |s|^2 = 1 \\ &\Leftrightarrow \ s_1^2 + s_2^2 + s_3^2 = 1 \ \text{ and } \ s_0 = 0. \end{aligned}$$

Therefore the S-spectrum is

$$\sigma_S(T_1) = \{1\} \cup \mathbb{S}.$$

Example 4.9.2. We determine the S-spectrum of the matrix

$$T_2 = \begin{bmatrix} i & 0 \\ 0 & j \end{bmatrix}.$$

We have

$$\begin{bmatrix} -1 - 2is_0 + |s|^2 & 0 \\ 0 & -1 - 2js_0 + |s|^2 \end{bmatrix} \begin{bmatrix} q_1 \\ q_2 \end{bmatrix} = 0$$

and therefore

$$\sigma_S(T_2) = \mathbb{S}.$$

Example 4.9.3. We determine the S-spectrum of the real diagonal matrix

$$T_3 = \begin{bmatrix} a & 0 \\ 0 & b \end{bmatrix},$$

which gives

$$\begin{bmatrix} a^2 - 2as_0 + |s|^2 & 0 \\ 0 & b^2 - 2bs_0 + |s|^2 \end{bmatrix} \begin{bmatrix} q_1 \\ q_2 \end{bmatrix} = 0$$

from which we have

$$(s_0 - a)^2 + s_1^2 + s_2^2 + s_3^2 = 0$$

and

$$(s_0 - b)^2 + s_1^2 + s_2^2 + s_3^2 = 0.$$

Therefore $\sigma_S(T_3) = \{a, b\}$.

Example 4.9.4. We determine the S-spectrum of

$$T_4 = \begin{bmatrix} 0 & i \\ -i & 0 \end{bmatrix}.$$

We consider

$$(T_4^2 - 2s_0T_4 + |s|^2) \begin{bmatrix} q_1 \\ q_2 \end{bmatrix} = \begin{bmatrix} 1 + |s|^2 & -2s_0 i \\ 2s_0 i & 1 + |s|^2 \end{bmatrix} \begin{bmatrix} q_1 \\ q_2 \end{bmatrix} = 0$$

and get

$$(1 + |s|^2)q_1 - 2s_0 i q_2 = 0, \qquad 2s_0 i q_1 + (1 + |s|^2)q_2 = 0.$$

Observe that $1+|s|^2 \neq 0$ for every $s \in \mathbb{H}$, so

$$q_2 = -\frac{2s_0 i}{1+|s|^2} q_1,$$

and replacing it in the first equation we obtain

$$(1+|s|^2)q_1 + 2s_0 i \frac{2s_0 i}{1+|s|^2} q_1 = 0.$$

Since it must be $q_1 \neq 0$, it is

$$(1+|s|^2)^2 - 4s_0^2 = 0$$

whose solutions are

$$1+|s|^2 \pm 2s_0 = 0,$$

which gives

$$(1 \pm s_0)^2 + s_1^2 + s_2^2 + s_3^2 = 0.$$

The S-spectrum is therefore

$$\sigma_S(T_4) = \{s_0 = \pm 1, \quad s_1 = s_2 = s_3 = 0\}.$$

4.10 The quaternionic functional calculus

We begin by recalling the quaternionic version of the Hahn-Banach theorem, originally proved in [97], and one of its corollaries which, with the Cauchy integral formula II, are the main tools to prove that the definition of the quaternionic functional calculus is well posed.

Theorem 4.10.1 (The quaternionic version of the Hahn-Banach theorem). *Let V_0 be a right subspace of a right vector space V on $\mathbb{H}$. Suppose that p is a seminorm on V and let ϕ be a linear and continuous functional on V_0 such that*

$$|\langle \phi, v\rangle| \leq p(v), \qquad \forall v \in V_0. \tag{4.52}$$

Then it is possible to extend ϕ to a linear and continuous functional Φ on V satisfying the estimate (4.52) *for all $v \in V$.*

Proof. Note that, for any quaternion q we have $q = q_0 + q_1 i + q_2 j + q_3 k = z_1(q) + z_2(q)j$, where $z_1, z_2 \in \mathbb{C} = \mathbb{R} + \mathbb{R}i$ and $qj = -z_2(q) + z_1(q)j$, so $q = z_1(q) - z_1(qj)j$. The functional ϕ can be written as $\phi = \phi_0 + \phi_1 i + \phi_2 j + \phi_3 k = \psi_1(\phi) + \psi_2(\phi)j$, with $\psi_1(\phi) = \phi_0 + \phi_1 i$ and $\psi_2(\phi) = \phi_2 + \phi_3 i$ which are complex functionals. It is immediate that

$$\langle \phi, v\rangle = \langle \psi_1, v\rangle - \langle \psi_1, vj\rangle j, \qquad \forall v \in V_0$$

where ψ_1 is a $\mathbb{C}$–linear functional. So we can apply the complex version of the Hahn–Banach theorem to deduce the existence of a functional $\tilde{\psi}_1$ that extends ψ_1 to the whole of V (as a complex vector space). The functional Ψ given by

$$\langle \Psi, v\rangle = \langle \tilde{\psi}_1, v\rangle - \langle \tilde{\psi}_1, vj\rangle j$$

is defined on V and it is the extension that satisfies estimate (4.52) for all $v \in V$. □

The following result is an immediate consequence of the quaternionic version of the Hahn-Banach theorem. Its proof mimics the analog proof in the complex case.

Corollary 4.10.2. *Let V be a right vector space on $\mathbb{H}$ and let $v \in V$. If $\langle \phi, v\rangle = 0$ for every linear and continuous functional ϕ in V', then $v = 0$.*

Theorem 4.10.3. *Let V be a two-sided quaternionic Banach space and $T \in \mathcal{B}(V)$. Let $U \subset \mathbb{H}$ be a T-admissible domain and set $ds_I = -dsI$. Then the integrals*

$$\frac{1}{2\pi}\int_{\partial(U\cap\mathbb{C}_I)} S_L^{-1}(s,T)\, ds_I\, f(s), \quad f \in \mathcal{R}^L_{\sigma_S(T)} \tag{4.53}$$

and

$$\frac{1}{2\pi}\int_{\partial(U\cap\mathbb{C}_I)} f(s)\, ds_I\, S_R^{-1}(s,T), \quad f \in \mathcal{R}^R_{\sigma_S(T)} \tag{4.54}$$

do not depend on the choice of the imaginary unit $I \in \mathbb{S}$ and on U.

Proof. Let us prove that (4.53) does not depend on the choice of the imaginary unit $I \in \mathbb{S}$ and on U. Let us consider $T \in \mathcal{B}^R(V)$ (the case $T \in \mathcal{B}^L(V)$ works with obvious different interpretations of the action of the operators involved). We first observe that the function $S_L^{-1}(s,q)$ is right s-regular in the variable s in its domain of definition by Proposition 4.4.9. Now observe that we can replace q with an operator $T \in \mathcal{B}(V)$ in the Cauchy formula (4.28), thanks to Theorem 4.7.4. For every linear and continuous functional $\phi \in V'$, consider the duality $\langle \phi, S_L^{-1}(s,T)v\rangle$, for $v \in V$ and define the function

$$g(s) := \langle \phi, S_L^{-1}(s,T)v\rangle, \quad \text{for} \quad v \in V, \quad \phi \in V'.$$

The function g remains right s-regular in the variable s on the complement of $\sigma_S(T)$ and since $g(s) \to 0$ as $s \to \infty$ we have that g is s-regular also at infinity. Suppose that U is a T-admissible open set such that $\partial(U \cap \mathbb{C}_I)$ does not cross the S-spectrum of T for every $I \in \mathbb{S}$. The fact that, for fixed $I \in \mathbb{S}$, the integral

$$\frac{1}{2\pi}\int_{\partial(U\cap\mathbb{C}_I)} g(s)ds_I\, f(s) \tag{4.55}$$

does not depend on U follows from the Cauchy theorem. By Corollary 4.10.2 also the integral (4.53) does not depend on U. We now prove that the integral (4.55) does not depend on $I \in \mathbb{S}$. Since g is a right s-regular function on the complement of the S-spectrum of T, we can consider an open set U' such that $\overline{U}' \subset \rho_S(T)$, $U' \cap \mathbb{R} \neq \emptyset$ and $[q] \subset U'$ whenever $q \in U'$. We assume that $\partial(U' \cap \mathbb{C}_I)$ consists of a finite number of continuously differentiable Jordan curves $\forall I \in \mathbb{S}$ and that $\partial U \subset U'$ where U is an open set as above so, in particular, U contains $[s]$ whenever $s \in U$. Choose $J \in \mathbb{S}$, $J \neq I$ and represent $g(s)$ by the Cauchy integral formula (4.30) as

$$g(s) = -\frac{1}{2\pi} \int_{\partial(U' \cap \mathbb{C}_J)^-} g(t)\, dt_J\, S_L^{-1}(s,t) \tag{4.56}$$

where the boundary $\partial(U' \cap \mathbb{C}_J)^-$ is oriented clockwise to include the points $[s] \in \partial(U \cap \mathbb{C}_J)$ (recalling that the singularities of $S_L^{-1}(s,t)$ correspond to the 2-sphere $[s]$) and to exclude the points belonging to the S-spectrum of T.

Let us now plug the expression of $g(s)$ in (4.56) into the integral (4.55) and taking into account the orientation of $\partial(U' \cap \mathbb{C}_J)^-$ we obtain

$$\begin{aligned} &\frac{1}{2\pi} \int_{\partial(U \cap \mathbb{C}_I)} g(s)\, ds_I\, f(s) \\ &= \frac{1}{2\pi} \int_{\partial(U \cap \mathbb{C}_I)} \left[\frac{1}{2\pi} \int_{\partial(U' \cap \mathbb{C}_J)} g(t)\, dt_J\, S_L^{-1}(s,t)\right] ds_I\, f(s) \\ &= \frac{1}{2\pi} \int_{\partial(U' \cap \mathbb{C}_J)} g(t)\, dt_J \Big[\frac{1}{2\pi} \int_{\partial(U \cap \mathbb{C}_I)} S_L^{-1}(s,t)\, ds_I\, f(s)\Big] \end{aligned} \tag{4.57}$$

where we have used the Fubini theorem. Now observe that $\partial(U' \cap \mathbb{C}_J)$ consists of a finite number of Jordan curves inside and outside $U \cap \mathbb{C}_J$, but the integral

$$\frac{1}{2\pi} \int_{\partial(U \cap \mathbb{C}_I)} S_L^{-1}(s,t)\, ds_I\, f(s)$$

equals $f(t)$ for those $t \in \partial(U' \cap \mathbb{C}_J)$ belonging to $U \cap \mathbb{C}_J$. Thus we obtain:

$$\begin{aligned} &\frac{1}{2\pi} \int_{\partial(U' \cap \mathbb{C}_J)} g(t)\, dt_J \Big[\frac{1}{2\pi} \int_{\partial(U \cap \mathbb{C}_I)} S_L^{-1}(s,t)\, ds_I\, f(s)\Big] \\ &= \frac{1}{2\pi} \int_{\partial(U' \cap \mathbb{C}_J)} g(t)\, dt_J f(t). \end{aligned} \tag{4.58}$$

So from (4.57) and (4.58) we can write

$$\frac{1}{2\pi} \int_{\partial(U \cap \mathbb{C}_I)} g(s)\, ds_I\, f(s) = \frac{1}{2\pi} \int_{\partial(U' \cap \mathbb{C}_J)} g(t)\, dt_J f(t). \tag{4.59}$$

Now observe that $\partial(U' \cap \mathbb{C}_J)$ is positively oriented and surrounds the S-spectrum of T. By the independence of the integral on the open set, we can substitute $\partial(U' \cap \mathbb{C}_J)$ by $\partial(U \cap \mathbb{C}_J)$ in (4.59) and we get:

$$\frac{1}{2\pi} \int_{\partial(U \cap \mathbb{C}_I)} g(s)\, ds_I\, f(s) = \frac{1}{2\pi} \int_{\partial(U \cap \mathbb{C}_J)} g(t)\, dt_J\, f(t),$$

that is

$$\frac{1}{2\pi}\int_{\partial(U\cap\mathbb{C}_I)}\langle\phi,S_L^{-1}(s,T)v\rangle ds_I\ f(s)$$
$$=\frac{1}{2\pi}\int_{\partial(U\cap\mathbb{C}_J)}\langle\phi,S_L^{-1}(t,T)v\rangle dt_J\ f(t),\quad \text{for all}\ \ v\in V,\ \phi\in V',\ \ I,J\in\mathbb{S}.$$

Thus by Corollary 4.10.2 the integral (4.53) does not depend on $I\in\mathbb{S}$.

Let us prove with analogous arguments that the integral (4.54) does not depend on the choice of U and $I\in\mathbb{S}$. Also in this case let us consider $T\in\mathcal{B}^R(V)$ (the case $T\in\mathcal{B}^L(V)$ works with obvious different interpretations of the action of the operators involved).

For every linear and continuous functional $\phi\in V'$, consider the duality $\langle\phi,S_R^{-1}(s,T)v\rangle$, for $v\in V$ and define the function

$$g(s):=\langle\phi,S_R^{-1}(s,T)v\rangle,\quad \text{for}\quad v\in V,\quad \phi\in V'.$$

The function g is left regular in the variable s on the complement of $\sigma_S(T)$ – recall that the S-spectrum of a bounded linear quaternionic operator T is a compact nonempty set – and since $g(s)\to 0$ as $s\to\infty$ we have that g is regular also at infinity. Suppose that U is a T-admissible open set such that $\partial(U\cap\mathbb{C}_I)$ does not cross the S-spectrum of T for every $I\in\mathbb{S}$. The fact that, for fixed $I\in\mathbb{S}$, the integral

$$\frac{1}{2\pi}\int_{\partial(U\cap\mathbb{C}_I)}f(s)ds_I\ g(s)\tag{4.60}$$

does not depend on U follows from the Cauchy theorem and on Corollary 4.10.2, so also the integral (4.54) does not depend on U. We now prove that the integral (4.54) does not depend on $I\in\mathbb{S}$. Let $\varepsilon>0$ and set

$$W_\varepsilon=\{q\in U\mid \operatorname{dist}(q,\partial U)>\varepsilon\}.$$

We have that $\overline{W_\varepsilon}\subset U$, moreover W_ε is axially symmetric since U is axially symmetric. Let U' be the complement of $\overline{W_\varepsilon}$. Then U' is axially symmetric, $\overline{U}'\subset\rho_S(T)$ and $\partial U'=\partial W_\varepsilon\subset U$. Note that $\partial(U'\cap\mathbb{C}_I)$ consists of a finite number of continuously differentiable Jordan curves $\forall I\in\mathbb{S}$. Since g is a left regular function on the complement of the S-spectrum of T, function g is regular on U'. Choose $J\in\mathbb{S}$, $J\neq I$ and represent $g(s)$ by the Cauchy integral formula (4.29) as

$$g(s)=\frac{1}{2\pi}\int_{\partial(U'\cap\mathbb{C}_J)^-}S_L^{-1}(t,s)\ dt_Jg(t)\tag{4.61}$$
$$=-\frac{1}{2\pi}\int_{\partial(U'\cap\mathbb{C}_J)^-}S_R^{-1}(s,t)\ dt_Jg(t)$$

where the boundary $\partial(U'\cap\mathbb{C}_J)^-$ is oriented clockwise to include the points $[s]\in\partial(U\cap\mathbb{C}_J)$ (recall that the singularities of $S^{-1}(s,t)$ correspond to the 2-sphere $[s]$) and to exclude the points belonging to the S-spectrum of T.

Let us now plug the expression of $g(s)$ in (4.61) into the integral (4.60) and, taking into account the orientation of $\partial(U'\cap\mathbb{C}_J)^-$, we obtain

$$\begin{aligned}&\frac{1}{2\pi}\int_{\partial(U\cap\mathbb{C}_I)} f(s)\,ds_I\,g(s) \\ &=\frac{1}{2\pi}\int_{\partial(U\cap\mathbb{C}_I)} f(s)\,ds_I\Big[\frac{1}{2\pi}\int_{\partial(U'\cap\mathbb{C}_J)} S_R^{-1}(s,t)\,dt_J g(t)\Big] \\ &=\frac{1}{2\pi}\int_{\partial(U'\cap\mathbb{C}_J)}\frac{1}{2\pi}\int_{\partial(U\cap\mathbb{C}_I)}\Big[f(s)\,ds_I S_R^{-1}(s,t)\Big]\,dt_J g(t)\end{aligned} \tag{4.62}$$

where we have used the Fubini theorem. Since $\partial(U'\cap\mathbb{C}_J)\subset U\cap\mathbb{C}_J$, the integral

$$\frac{1}{2\pi}\int_{\partial(U\cap\mathbb{C}_I)} f(s)\,ds_I\,S_R^{-1}(s,t)$$

equals $f(t)$. Thus we obtain:

$$\frac{1}{2\pi}\int_{\partial(U\cap\mathbb{C}_I)} g(s)\,ds_I\,f(s)=\frac{1}{2\pi}\int_{\partial(U'\cap\mathbb{C}_J)} g(t)\,dt_J f(t). \tag{4.63}$$

Now observe that $\partial(U'\cap\mathbb{C}_J)$ is positively oriented and surrounds the S-spectrum of T. By the independence of the integral on the open set, we can substitute $\partial(U'\cap\mathbb{C}_J)$ by $\partial(U\cap\mathbb{C}_J)$ in (4.63) and we get:

$$\frac{1}{2\pi}\int_{\partial(U\cap\mathbb{C}_I)} g(s)\,ds_I\,f(s)=\frac{1}{2\pi}\int_{\partial(U\cap\mathbb{C}_J)} g(t)\,dt_J\,f(t),$$

that is

$$\begin{aligned}&\frac{1}{2\pi}\int_{\partial(U\cap\mathbb{C}_I)}\langle\phi,S_L^{-1}(s,T)v\rangle ds_I\,f(s)\\ &=\frac{1}{2\pi}\int_{\partial(U\cap\mathbb{C}_J)}\langle\phi,S_L^{-1}(t,T)v\rangle dt_J\,f(t),\quad\text{for all }\ v\in V,\ \phi\in V',\ \ I,J\in\mathbb{S}.\end{aligned}$$

Thus by Corollary 4.10.2 the integral (4.54) does not depend on $I\in\mathbb{S}$. □

Definition 4.10.4 (The quaternionic functional calculus). *Let V be a two-sided quaternionic Banach space and $T\in\mathcal{B}(V)$. Let $U\subset\mathbb{H}$ be a T-admissible domain and set $ds_I=-dsI$. We define*

$$f(T)=\frac{1}{2\pi}\int_{\partial(U\cap\mathbb{C}_I)} S_L^{-1}(s,T)\,ds_I\,f(s),\quad\textit{for}\quad f\in\mathcal{R}^L_{\sigma_S(T)}, \tag{4.64}$$

and

$$f(T) = \frac{1}{2\pi} \int_{\partial(U \cap \mathbb{C}_I)} f(s)\, ds_I\, S_R^{-1}(s,T), \quad \textit{for} \quad f \in \mathcal{R}^R_{\sigma_S(T)}. \tag{4.65}$$

Remark 4.10.5. By Remark 4.6.4, it follows that when $T \in \mathcal{B}^L(V)$ we have $f(T)(v) = vf(T)$ while if $T \in \mathcal{B}^R(V)$ we have $f(T)(v) = f(T)v$.

An immediate consequence is the following.

Theorem 4.10.6. *Let V be a quaternionic Banach space and let $T \in \mathcal{B}(V)$. Assume that $f_n \in \mathcal{R}^L_{\sigma_S(T)}$ (resp. $f_n \in \mathcal{R}^R_{\sigma_S(T)}$), for all $n \in \mathbb{N}$ and let U be a T-admissible open set. If f_n converges uniformly to f on $U \cap \mathbb{C}_I$, $I \in \mathbb{S}$, then $f_n(T)$ converges to $f(T)$ in $\mathcal{B}(V)$.*

Proof. Let W be a T-admissible domain such that $\sigma_S(T) \subset \overline{W} \subset U$. Then $f_n \to f$ converges uniformly on $\partial(W \cap \mathbb{C}_I)$ and consequently

$$f_n(T) = \frac{1}{2\pi} \int_{\partial(W \cap \mathbb{C}_I)} S_L^{-1}(s,T)\, ds_I\, f_n(s)$$

converges in the uniform topology of operators to

$$f(T) = \frac{1}{2\pi} \int_{\partial(W \cap \mathbb{C}_I)} S_L^{-1}(s,T)\, ds_I\, f(s).$$

The case $f_n \in \mathcal{R}^R_{\sigma_S(T)}$ follows by the functional calculus in (4.65). □

4.11 Algebraic properties of the quaternionic functional calculus

An immediate consequence of Definition 4.10.4 are the linearity properties of the functional calculus.

Proposition 4.11.1. *Let V be a two-sided quaternionic Banach space and $T \in \mathcal{B}(V)$.*

(a) *If $f, g \in \mathcal{R}^L_{\sigma_S(T)}$, then*

$$(f+g)(T) = f(T) + g(T), \qquad (fp)(T) = f(T)p, \qquad \textit{for all} \quad p \in \mathbb{H}.$$

(b) *If $f, g \in \mathcal{R}^R_{\sigma_S(T)}$, then*

$$(f+g)(T) = f(T) + g(T), \qquad (pf)(T) = pf(T), \qquad \textit{for all} \quad p \in \mathbb{H}.$$

For the definition of the product of s-regular functions we need the following subclass of regular functions.

Definition 4.11.2. *Let $U \subset \mathbb{H}$ be an open set. We define*

$$\mathcal{N}^j(U) = \{f \in \mathcal{R}^j(U) \mid f(U \cap \mathbb{C}_I) \subseteq \mathbb{C}_I, \quad \text{for all } I \in \mathbb{S}\}, \quad \text{for} \quad j = L, R.$$

Proposition 4.11.3. *Let $U \subset \mathbb{H}$ be an open set. Then we have $\mathcal{N}^L(U) = \mathcal{N}^R(U)$.*

Proof. Since $f_I : \mathbb{C}_I \to \mathbb{C}_I$ we have that f_I commutes with $I \in \mathbb{S}$ thus f is left regular if and only if it is right regular. □

Thanks to Proposition 4.11.3 we set

$$\mathcal{N}^L(U) = \mathcal{N}^R(U) := \mathcal{N}(U).$$

Definition 4.11.4. *Let V be a two-sided quaternionic Banach space and let $T \in \mathcal{B}(V)$. We will denote by $\mathcal{N}_{\sigma_S(T)}$ the set of slice regular functions for which there exists a T-admissible open set $U \subset \mathbb{H}$ such that $f \in \mathcal{N}(U)$, and where $\overline{U}$ is contained in the set of slice regularity of f.*

The following result on the s-regular functions will be useful to study some of the properties of the quaternionic functional calculus for bounded linear operators.

Proposition 4.11.5. *Let $U \subset \mathbb{H}$ be an open set.*

(1) *Let $f \in \mathcal{N}(U)$, $g \in \mathcal{R}^L(U)$, then $fg \in \mathcal{R}^L(U)$.*

(2) *Let $f \in \mathcal{N}(U)$, $g \in \mathcal{R}^R(U)$, then $gf \in \mathcal{R}^R(U)$.*

(3) *Let $f, g \in \mathcal{N}(U)$, then $fg = gf$ and $fg \in \mathcal{N}(U)$.*

Proof. Point (1): Consider $I \in \mathbb{S}$ and set $z = x + Iy$. The restriction $f_I(z)$ of f equals $F(z)$ with $F : U \cap \mathbb{C}_I \to \mathbb{C}_I$ holomorphic and we have that

$$\left(\frac{\partial}{\partial x} + I\frac{\partial}{\partial y}\right)(fg)(z) = \frac{\partial F}{\partial x}(z)g(z) + F(z)\frac{\partial g}{\partial x}(z) + I\frac{\partial F}{\partial y}(z)g(z) + IF(z)\frac{\partial g}{\partial y}(z)$$

and since I commutes with $F(z)$ we obtain:

$$\left(\frac{\partial}{\partial x} + I\frac{\partial}{\partial y}\right)(fg)(z) = \Big(\frac{\partial F}{\partial x}(z) + I\frac{\partial F}{\partial y}(z)\Big)g(z) + F(z)\Big(\frac{\partial g}{\partial x}(z) + I\frac{\partial g}{\partial y}(z)\Big) = 0.$$

Point (2) is analogous. Point (3) follows from the fact that both f and g take $\mathbb{C}_I$ to itself for all $I \in \mathbb{S}$. □

Theorem 4.11.6. *Let V be a two-sided quaternionic Banach space and $T \in \mathcal{B}(V)$.*

(1) *If $\phi \in \mathcal{N}_{\sigma_S(T)}$ and $g \in \mathcal{R}^L_{\sigma_S(T)}$, then $(\phi g)(T) = \phi(T)g(T)$.*

(2) *If $\phi \in \mathcal{N}_{\sigma_S(T)}$ and $g \in \mathcal{R}^R_{\sigma_S(T)}$, then $(g\phi)(T) = g(T)\phi(T)$.*

Proof. We prove point (1), since the proof of point (2) is analogous. Denote by U a T-admissible open set on which both ϕ and g are s-regular. Observe that ϕg is s-regular on U thanks to Lemma 4.11.5. Let G_1 and G_2 be two T-admissible open sets such that $G_1 \cup \partial G_1 \subset G_2$ and $G_2 \cup \partial G_2 \subset U$. Take $s \in \partial(G_1 \cap \mathbb{C}_I)$ and $t \in \partial(G_2 \cap \mathbb{C}_I)$ and observe that, for $I \in \mathbb{S}$, we have

$$g(s) = \frac{1}{2\pi} \int_{\partial(G_2 \cap \mathbb{C}_I)} S_L^{-1}(t,s)\, dt_I\, g(t).$$

Now consider

$$\begin{aligned}(\phi g)(T) &= \frac{1}{2\pi} \int_{\partial(G_1 \cap \mathbb{C}_I)} S_L^{-1}(s,T)\, ds_I\, \phi(s)\, g(s) \\ &= \frac{1}{2\pi} \int_{\partial(G_1 \cap \mathbb{C}_I)} S_L^{-1}(s,T)\, ds_I\, \phi(s) \left[\frac{1}{2\pi} \int_{\partial(G_2 \cap \mathbb{C}_I)} S_L^{-1}(t,s)\, dt_I\, g(t)\right].\end{aligned}$$

By the vectorial version of the Fubini theorem we have

$$(\phi g)(T) = \frac{1}{(2\pi)^2} \int_{\partial(G_2 \cap \mathbb{C}_I)} \int_{\partial(G_1 \cap \mathbb{C}_I)} S_L^{-1}(s,T)\, ds_I\, \phi(s) S_L^{-1}(t,s)\, dt_I\, g(t).$$

Finally observe that $S_L^{-1}(t,s)$ is s-regular in the variable s on the S-spectrum of T and $\phi(s)S_L^{-1}(t,s)$ is s-regular in the variable s thanks to Lemma 4.11.5, so we have

$$(\phi g)(T) = \frac{1}{2\pi} \int_{\partial(G_2 \cap \mathbb{C}_I)} \phi(T) S_L^{-1}(t,T) dt_I g(t) = \phi(T) g(T). \qquad \square$$

4.12 The S-spectral radius

In this section we give the definition of the S-spectral radius which is the analog of the spectral radius for the Riesz–Dunford case. The main result of this section is Theorem 4.12.6. This theorem is based on the S-spectral mapping theorem for the powers T^n, $n \in \mathbb{N}$, of a quaternionic bounded linear operator T, and it can be proved using some algebraic properties of quaternionic polynomials. In Section 4.13 we will generalize the S-spectral mapping theorem to a wider class of s-regular functions.

Definition 4.12.1 (The S-spectral radius of T). *Let V be a two-sided quaternionic Banach space and $T \in \mathcal{B}(V)$. We call S-spectral radius of T the nonnegative real number*

$$r_S(T) := \sup\{\, |s| \; : \; s \in \sigma_S(T)\,\}.$$

Before we can state and prove the S-spectral radius theorem, we need two preliminary lemmas on quaternionic polynomials. For the sequel, it is useful to recall that any quaternion $q = \mathrm{Re}[q] + I_q|\mathrm{Im}[q]|$ is associated to the 2-sphere defined by $[q]$ which reduces to q only when q is real.

Remark 4.12.2. We recall that quaternionic polynomials with real coefficients are both left and right s-regular functions.

Lemma 4.12.3. *Let $n \in \mathbb{N}$ and $q,\ s \in \mathbb{H}$. Let*

$$P_{2n}(q) := q^{2n} - 2\,\mathrm{Re}[s^n]\, q^n + |s^n|^2.$$

Then

$$\begin{aligned} P_{2n}(q) &= Q_{2n-2}(q)(q^2 - 2\,\mathrm{Re}[s]\, q + |s|^2) \\ &= (q^2 - 2\,\mathrm{Re}[s]\, q + |s|^2)Q_{2n-2}(q), \end{aligned} \tag{4.66}$$

where $Q_{2n-2}(q)$ is a polynomial of degree $2n-2$ in q.

Proof. First of all we observe that

$$P_{2n}(s) = s^{2n} - 2\mathrm{Re}[s^n]s^n + |s^n|^2 = s^{2n} - (s^n + \bar{s}^n)s^n + s^n\bar{s}^n = 0.$$

Moreover, the substitution of s by any s' on the same 2-sphere leaves the coefficients of the polynomial $P_{2n}(q)$ unchanged, and $P_{2n}(s') = 0$. We conclude that the whole 2-sphere defined by s is a solution to the equation $P_{2n}(q) = 0$. The statement follows from the factorization theorem, see [71], and the fact that the second degree polynomial $q^2 - 2\mathrm{Re}[s]q + |s|^2$ has real coefficients. □

Lemma 4.12.4. *Let $n \in \mathbb{N}$ and $q,\ p \in \mathbb{H}$. Let λ_j, $j = 0, 1, \ldots, n-1$ be the solutions of $\lambda^n = p$ in the complex plane $\mathbb{C}_{I_p}$. Then*

$$q^{2n} - 2\,\mathrm{Re}[p]\, q^n + |p|^2 = \prod_{j=0}^{n-1} (q^2 - 2\,\mathrm{Re}[\lambda_j]\, q + |\lambda_j|^2). \tag{4.67}$$

Proof. The equation $\lambda^n = p$ can be solved in the complex plane $x + I_p y$ containing $p = p_0 + I_p p_1$ where it admits n solutions $\lambda_j = \lambda_{j0} + I_p \lambda_{j1}$, $j = 0, 1, \ldots, n-1$. By reason of degree, these are the only solutions to the equation in the complex plane $\mathbb{C}_{I_p}$. Note that if we take any $p' = p_0 + Ip_1$, $I \in \mathbb{S}$ in the 2-sphere of p, then the solutions to the equation $\lambda^n = p'$ are $\lambda'_j = \lambda_{j0} + I\lambda_{j1}$, $j = 0, 1, \ldots, n-1$, $I \in \mathbb{S}$. We consider the polynomial

$$P_{2n}(q) = q^{2n} - 2\,\mathrm{Re}[p]\, q^n + |p|^2$$

and we observe that $q = \lambda_j$ is a root of $P_{2n}(q) = 0$, in fact

$$P_{2n}(\lambda_j) = \lambda_j^{2n} - 2\,\mathrm{Re}[p]\, \lambda_j^n + |p|^2 = p^2 - 2\,\mathrm{Re}[p]\, p + |p|^2 = 0.$$

The substitution of p by p' on the same 2-sphere leaves P_{2n} unchanged and it is immediate that $P_{2n}(\lambda'_j) = 0$ when I varies in $\mathbb{S}$. This proves that the roots of $P_{2n}(q) = 0$ lie on the 2-spheres of λ_j, $j = 0, \ldots, n-1$. The statement follows from the factorization theorem. □

Theorem 4.12.5 (A particular case of the S-spectral mapping theorem). *Let V be a two-sided quaternionic Banach space and let $T \in \mathcal{B}(V)$. Then*

$$\sigma_S(T^n) = (\sigma_S(T))^n = \{s^n \in \mathbb{H} : s \in \sigma_S(T)\}.$$

Proof. Recall that

$$\sigma_S(T) = \{s \in \mathbb{H} \quad : \quad T^2 - 2\ \mathrm{Re}[s]T + |s|^2\mathcal{I} \quad \text{is not invertible}\}$$

and

$$\sigma_S(T^n) = \{p \in \mathbb{H} \quad : \quad T^{2n} - 2\ \mathrm{Re}[p]T^n + |p|^2\mathcal{I} \quad \text{is not invertible}\}.$$

Note that the operator $T^{2n} - 2\mathrm{Re}[s^n]T^n + |s^n|^2\mathcal{I}$, thanks to Lemma 4.12.3 and Theorem 4.11.6, can be factorized as

$$T^{2n} - 2\mathrm{Re}[s^n]T^n + |s^n|^2\mathcal{I} = Q_{2n-2}(T)(T^2 - 2\mathrm{Re}[s]T + |s|^2\mathcal{I}).$$

So we deduce that if $T^2 - 2\mathrm{Re}[s]T + |s|^2\mathcal{I}$ is not injective also $T^{2n} - 2\mathrm{Re}[s^n]T^n + |s^n|^2\mathcal{I}$ is not injective. This proves that $(\sigma_S(T))^n \subseteq \sigma_S(T^n)$. Let us now consider $p \in \sigma_S(T^n)$. By Lemma 4.12.4 and Theorem 4.11.6 we can write

$$T^{2n} - 2\mathrm{Re}[p]T^n + |p|^2\mathcal{I} = \prod_{j=0}^{n-1}(T^2 - 2\mathrm{Re}[\lambda_j]T + |\lambda_j|^2\mathcal{I}).$$

Since if $T^{2n} - 2\mathrm{Re}[p]T^n + |p|^2\mathcal{I}$ is not invertible, then at least one of the operators $T^2 - 2\mathrm{Re}[\lambda_j]T + |\lambda_j|^2\mathcal{I}$ for some j is not invertible. This proves that $\sigma_S(T^n) \subseteq (\sigma_S(T))^n$. □

We can now conclude this section with the S-spectral radius theorem.

Theorem 4.12.6 (The S-spectral radius theorem). *Let V be a two-sided quaternionic Banach space, let $T \in \mathcal{B}(V)$, and let $r_S(T)$ be its S-spectral radius. Then*

$$r_S(T) = \lim_{n\to\infty} \|T^n\|^{1/n}.$$

Proof. For every $s \in \mathbb{H}$ such that $|s| > r_S(T)$ the series $\sum_{n\geq 0} T^n s^{-1-n}$ converges in $\mathcal{B}(V)$ to the S-resolvent operator $S_L^{-1}(s,T)$ (we reason analogously for $\sum_{n\geq 0} s^{-1-n}T^n$). So the sequence $T^n s^{-1-n}$ is bounded in the norm of $\mathcal{B}(V)$ and

$$\limsup_{n\to\infty} \|T^n\|^{1/n} \leq r_S(T). \tag{4.68}$$

Theorem 4.12.5 implies $\sigma_S(T^n) = (\sigma_S(T))^n$, so we have

$$(r_S(T))^n = r_S(T^n) \leq \|T^n\|,$$

from which we get

$$r_S(T) \leq \liminf \|T^n\|^{1/n}. \tag{4.69}$$

From (4.68), (4.69) we obtain

$$r_S(T) \leq \liminf_{n\to\infty} \|T^n\|^{1/n} \leq \limsup_{n\to\infty} \|T^n\|^{1/n} \leq r_S(T). \tag{4.70}$$

The chain of inequalities (4.70) also proves the existence of the limit. □

4.13 The S-spectral mapping and the composition theorems

We collect in the following lemma some useful properties of s-regular functions that will be used to prove the main results of this section.

Lemma 4.13.1. *Let $U \subset \mathbb{H}$ be an open set.*

(a) *Suppose that $P(q), Q(q)$ are polynomials in the quaternionic variable q with real coefficients and assume that $Q(q)$ has no zeros in U.*
If $F(q) = (Q(q))^{-1}P(q)$ (or $F(q) = P(q)(Q(q))^{-1}$), then $F \in \mathcal{N}(U)$.

(b) *If $f \in \mathcal{N}(U)$, then $f^2 \in \mathcal{N}(U)$.*

(c) *Let U, U' be two open sets in $\mathbb{H}$ and $f \in \mathcal{N}(U')$, $g \in \mathcal{N}(U)$ with $g(U) \subseteq U'$. Then $f(g(q))$ is s-regular in U.*

Proof. Part a) trivially follows by replacing q by $z = x + Iy$ and observing that

$$\left(\frac{\partial}{\partial x} + I\frac{\partial}{\partial y}\right) F(x + Iy) = 0$$

for all $I \in \mathbb{S}$. To prove b), consider $\mathbb{C}_I$ for any $I \in \mathbb{S}$ and the restriction $f_I(z) = F(z)$, where $F : U \cap \mathbb{C}_I \to \mathbb{C}_I$ is a holomorphic function. This implies that also the function f^2 belongs to $\mathcal{N}(U)$. Finally, to prove c) set $q = x + Iy$. By hypothesis, $g(x + Iy) = \alpha(x, y) + I\beta(x, y)$, where α and β are real-valued functions and

$$f(g(x + Iy)) = f(\alpha(x, y) + I\beta(x, y)) \subseteq \mathbb{C}_I.$$

The function $f(g(x + Iy))$ is holomorphic on each plane $\mathbb{C}_I$ thus it satisfies the condition

$$\left(\frac{\partial}{\partial x} + I\frac{\partial}{\partial y}\right) f(g(x + Iy)) = 0$$

for all $I \in \mathbb{S}$ and so $f(g(q))$ is s-regular. □

Theorem 4.13.2 (The S-spectral mapping theorem). *Let V be a two-sided quaternionic Banach space, $T \in \mathcal{B}(V)$ and $f \in \mathcal{N}_{\sigma_S(T)}$. Then*

$$\sigma_S(f(T)) = f(\sigma_S(T)) = \{f(s) : s \in \sigma_S(T)\}.$$

Proof. Since $f \in \mathcal{N}_{\sigma_S(T)}$, there exists a T-admissible open set $U \subset \mathbb{H}$ containing $\sigma_S(T)$, and such that $f \in \mathcal{N}(U)$. Let us fix $\lambda \in \sigma_S(T)$. For $q \notin [\lambda]$, let us define the function $\tilde{g}(q)$ by

$$\tilde{g}(q) = (q^2 - 2\mathrm{Re}[\lambda]q + |\lambda|^2)^{-1}(f^2(q) - 2\mathrm{Re}[f(\lambda)]f(q) + |f(\lambda)|^2).$$

Observe that the assumption $f \in \mathcal{N}(U)$ implies that $f^2 \in \mathcal{N}(U)$ by Lemma 4.13.1 (b), so also $f^2(q) - 2\mathrm{Re}[f(\lambda)]f(q) + |f(\lambda)|^2 \in \mathcal{N}(U)$. The function $(q^2 - 2\mathrm{Re}[\lambda]q + |\lambda|^2)^{-1} \in \mathcal{N}(U \setminus \{[\lambda]\})$, by Lemma 4.13.1 (a), thus $\tilde{g}(q) \in \mathcal{N}(U \setminus \{[\lambda]\})$ by Lemma 4.11.5. We can extend $\tilde{g}(q)$ to an s-regular function whose domain is U: if the 2-sphere $[\lambda]$ is not reduced to a real point, then we define

$$g(q) = \begin{cases} \tilde{g}(q) & \text{if } q \notin [\lambda], \\ \dfrac{\partial}{\partial x} f(\mu)\, \dfrac{f(\mu) - \overline{f(\mu)}}{\mu - \overline{\mu}} & \text{if } q = \mu = \lambda_0 + I\lambda_1 \in [\lambda],\ I \in \mathbb{S}. \end{cases}$$

Now, the auxiliary function g, is defined on U and is s-regular, see the proof of Theorem 3.5.9, with suitable variations. Thanks to Theorem 4.11.6 we can write

$$f^2(T) - 2\mathrm{Re}[f(\lambda)]f(T) + |f(\lambda)|^2\mathcal{I} = (T^2 - 2\mathrm{Re}[\lambda]T + |\lambda|^2\mathcal{I})g(T).$$

If $f^2(T) - 2\mathrm{Re}[f(\lambda)]f(T) + |f(\lambda)|^2\mathcal{I}$ admits a bounded inverse

$$B := (f^2(T) - 2\mathrm{Re}[f(\lambda)]f(T) + |f(\lambda)|^2\mathcal{I})^{-1} \in \mathcal{B}(V),$$

then we have

$$(T^2 - 2\mathrm{Re}[\lambda]T + |\lambda|^2\mathcal{I})g(T)B = \mathcal{I},$$

i.e., $g(T)B$ is the inverse of $T^2 - 2\mathrm{Re}[\lambda]T + |\lambda|^2\mathcal{I}$. Thus $f(\sigma_S(T)) \subseteq \sigma_S(f(T))$. Now we take $p \in \sigma_S(f(T))$ such that $p \notin f(\sigma_S(T))$. We define the function

$$h(q) := (f^2(q) - 2\mathrm{Re}[p]f(q) + |p|^2)^{-1}$$

which is s-regular on $\sigma_S(T)$. By Theorem 4.11.6 we obtain

$$h(T)(f^2(T) - 2\mathrm{Re}[p]f(T) + |p|^2\mathcal{I}) = \mathcal{I},$$

which means that $p \notin \sigma_S(f(T))$, but this contradicts the assumption. So $p \in f(\sigma_S(T))$. □

Theorem 4.13.3. *Let V be a two-sided quaternionic Banach space and let $T \in \mathcal{B}(V)$. Suppose that $f \in \mathcal{N}_{\sigma_S(T)}$, $\phi \in \mathcal{N}_{f(\sigma_S(T))}$ and define $F(s) = \phi(f(s))$. Then $F \in \mathcal{R}_{\sigma_S(T)}$ and $F(T) = \phi(f(T))$.*

Proof. The statement $F \in \mathcal{R}_{\sigma_S(T)}$ follows from Lemma 4.13.1 (c). Let $U \supset \sigma_S(f(T))$ be a T-admissible open set whose boundary is denoted by ∂U. Suppose that $U \cup \partial U$ is contained in the domain in which ϕ is s-regular. Let W be a T-admissible neighborhood of $\sigma_S(T)$ whose boundary is denoted by ∂W and suppose that $W \cup \partial W$ is contained in the domain where f is s-regular. Finally suppose that $f(W \cup \partial W) \subset U$. Let $I \in \mathbb{S}$ and define the operator

$$S^{-1}(\lambda, f(T)) = \frac{1}{2\pi} \int_{\partial(U \cap \mathbb{C}_I)} S^{-1}(s,T)\, ds_I\, S^{-1}(\lambda, f(s))$$

where

$$S^{-1}(\lambda, f(s)) = -(f(s)^2 - 2\mathrm{Re}[\lambda] f(s) + |\lambda|^2)^{-1}(f(s) - \overline{\lambda}). \tag{4.71}$$

By applying Lemmas 4.13.1 and 4.11.5 and with some easy calculations it follows that $S^{-1}(\lambda, f(s))$ is s-regular in the variable s and it is right s-regular in the variable λ.

Take $\lambda \in \mathbb{R}$, so that also $S(\lambda, f(s))$ is an s-regular function and observe that

$$S^{-1}(\lambda, f(s))S(\lambda, f(s)) = S(\lambda, f(s))S^{-1}(\lambda, f(s)) = 1$$

so by Theorem 4.11.6 the operator $S^{-1}(\lambda, f(T))$ satisfy the equation:

$$S(\lambda, f(T))S^{-1}(\lambda, f(T)) = S^{-1}(\lambda, f(T))S(\lambda, f(T)) = \mathcal{I}. \tag{4.72}$$

Observe now that also when λ is not necessarily a real number, identity (4.72) remains valid as it can be easily shown by replacing $S^{-1}(\lambda, f(T))$ and $S(\lambda, f(T))$ by their explicit expressions

$$S^{-1}(\lambda, f(T)) = -(f(T)^2 - 2\mathrm{Re}[\lambda] f(T) + |\lambda|^2)^{-1}(f(T) - \overline{\lambda})$$

and

$$S(\lambda, f(T)) = -(f(T) - \overline{\lambda})^{-1}(f(T)^2 - 2\mathrm{Re}[\lambda] f(T) + |\lambda|^2)$$

in (4.72) and verifying that we get an identity. Consequently we obtain

$$\begin{aligned}
\phi(f(T)) &= \frac{1}{2\pi} \int_{\partial(W \cap \mathbb{C}_I)} S^{-1}(\lambda, f(T))\, d\lambda_I\, \phi(\lambda) \\
&= \frac{1}{2\pi} \int_{\partial(W \cap \mathbb{C}_I)} \Big(\frac{1}{2\pi} \int_{\partial(U \cap \mathbb{C}_I)} S^{-1}(s,T)\, ds_I\, S^{-1}(\lambda, f(s))\Big)\, d\lambda_I\, \phi(\lambda) \\
&= \frac{1}{2\pi} \int_{\partial(U \cap \mathbb{C}_I)} S^{-1}(s,T)\, ds_I \Big(\frac{1}{2\pi} \int_{\partial(W \cap \mathbb{C}_I)} S^{-1}(\lambda, f(s))\, d\lambda_I\, \phi(\lambda)\Big) \\
&= \frac{1}{2\pi} \int_{\partial(U \cap \mathbb{C}_I)} S^{-1}(s,T)\, ds_I\, \phi(f(s)) \\
&= \frac{1}{2\pi} \int_{\partial(U \cap \mathbb{C}_I)} S^{-1}(s,T)\, ds_I\, F(s) = F(T),
\end{aligned}$$

so this concludes the proof. □

4.14 Bounded perturbations of the S-resolvent operator

In this section we prove that, as in the case of the Riesz–Dunford functional calculus, bounded perturbations of the S-resolvent operator produce bounded perturbations of the function $f(T)$. The main result of this section is Theorem 4.14.14.

To start with, we introduce the notions of left and right spectrum of a quaternionic linear operator. In the literature, these notions are usually introduced by specifying the side on which the multiplication by a quaternion is done. Thus we can state the following definition:

Definition 4.14.1. *Let $T : V \to V$ be a right linear quaternionic operator on a quaternionic Banach space V. We denote by $\sigma_L(T)$ the left spectrum of T related to the resolvent operator $(s\mathcal{I} - T)^{-1}$, that is*

$$\sigma_L(T) = \{s \in \mathbb{H} : \quad s\mathcal{I} - T \quad \text{is not invertible in } \mathcal{B}^R(V)\},$$

where the notation $s\mathcal{I}$ in $\mathcal{B}^R(V)$ means that $(s\mathcal{I})(v) = sv$. Let $T : V \to V$ be a left linear quaternionic operator on a quaternionic Banach space V. We denote by $\sigma_R(T)$ the right spectrum of T related to the resolvent operator $(\mathcal{I}s - T)^{-1}$, that is

$$\sigma_R(T) = \{s \in \mathbb{H} : \quad \mathcal{I}s - T \quad \text{is not invertible in } \mathcal{B}^L(V)\},$$

where in $\mathcal{B}^L(V)$ the notation $\mathcal{I}s$ means that $(\mathcal{I}s)(v) = vs$.

We recall that the multiplication operator by an element $s \in \mathbb{H}$ is not a left (resp. right) linear operator unless the multiplication is performed on the right (resp. left).

The S-spectrum and the left spectrum, called from now on, L-spectrum are not, in general, related. To convince ourselves, we will rework the examples given in Section 4.9 and compute their L-spectra.

Example 4.14.2. Consider the matrix

$$T_1 = \begin{bmatrix} 1 & 0 \\ 0 & j \end{bmatrix}.$$

Recall that the S-spectrum was given by $\{1\} \cup \mathbb{S}$ while an immediate computation shows that $\sigma_L(T_1) = \{1, j\}$.

Example 4.14.3. Consider the matrix

$$T_2 = \begin{bmatrix} i & 0 \\ 0 & j \end{bmatrix}.$$

Recall that $\sigma_S(T_2) = \mathbb{S}$, while, with simple computations, we obtain that the L-spectrum is $\sigma_L(T_2) = \{i, j\}$.

Example 4.14.4. When we consider a real diagonal matrix

$$T_3 = \begin{bmatrix} a & 0 \\ 0 & b \end{bmatrix},$$

we obtain $\sigma_L(T_3) = \sigma_S(T_3) = \{a, b\}$.

Example 4.14.5. Consider the matrix

$$T_4 = \begin{bmatrix} 0 & i \\ -i & 0 \end{bmatrix}.$$

Recall that the S-spectrum is $\sigma_S(T_4) = \{s_0 = \pm 1, \quad s_1 = s_2 = s_3 = 0\}$. We now determine the L-spectrum:

$$(sI - T_4)q = \begin{bmatrix} s & -i \\ i & s \end{bmatrix} \begin{bmatrix} q_1 \\ q_2 \end{bmatrix} = 0,$$

which gives

$$sq_1 - iq_2 = 0, \qquad iq_1 + sq_2 = 0.$$

If we replace from the first equation $sq_1 = iq_2$ into the second one, we get

$$(i - s\,i\,s)q_1 = 0$$

and therefore

$$i - s\,i\,s = 0.$$

Expanding s we obtain

$$i - (s_0 + is_1 + js_2 + ks_3)i(s_0 + is_1 + js_2 + ks_3) = 0,$$

i.e.,

$$i - i(s_0 + is_1 - js_2 - ks_3)(s_0 + is_1 + js_2 + ks_3) = 0,$$

which gives the system

$$1 - (s_0^2 - s_1^2 + s_2^2 + s_3^2) = 0, \quad s_0 s_1 = 0, \quad s_1 s_2 = 0, \qquad s_1 s_3 = 0$$

so that

$$\sigma_L(T) = \{s_1 = 0, \quad s_0^2 + s_2^2 + s_3^2 = 1\}.$$

The notion of L-spectrum is related to the existence of the algebraic inverse of the operator $s\mathcal{I} - T$. However, the operator $(s\mathcal{I} - T)^{-1}$ is not the analog of the resolvent operator for the Riesz–Dunford functional calculus. We describe it in the following result:

Theorem 4.14.6. *Let $T \in \mathcal{B}(V)$ and let $s \in \mathbb{H}$ be such that $\|T\| < |s|$. Then the operator*

$$\sum_{n\geq 0}(s^{-1}T)^n s^{-1}\mathcal{I}$$

is the right and left algebraic inverse of $s\mathcal{I} - T$ and the series converges in the operator norm.

Proof. Let us directly compute

$$\begin{aligned}
&(s\mathcal{I} - T)\sum_{n\geq 0}(s^{-1}T)^n s^{-1}\mathcal{I} \\
&= s\mathcal{I}\sum_{n\geq 0}(s^{-1}T)^n s^{-1}\mathcal{I} - T\sum_{n\geq 0}(s^{-1}T)^n s^{-1}\mathcal{I} \\
&= s\mathcal{I}s^{-1}\mathcal{I} + Ts^{-1}\mathcal{I} + T(s^{-1}T)s^{-1}\mathcal{I} + T(s^{-1}T)^2 s^{-1}\mathcal{I} + \dots \\
&\quad - Ts^{-1}\mathcal{I} - T(s^{-1}T)s^{-1}\mathcal{I} - T(s^{-1}T)^2 s^{-1}\mathcal{I} - T(s^{-1}T)^3 s^{-1}\mathcal{I} + \dots = \mathcal{I}.
\end{aligned}$$

Similarly, we can prove that

$$\sum_{n\geq 0}(s^{-1}T)^n s^{-1}\mathcal{I}(s\mathcal{I} - T) = \mathcal{I}.$$

Finally we consider

$$\begin{aligned}
\|\sum_{n\geq 0}(s^{-1}T)^n s^{-1}\mathcal{I}\| &\leq \sum_{n\geq 0}\|(s^{-1}T)^n s^{-1}\mathcal{I}\| \\
&\leq \sum_{n\geq 0}\|(s^{-1}T)\|^n |s^{-1}| \leq \sum_{n\geq 0}\|T\|^n |s^{-1}|^{n+1}
\end{aligned}$$

which converges for $\|T\| < |s|$. This completes the proof. □

The L-spectrum $\sigma_L(T)$ for bounded operators is bounded and, in particular, is contained in the same ball as the S-spectrum $\sigma_S(T)$, as shown in the next result:

Theorem 4.14.7. *Let $T \in \mathcal{B}(V)$. Then $\sigma_L(T)$ is contained in the set $\{s \in \mathbb{H} : |s| \leq \|T\|\}$.*

Proof. Since the series

$$\sum_{n\geq 0}(s^{-1}T)^n s^{-1}\mathcal{I},$$

converges if and only if $|s^{-1}|\|T\| < 1$, we get the statement. □

We now provide a simple relation between the L-spectrum and the S- spectrum.

Proposition 4.14.8. *Let $T \in \mathcal{B}(V)$ and $s \in \sigma_L(T)$ and let v be the corresponding L-eigenvector. Then $s \in \sigma_S(T)$ and v is the corresponding S-eigenvector if and only if*

$$(T - s\mathcal{I})(sv) = 0.$$

Proof. It follows from the relations:

$$\begin{aligned} T^2v - 2Re[s]Tv + \overline{s}s\mathcal{I}v &= T(sv) - 2Re[s](sv) + \overline{s}(sv) \\ &= (T - s\mathcal{I})(sv) = 0. \end{aligned}$$

□

Lemma 4.14.9. *The set $\mathcal{U}(V)$ of elements in $\mathcal{B}(V)$ which have inverse in $\mathcal{B}(V)$ is an open set in the uniform topology of $\mathcal{B}(V)$. If $\mathcal{U}(V)$ contains an element A, then it contains the ball*

$$\Sigma = \{B \in \mathcal{B}(V) \ : \ \|A - B\| < \|A^{-1}\|^{-1}\}.$$

If $B \in \Sigma$, its inverse is given by the series

$$B^{-1} = A^{-1} \sum_{n \geq 0} [(A - B)A^{-1}]^n. \tag{4.73}$$

Furthermore, the map $A \mapsto A^{-1}$ from $\mathcal{U}(V)$ onto $\mathcal{U}(V)$, is a homeomorphism in the uniform operator topology.

Proof. Let $\|\mathcal{I} - B\| < 1$, so the series

$$Q = \sum_{n \geq 0} (\mathcal{I} - B)^n$$

converges. Since

$$QB = BQ = [\mathcal{I} - (\mathcal{I} - B)]Q = \sum_{n \geq 0} (\mathcal{I} - B)^n - \sum_{n \geq 1} (\mathcal{I} - B)^n = \mathcal{I},$$

it follows that

$$\{B \in \mathcal{B}(V) \ : \ \|\mathcal{I} - B\| < 1\} \subset \mathcal{U}(V).$$

Now let $A \in \mathcal{U}(V)$ and let $\|A - B\| < \|A^{-1}\|^{-1}$. Then

$$\|\mathcal{I} - BA^{-1}\| = \|(A - B)A^{-1}\| < 1$$

hence BA^{-1} has an inverse in $\mathcal{B}(V)$, given by the series

$$\sum_{n \geq 0} (\mathcal{I} - BA^{-1})^n = \sum_{n \geq 0} [(A - B)A^{-1}]^n.$$

Thus B has an inverse in $\mathcal{B}(V)$, given by formula (4.73):

$$B^{-1} = A^{-1} \sum_{n \geq 0} [(A - B)A^{-1}]^n = A^{-1} + A^{-1} \sum_{n \geq 1} [(A - B)A^{-1}]^n$$

so that

$$\|B^{-1} - A^{-1}\| \leq \|A^{-1}\| \sum_{n\geq 1} \|[(A-B)A^{-1}]\|^n \leq \frac{\|A-B\|\|A^{-1}\|^2}{1 - \|A-B\|\|A^{-1}\|}$$

from which it follows that the map $B \mapsto B^{-1}$ from $\mathcal{U}(V)$ onto $\mathcal{U}(V)$ is a homeomorphism. □

Definition 4.14.10. *Let $\mathcal{W}$ be a subset of $\mathbb{H}$. We denote by $\mathcal{E}(\mathcal{W}, \varepsilon)$, for $\varepsilon > 0$, the ε-neighborhood of $\mathcal{W}$ defined as*

$$\mathcal{E}(\mathcal{W}, \varepsilon) := \{q \in \mathbb{H} \, : \, \inf_{s\in\mathcal{W}} |s - q| < \varepsilon\}.$$

Let $\sigma_L(T)$ be the left spectrum of T. We will use the notation

$$\overline{\sigma_L(T)} = \{q \in \mathbb{H} \, : \, \overline{s} \in \sigma_L(T)\}.$$

We now state and prove two lemmas in the case of right linear operators. The results can be stated and proved also in the case of left linear operators, with suitable modifications.

Lemma 4.14.11. *Let T, $Z \in \mathcal{B}^R(V)$ and let $s \notin \overline{\sigma_L(T)} \cup \overline{\sigma_L(Z)}$ and consider*

$$S_L(s,T) = (T - \overline{s}\mathcal{I})^{-1} s\,(T - \overline{s}\mathcal{I}) - T, \qquad S_L(s,Z) = (Z - \overline{s}\mathcal{I})^{-1} s\,(Z - \overline{s}\mathcal{I}) - Z$$

and

$$S_R(s,T) = (T - \mathcal{I}\overline{s})\, s\,(T - \mathcal{I}\overline{s})^{-1} - T, \qquad S_R(s,Z) = (Z - \mathcal{I}\overline{s})\, s\,(Z - \mathcal{I}\overline{s})^{-1} - Z.$$

Then there exist positive constants $K(s)$, $K'(s)$ depending on the operators T and Z, such that

$$\|S_L(s,T) - S_L(s,Z)\| \leq K(s)\|T - Z\|, \tag{4.74}$$

$$\|S_R(s,T) - S_R(s,Z)\| \leq K'(s)\|T - Z\|. \tag{4.75}$$

Proof. Consider the following chain of inequalities:

$$\begin{aligned}
&\|S_L(s,T) - S_L(s,Z)\| \\
&\leq \|(T - \overline{s}\mathcal{I})^{-1}\| \; |s| \; \|T - Z\| \\
&\quad + \|(T - \overline{s}\mathcal{I})^{-1}\| \; \|Z - T\| \; \|(Z - \overline{s}\mathcal{I})^{-1}\| \, |s| \, \|Z - \overline{s}\mathcal{I}\| + \|T - Z\| \\
&\leq \Big[\|(T - \overline{s}\mathcal{I})^{-1}\| \, |s| \, + \|(T - \overline{s}\mathcal{I})^{-1}\| \; \|(Z - \overline{s}\mathcal{I})^{-1}\| \; |s| \; \|Z - \overline{s}\mathcal{I}\| + 1\Big] \|T - Z\| \\
&\leq \Big[|s| \; \|(T - \overline{s}\mathcal{I})^{-1}\| \; \Big(1 + \|(Z - \overline{s}\mathcal{I})^{-1}\| \; \|Z - \overline{s}\mathcal{I}\|\Big) + 1\Big] \|T - Z\|.
\end{aligned}$$

We set

$$K(s) := |s| \; \|(T - \overline{s}\mathcal{I})^{-1}\| \; \Big(1 + \|(Z - \overline{s}\mathcal{I})^{-1}\| \; \|Z - \overline{s}\mathcal{I}\|\Big) + 1, \tag{4.76}$$

so we get the statement. The strategy to prove (4.75) follows the same lines. □

Lemma 4.14.12. *Let T, $Z \in \mathcal{B}^R(V)$, let $s \in \rho_S(T)$, $s \notin \overline{\sigma_L(T)} \cup \overline{\sigma_L(Z)}$ and suppose that*

$$\|T - Z\| < \frac{1}{K(s)} \|S_L^{-1}(s, T)\|^{-1}, \tag{4.77}$$

where $K(s)$ is defined in (4.76). Then $s \in \rho_S(Z)$ and

$$S_L^{-1}(s, Z) - S_L^{-1}(s, T) = S_L^{-1}(s, T) \sum_{n \geq 1} [(S_L(s, T) - S_L(s, Z)) S_L^{-1}(s, T)]^n. \tag{4.78}$$

Analogously, suppose that $s \in \rho_S(T)$, $s \notin \overline{\sigma_L(T)} \cup \overline{\sigma_L(Z)}$ and

$$\|T - Z\| < \frac{1}{K'(s)} \|S_R^{-1}(s, T)\|^{-1},$$

where $K'(s)$ is obtained with analogous calculations as for $K(s)$. Then $s \in \rho_S(Z)$ and

$$S_R^{-1}(s, Z) - S_R^{-1}(s, T) = S_R^{-1}(s, T) \sum_{n \geq 1} [(S_R(s, T) - S_R(s, Z)) S_R^{-1}(s, T)]^n. \tag{4.79}$$

Proof. Let us consider

$$S_L(s, T) = (T - \overline{s}\mathcal{I})^{-1} s (T - \overline{s}\mathcal{I}) - T, \qquad S_L(s, Z) = (Z - \overline{s}\mathcal{I})^{-1} s (Z - \overline{s}\mathcal{I}) - Z.$$

Using the estimate (4.74) and hypothesis (4.77), we get

$$\|S_L(s, T) - S_L(s, Z)\| \leq K(s) \|T - Z\| < \|S_L^{-1}(s, T)\|^{-1}.$$

If we apply Lemma 4.14.9 where we set

$$A := S_L(s, T), \qquad B := S_L(s, Z), \qquad A^{-1} = S_L^{-1}(s, T), \tag{4.80}$$

we obtain that $S_L(s, Z)$ is invertible, so we conclude that $s \in \rho_S(T)$. Moreover, its inverse $S_L^{-1}(s, Z)$ is given by formula (4.73), i.e.,

$$S_L^{-1}(s, Z) = S_L^{-1}(s, T) \sum_{n \geq 0} [(S_L(s, T) - S_L(s, Z)) S_L^{-1}(s, T)]^n, \tag{4.81}$$

and the series converges since

$$\|(S_L(s, T) - S_L(s, Z)) S_L^{-1}(s, T)\| \leq K(s) \|T - Z\| \|S_L^{-1}(s, T)\| < 1.$$

To prove (4.79) we follow an analogous argument. □

Theorem 4.14.13. *Let $T, Z \in \mathcal{B}^R(V)$, $s \in \rho_S(T)$, $s \notin \overline{\sigma_L(T)} \cup \overline{\sigma_L(Z)}$ and let $\varepsilon > 0$. Then there exists $\delta > 0$ such that, for $\|T - Z\| < \delta$, we have*

$$\sigma_S(Z) \subseteq \mathcal{E}(\sigma_S(T) \cup \overline{\sigma_L(T)}, \varepsilon), \tag{4.82}$$

$$\|S_L^{-1}(s,Z) - S_L^{-1}(s,T)\| < \varepsilon, \quad \textit{for} \quad s \notin \mathcal{E}(\sigma_S(T) \cup \overline{\sigma_L(T)}, \varepsilon), \tag{4.83}$$

and

$$\|S_R^{-1}(s,Z) - S_R^{-1}(s,T)\| < \varepsilon, \quad \textit{for} \quad s \notin \mathcal{E}(\sigma_S(T) \cup \overline{\sigma_L(T)}, \varepsilon). \tag{4.84}$$

A similar statement holds for $T \in \mathcal{B}^L(V)$ when $s \in \rho_S(T)$, $s \notin \overline{\sigma_R(T)} \cup \overline{\sigma_R(Z)}$.

Proof. Recall that we have assumed $T, Z \in \mathcal{B}^R(V)$. Let $\varepsilon > 0$; thanks to Lemma 4.14.9 there exists $\eta > 0$ such that if

$$\|T - Z\| < \eta,$$

then $\overline{\sigma_L(Z)} \subset \mathcal{E}(\overline{\sigma_L(T)}, \varepsilon)$, where $\mathcal{E}(\overline{\sigma_L(T)}, \varepsilon)$ is the ε-neighborhood of $\overline{\sigma_L(T)}$. So we can always choose η such that $\overline{\sigma_L(Z)} \subset \mathcal{E}(\sigma_S(T) \cup \overline{\sigma_L(T)}, \varepsilon)$. Consider the function $K(s)$ defined in (4.76) and observe that the constant K_ε defined by

$$K_\varepsilon = \sup_{s \notin \mathcal{E}(\sigma_S(T) \cup \overline{\sigma_L(T)}, \varepsilon)} K(s) \tag{4.85}$$

is finite since $s \notin \mathcal{E}(\sigma_S(T) \cup \overline{\sigma_L(T)}, \varepsilon)$, the set $\overline{\sigma_L(Z)}$ is contained in $\mathcal{E}(\sigma_S(T) \cup \overline{\sigma_L(T)}, \varepsilon)$ and because

$$\lim_{s \to \infty} \|(s\mathcal{I} - \overline{Z})^{-1}\| = \lim_{s \to \infty} \|(s\mathcal{I} - \overline{T})^{-1}\| = 0.$$

Observe that $s \in \rho_S(T)$ implies that the map $s \mapsto \|S_L^{-1}(s,T)\|$ is continuous and

$$\lim_{s \to \infty} \|S_L^{-1}(s,T)\| = \lim_{s \to \infty} \|(T^2 - 2\mathrm{Re}[s]T + |s|^2\mathcal{I})^{-1}(T - \overline{s}\mathcal{I})\| = 0,$$

and so, for s in the complement set of $\mathcal{E}(\sigma_S(T), \varepsilon)$, we have that there exists a positive constant N_ε such that

$$\|S_L^{-1}(s,T)\| \leq N_\varepsilon.$$

From Lemma 4.14.12 if $\delta_1 > 0$ is such that

$$\|Z - T\| < \frac{1}{K_\varepsilon N_\varepsilon} := \delta_1,$$

where K_ε is defined in (4.85), then $s \in \rho_S(Z)$ and

$$\begin{aligned} \|S_L^{-1}(s,Z) - S_L^{-1}(s,T)\| &\leq \frac{\|S_L^{-1}(s,T)\|^2 \; \|S_L(s,T) - S_L(s,Z)\|}{1 - \|S_L^{-1}(s,T)\| \; \|S_L(s,T) - S_L(s,Z)\|} \\ &\leq \frac{N_\varepsilon^2 K_\varepsilon \|Z - T\|}{1 - N_\varepsilon K \|Z - T\|} < \varepsilon \end{aligned}$$

if we take

$$\|Z - T\| < \delta_2 := \frac{\varepsilon}{K_\varepsilon(N_\varepsilon^2 + \varepsilon N_\varepsilon)}.$$

To get the statement it suffices to set $\delta = \min\{\eta, \delta_1, \delta_2\}$. So we have shown (4.82) and (4.83). To prove (4.84) we reason in a similar way. □

Theorem 4.14.14. *Let $T, Z \in \mathcal{B}(V)$, $f \in \mathcal{R}^L_{\sigma_S(T)}$ and let $\varepsilon > 0$. Then there exists $\delta > 0$ such that, for $\|Z - T\| < \delta$, we have $f \in \mathcal{R}^L_{\sigma_S(Z)}$ and*

$$\|f(Z) - f(T)\| < \varepsilon,$$

where

$$f(T) = \frac{1}{2\pi} \int_{\partial(U \cap L_I)} S_L^{-1}(s, T)\, ds_I\, f(s)$$

and $U \subset \mathbb{H}$ is a T-admissible domain, $ds_I = ds/I$ for $I \in \mathbb{S}$.

Proof. Suppose that U is an ε-neighborhood of $\sigma_S(T) \cup \overline{\sigma_L(T)}$ that is contained in the domain in which f is left s-regular. By Lemma 4.14.13 there is a $\delta_1 > 0$ such that $\sigma_S(Z) \subset U$ for $\|Z - T\| < \delta_1$. Consequently $f \in \mathcal{R}^L_{\sigma_S(Z)}$ for $\|Z - T\| < \delta_1$. By Lemma 4.14.13, the operator $S_L^{-1}(s, T)$ is uniformly near to $S_L^{-1}(s, Z)$ with respect to $s \in \partial(U \cap L_I)$ for $I \in \mathbb{S}$ if $\|Z - T\|$ is small enough, so for some positive $\delta \leq \delta_1$ we get

$$\|f(T) - f(Z)\| = \frac{1}{2\pi} \| \int_{\partial(U \cap L_I)} [S_L^{-1}(s, T) - S_L^{-1}(s, Z)]\, ds_I\, f(s)\| < \varepsilon. \qquad \square$$

Remark 4.14.15. Theorem 4.14.14 can be stated and proved also when $f \in \mathcal{R}^R_{\sigma_S(T)}$ with minor changes in the proof.

4.15 Linear closed quaternionic operators

Let V be a two-sided quaternionic Banach space. In analogy with the complex case, we say that a linear operator, whose domain is a linear manifold $\mathcal{D}(T)$, is said to be closed if its graph is closed. For the powers of an operator T, we have

$$\mathcal{D}(T^n) = \{v \ : \ v \in \mathcal{D}(T^{n-1}), \ \ T^{n-1} v \in \mathcal{D}(T) \ \}.$$

A quaternionic linear operator T can be written in the form

$$T = T_0 + iT_1 + jT_2 + kT_3.$$

Recalling the definitions in Section 4.6, the operators T_ℓ, for $\ell = 0, \dots, 3$, are given by the rules:

$$T_0 = +\frac{1}{4}(T - iTi - jTj - kTk),$$
$$T_1 = -\frac{1}{4}(iT + Ti - jTk + kTj),$$
$$T_2 = -\frac{1}{4}(jT + Tj - kTi + iTk),$$
$$T_3 = -\frac{1}{4}(kT + Tk - iTj + jTi).$$

Moreover we have $[T_\ell, i\mathcal{I}] = [T_\ell, j\mathcal{I}] = [T_\ell, k\mathcal{I}] = 0$, $\ell = 0, \dots, 3$, where $[\cdot,\cdot]$ denotes the commutator. These properties justify the fact that we will refer to T_ℓ, $\ell = 0, \dots, 3$, as the formal real components of the operator T. As we have already pointed out, if we consider a right regular polynomial $P_m(q) = \sum_{j=0}^m a_j q^j$, where $a_j \in \mathbb{H}$ for $j = 0, \dots, n$ of degree $m \in \mathbb{N}$, the right (resp. left) linear quaternionic operator

$$P_m(T) = \sum_{j=0}^{m} a_j T^j : \mathcal{D}(T^m) \to V$$

is obtained replacing q by a right (resp. left) linear quaternionic operator T. Analogously, if we consider a left regular polynomial of degree $m \in \mathbb{N}$, $P_m(q) = \sum_{j=0}^m q^j a_j$, where $a_j \in \mathbb{H}$ for $j = 0, \dots, n$ the right (resp. left) linear quaternionic operator

$$P_m(T) = \sum_{j=0}^{m} T^j a_j : \mathcal{D}(T^m) \to V$$

is obtained replacing q by a right (resp. left) linear quaternionic operator T. Let $T = T_0 + iT_1 + jT_2 + kT_3$ where $T_\ell : \mathcal{D}(T_\ell) \to V$, $\ell = 0, 1, 2, 3$ are linear operators and $\mathcal{D}(T_\ell)$ denotes the domain of T_ℓ. The domain of T is defined as $\mathcal{D}(T) = \bigcap_{\ell=0}^3 \mathcal{D}(T_\ell)$. When at least one of the T_ℓ's is an unbounded operator, we define the extended S-spectrum of T as

$$\overline{\sigma}_S(T) := \sigma_S(T) \cup \{\infty\}.$$

Let us consider $\overline{\mathbb{H}} = \mathbb{H} \cup \{\infty\}$ endowed with the natural topology. Precisely, a set is open if and only if it is the union of open discs $D(q, r)$ with center at points in $q \in \mathbb{H}$ and radius r, for some r, and/or the union of sets of the form $D'(\infty, r) \cup \{\infty\}$, for some r, where $D'(\infty, r) = \{q \in \mathbb{H} \mid |q| > r\}$.

We recall that $f(q)$ is an s-regular function at ∞ if $f(q)$ is an s-regular function in a set $D'(\infty, r)$ and $\lim_{q \to \infty} f(q)$ exists and it is finite. We define $f(\infty)$ to be the value of this limit.

Remark 4.15.1. If T is a right (resp. left) linear and bounded quaternionic operator, then $\sigma_S(T)$ is a compact nonempty set, but for unbounded operators, as in the classical case, the S-spectrum $\sigma_S(T)$ can be empty, bounded or unbounded and it can also be $\sigma_S(T) = \mathbb{H}$. In the sequel, we will assume that the S-resolvent set $\rho_S(T)$ is nonempty.

Definition 4.15.2. *Let V be a two-sided quaternionic Banach space.*

(i) *We denote by $\mathcal{K}^R(V)$ the set of right linear closed operators $T : \mathcal{D}(T) \subset V \to V$, such that*

   (1) *$\mathcal{D}(T)$ is dense in V,*

   (2) *$\mathcal{D}(T^2) \subset \mathcal{D}(T)$ is dense in V,*

(3) *$T - \overline{s}\mathcal{I}$ is densely defined in V.*

(ii) *We denote by $\mathcal{K}^L(V)$ the set of left linear closed operators satisfying* (1) *and* (2) *and such that $T - \mathcal{I}\overline{s}$ is densely defined in V.*

(iii) *We use the symbol $\mathcal{K}(V)$ when we do not distinguish between $\mathcal{K}^L(V)$ and $\mathcal{K}^R(V)$.*

Since T is a closed operator, then $T^2 - 2T\text{Re}[s] + |s|^2\mathcal{I} : \mathcal{D}(T^2) \subset V \to V$ is a closed operator. In analogy with the case of bounded operators, we define the S-spectrum and the S-resolvent sets of T.

Definition 4.15.3. *Let V be a two-sided quaternionic Banach space and let $T \in \mathcal{K}(V)$. We denote by $\rho_S(T)$ the S-resolvent set of T as*

$$\rho_S(T) = \{s \in \mathbb{H} \ : \ (T^2 - 2T\text{Re}[s] + |s|^2\mathcal{I})^{-1} \in \mathcal{B}(V) \ \}.$$

We define the S-spectrum $\sigma_S(T)$ of T as

$$\sigma_S(T) = \mathbb{H} \setminus \rho_S(T).$$

In the sequel, we will use the notation introduced in the following definition:

Definition 4.15.4. *Let $T \in \mathcal{K}(V)$ and $s \in \rho_S(T)$. We denote by $Q_s(T)$ the operator:*

$$Q_s(T) := (T^2 - 2T\text{Re}[s] + |s|^2\mathcal{I})^{-1} : \quad V \to \mathcal{D}(T^2). \tag{4.86}$$

Remark 4.15.5. The S-resolvent operators S_L^{-1}, S_R^{-1} present a deep difference when considering the case of left or of right linear operators. Consider, for $s \in \rho_S(T)$, the left S-resolvent operator used in the bounded case, that is:

$$S_L^{-1}(s,T) = -Q_s(T)(T - \overline{s}\mathcal{I}), \tag{4.87}$$

and observe that in the case of right linear unbounded operators, this resolvent turns out to be defined only on $\mathcal{D}(T)$ while in the case of left linear unbounded operators it is defined on all of V . This fact is due to the presence of the term $Q_s(T)T$. However, for $T \in \mathcal{K}^R(V)$, observe that the operator $Q_s(T)T$ is the restriction to the dense subspace $\mathcal{D}(T)$ of V of a bounded linear operator defined on V. This fact follows by the commutation relation $Q_s(T)Tv = TQ_s(T)v$ which holds for all $v \in \mathcal{D}(T)$ since the polynomial operator

$$T^2 - 2\text{Re}[s]T + |s|^2\mathcal{I} : \quad \mathcal{D}(T^2) \to V$$

has real coefficients. More precisely, for $T \in \mathcal{K}^R(V)$, we have

$$TQ_s(T) : \quad V \to \mathcal{D}(T)$$

and this operator is continuous for $s \in \rho_S(T)$.

Definition 4.15.6 (The S-resolvent operators for unbounded right linear operators). *Let V be a two-sided quaternionic Banach space, let $T \in \mathcal{K}^R(V)$ and $s \in \rho_S(T)$. We define the left S-resolvent operator as*

$$S_L^{-1}(s,T)v := -Q_s(T)(T - \overline{s}\mathcal{I})v, \quad \textit{for all} \quad v \in \mathcal{D}(T),$$

and we will call

$$\hat{S}_L^{-1}(s,T)v = Q_s(T)\overline{s}v - TQ_s(T)v, \quad \textit{for all} \quad v \in V, \tag{4.88}$$

the extended left S-resolvent operator. We define the right S-resolvent operator as

$$S_R^{-1}(s,T)v := -(T - \mathcal{I}\overline{s})Q_s(T)v, \quad \textit{for all} \quad v \in V. \tag{4.89}$$

Remark 4.15.7. Observe that for $s \in \rho_S(T)$ the operator $Q_s(T) : V \to \mathcal{D}(T^2)$ is bounded and so also

$$S_R^{-1}(s,T) = -(T - \mathcal{I}\overline{s})Q_s(T) : V \to \mathcal{D}(T)$$

is bounded.

Definition 4.15.8 (The S-resolvent operators for unbounded left linear operators). *Let V be a two-sided quaternionic Banach space, let $T \in \mathcal{K}^L(V)$ and $s \in \rho_S(T)$. We define the left S-resolvent operator as*

$$vS_L^{-1}(s,T) := -vQ_s(T)(T - \overline{s}\mathcal{I}), \quad \textit{for all} \quad v \in V. \tag{4.90}$$

We define the right S-resolvent operator as

$$vS_R^{-1}(s,T) := -v(T - \mathcal{I}\overline{s})Q_s(T), \quad \textit{for all} \quad v \in \mathcal{D}(T), \tag{4.91}$$

and we will call

$$v\hat{S}_R^{-1}(s,T) = vQ_s(T)\overline{s} - vQ_s(T)T, \quad \textit{for all} \quad v \in V, \tag{4.92}$$

the extended right S-resolvent operator.

This motivates the following definition.

Definition 4.15.9. *Let A be an operator containing the term $Q_s(T)T$ (resp. $TQ_s(T)$). We define $\hat{A}$ to be the operator obtained from A by substituting each occurrence of $Q_s(T)T$ (resp. $TQ_s(T)$) by $TQ_s(T)$ (resp. $Q_s(T)T$).*

A second difference between the left and the right functional calculus are the S-resolvent equations which, in order to hold on V, need different extensions of the operators involved.

Theorem 4.15.10 (The S-resolvent equations). *Let V be a two-sided quaternionic Banach space.*

(i) *If $T \in \mathcal{K}^R(V)$ and $s \in \rho_S(T)$, then the left S-resolvent operator satisfies the equations*

$$S_L^{-1}(s,T)sv - TS_L^{-1}(s,T)v = \mathcal{I}v, \quad \text{for all} \quad v \in \mathcal{D}(T), \tag{4.93}$$

$$\hat{S}_L^{-1}(s,T)sv - T\hat{S}_L^{-1}(s,T)v = \mathcal{I}v, \quad \text{for all} \quad v \in V. \tag{4.94}$$

Moreover, the right S-resolvent operator satisfies the equations

$$sS_R^{-1}(s,T)v - S_R^{-1}(s,T)Tv = \mathcal{I}v, \quad \text{for all} \quad v \in \mathcal{D}(T), \tag{4.95}$$

$$sS_R^{-1}(s,T)v - (\widehat{S_R^{-1}(s,T)T})v = \mathcal{I}v, \quad \text{for all} \quad v \in V. \tag{4.96}$$

(ii) *If $T \in \mathcal{K}^L(V)$ and $s \in \rho_S(T)$, then the left S-resolvent operator satisfies the equation*

$$vS_L^{-1}(s,T)s - vTS_L^{-1}(s,T) = v\mathcal{I}, \quad \text{for all} \quad v \in \mathcal{D}(T), \tag{4.97}$$

$$v\hat{S}_L^{-1}(s,T)s - \widehat{vTS_L^{-1}}(s,T) = v\mathcal{I}, \quad \text{for all} \quad v \in V. \tag{4.98}$$

Moreover, the right S-resolvent operator satisfies the equations

$$vsS_R^{-1}(s,T) - vS_R^{-1}(s,T)T = v\mathcal{I}, \quad \text{for all} \quad v \in \mathcal{D}(T), \tag{4.99}$$

$$vs\hat{S}_R^{-1}(s,T) - v(\hat{S}_R^{-1}(s,T)T) = v\mathcal{I}, \quad \text{for all} \quad v \in V. \tag{4.100}$$

Proof. To prove (4.93) we consider its left-hand side where we replace $S_L^{-1}(s,T)$ by $-Q_s(T)(T - \bar{s}\mathcal{I})$ and we obtain, for $v \in \mathcal{D}(T)$:

$$\begin{aligned}
&- Q_s(T)Tsv + Q_s(T)\bar{s}sv + TQ_s(T)Tv - TQ_s(T)\bar{s}v \\
&= Q_s(T)|s|^2v - 2s_0TQ_s(T)v + T^2Q_s(T)v \\
&= (|s|^2\mathcal{I} - 2s_0T + T^2)Q_s(T)v = \mathcal{I}v.
\end{aligned}$$

Equation (4.94) can be verified as

$$[Q_s(T)\bar{s} - TQ_s(T)]sv - T[Q_s(T)\bar{s} - TQ_s(T)]v = \mathcal{I}v, \quad \text{for all } v \in V.$$

Observe that $T[Q_s(T)\bar{s} - TQ_s(T)]v \in V$ since $Q_s(T) : V \to \mathcal{D}(T^2)$ and by trivial computations we get the identity

$$(T^2 - 2T\mathrm{Re}[s] + |s|^2\mathcal{I})Q_s(T)v = v, \quad \text{for all } v \in V,$$

which proves the statement. Equations (4.95)–(4.100) can be verified in the same way with obvious meaning of the symbols. □

In the classical case of a complex unbounded linear operator

$$B : \mathcal{D}(B) \subset X \to X,$$

where X is a complex Banach space, the resolvent operator

$$R(\lambda, B) := (\lambda\mathcal{I} - B)^{-1}, \quad \text{for} \quad \lambda \in \rho(B),$$

satisfies the following relations:

$$(\lambda\mathcal{I} - B)R(\lambda, B)x = x, \quad \text{for all} \quad x \in X,$$
$$R(\lambda, B)(\lambda\mathcal{I} - B)x = x, \quad \text{for all} \quad x \in \mathcal{D}(B).$$

We study what happens in the quaternionic case for unbounded operators. The analog of $\lambda\mathcal{I} - B$, associated to the left S-resolvent operator, is defined by

$$S_L(s,T) = (T - \overline{s}\mathcal{I})^{-1}\, s\,(T - \overline{s}\mathcal{I}) - T$$

for those $\overline{s} \in \mathbb{H}$ such that $(T - \overline{s}\mathcal{I})^{-1}$ is a bounded operator. Observe that for the operator $S_L(s,T)$ the following identity

$$(T - \overline{s}\mathcal{I})^{-1}\, s\,(T - \overline{s}\mathcal{I}) - T = -(T - \overline{s}\mathcal{I})^{-1}(T^2 - 2s_0T + |s|^2\mathcal{I}) \tag{4.101}$$

holds for bounded operators. Suppose now that $T \in \mathcal{K}^R(V)$. It is easy to see that the left-hand side of (4.101) is defined on $\mathcal{D}(T)$ while the right-hand side of (4.101) is defined on $\mathcal{D}(T^2)$. This fact motivates the following definition.

Definition 4.15.11. *Let V be a two-sided quaternionic Banach space. Take $\overline{s} \in \mathbb{H}$ such that $(T - \overline{s}\mathcal{I})^{-1}$ is a bounded operator.*

(i) *Let $T \in \mathcal{K}^R(V)$. Then we define*

$$S_L(s,T)v := -(T - \overline{s}\mathcal{I})^{-1}(T^2 - 2s_0T + |s|^2\mathcal{I})v : \quad v \in \mathcal{D}(T^2),$$
$$\hat{S}_L(s,T)v := [(T - \overline{s}\mathcal{I})^{-1}\, s\,(T - \overline{s}\mathcal{I}) - T]v : \quad v \in \mathcal{D}(T),$$

where, with an abuse of notation, we have denoted by $\hat{S}_L(s,T)$ the extension of $S_L(s,T)$ on $\mathcal{D}(T)$. Moreover, we set

$$S_R(s,T)v := [(T - \mathcal{I}\overline{s})\, s\,(T - \mathcal{I}\overline{s})^{-1} - T]v$$
$$\Big(= -(T^2 - 2s_0T + |s|^2\mathcal{I})(T - \overline{s}\mathcal{I})^{-1}v \Big), \quad v \in \mathcal{D}(T).$$

(ii) *Let $T \in \mathcal{K}^L(V)$. Then we define*

$$vS_L(s,T) := v[(T - \overline{s}\mathcal{I})^{-1}\, s\,(T - \overline{s}\mathcal{I}) - T]$$
$$\Big(= -v(T - \overline{s}\mathcal{I})^{-1}(T^2 - 2s_0T + |s|^2\mathcal{I}) \Big), \quad v \in \mathcal{D}(T).$$

Moreover, we set

$$vS_R(s,T) := -v(T^2 - 2s_0T + |s|^2\mathcal{I})(T - \overline{s}\mathcal{I})^{-1}, \quad v \in \mathcal{D}(T^2),$$
$$v\hat{S}_R(s,T) := v[(T - \mathcal{I}\overline{s})\, s\,(T - \mathcal{I}\overline{s})^{-1} - T], \quad v \in \mathcal{D}(T),$$

where, with an abuse of notation, we have denoted by $\hat{S}_R(s,T)$ the extension of $S_R(s,T)$ on $\mathcal{D}(T)$.

The abuse of notation in the previous definition is motivated by the following result:

Theorem 4.15.12. *Let V be a two-sided quaternionic Banach space. Take $\overline{s} \in \mathbb{H}$ such that $(T - \overline{s}\mathcal{I})^{-1}$ is a bounded operator and $s \in \rho_S(T)$.*

(i) *Let $T \in \mathcal{K}^R(V)$. Then we have*

$$\hat{S}_L(s,T)\hat{S}_L^{-1}(s,T)v = \mathcal{I}v, \quad \textit{for all} \quad v \in V, \tag{4.102}$$

$$\hat{S}_L^{-1}(s,T)\hat{S}_L(s,T)v = \mathcal{I}v, \quad \textit{for all} \quad v \in \mathcal{D}(T), \tag{4.103}$$

and

$$S_R(s,T)S_R^{-1}(s,T)v = \mathcal{I}v, \quad \textit{for all} \quad v \in V, \tag{4.104}$$

$$S_R^{-1}(s,T)S_R(s,T)v = \mathcal{I}v, \quad \textit{for all} \quad v \in \mathcal{D}(T). \tag{4.105}$$

(ii) *Let $T \in \mathcal{K}^L(V)$. Then we have*

$$vS_L(s,T)S_L^{-1}(s,T) = v\mathcal{I}, \quad \textit{for all} \quad v \in V, \tag{4.106}$$

$$vS_L^{-1}(s,T)S_L(s,T) = v\mathcal{I}, \quad \textit{for all} \quad v \in \mathcal{D}(T), \tag{4.107}$$

and

$$v\hat{S}_R(s,T)\hat{S}_R^{-1}(s,T) = v\mathcal{I}, \quad \textit{for all} \quad v \in V, \tag{4.108}$$

$$v\hat{S}_R^{-1}(s,T)\hat{S}_R(s,T) = v\mathcal{I}, \quad \textit{for all} \quad v \in \mathcal{D}(T). \tag{4.109}$$

Proof. Let us verify that (4.102) holds, i.e.,

$$[(T - \overline{s}\mathcal{I})^{-1}s(T - \overline{s}\mathcal{I}) - T]\hat{S}_L^{-1}(s,T)v = \mathcal{I}v, \quad \text{for all } v \in V,$$

from which we get

$$\begin{aligned}&(T - \overline{s}\mathcal{I})^{-1}[sT\hat{S}_L^{-1}(s,T)v - |s|^2\hat{S}_L^{-1}(s,T)v]\\ &= T\hat{S}_L^{-1}(s,T)v + \mathcal{I}v, \quad \text{for all } v \in V;\end{aligned}$$

using (4.94) on the right-hand side we obtain

$$\begin{aligned}&(T - \overline{s}\mathcal{I})^{-1}[s\hat{S}_L^{-1}(s,T)sv - sv - |s|^2\hat{S}_L^{-1}(s,T)v]\\ &= \hat{S}_L^{-1}(s,T)sv, \quad \text{for all } v \in V\end{aligned}$$

and

$$\begin{aligned}&(T - \overline{s}\mathcal{I})^{-1}[s\hat{S}_L^{-1}(s,T)sv - sv - \hat{S}_L^{-1}(s,T)\overline{s}sv]\\ &= \hat{S}_L^{-1}(s,T)sv, \quad \text{for all } v \in V\\ &(T - \overline{s}\mathcal{I})^{-1}[s\hat{S}_L^{-1}(s,T) - \mathcal{I} - \hat{S}_L^{-1}(s,T)\overline{s}]sv\\ &= \hat{S}_L^{-1}(s,T)sv, \quad \text{for all } v \in V;\end{aligned}$$

by (4.94) we have:

$$(T - \overline{s}\mathcal{I})^{-1}[s\hat{S}_L^{-1}(s,T) - (\hat{S}_L^{-1}(s,T)s - T\hat{S}_L^{-1}(s,T)) - \hat{S}_L^{-1}(s,T)\overline{s}]sv$$
$$= \hat{S}_L^{-1}(s,T)sv, \quad \text{for all } v \in V$$

from which we get

$$(T - \overline{s}\mathcal{I})^{-1}[T\hat{S}_L^{-1}(s,T) - \overline{s}\hat{S}_L^{-1}(s,T)]sv = \hat{S}_L^{-1}(s,T)sv, \quad \text{for all } v \in V,$$

that is

$$(T - \overline{s}\mathcal{I})^{-1}(T - \overline{s}\mathcal{I})\hat{S}_L^{-1}(s,T)(sv) = \hat{S}_L^{-1}(s,T)(sv), \quad \text{for all } v \in V,$$

which proves (4.102). To verify (4.103) observe that from (4.102) we get

$$(\hat{S}_L^{-1}(s,T)S_L(s,T))\hat{S}_L^{-1}(s,T)v = \hat{S}_L^{-1}(s,T)v, \quad \text{for all} \quad v \in V,$$

but since $\hat{S}_L^{-1}(s,T)v \in \mathcal{D}(T)$ for $v \in V$ we have that

$$\hat{S}_L^{-1}(s,T)S_L(s,T)w = \mathcal{I}w \quad \text{for all } w \in \mathcal{D}(T).$$

The proofs of (4.104)–(4.109) can be treated with analogous considerations. □

Remark 4.15.13. Let $T \in \mathcal{K}^R(V)$. Take $\overline{s} \in \mathbb{H}$ such that $(T - \overline{s}\mathcal{I})^{-1}$ is a bounded operator and $s \in \rho_S(T)$. Then it is easy to show that

$$S_L(s,T)S_L^{-1}(s,T)v = \mathcal{I}v, \quad \text{for all} \quad v \in \mathcal{D}(T), \tag{4.110}$$
$$S_L^{-1}(s,T)S_L(s,T)v = \mathcal{I}v, \quad \text{for all} \quad v \in \mathcal{D}(T^2). \tag{4.111}$$

Similar considerations can be applied for (4.108)–(4.109).

4.16 The functional calculus for unbounded operators

Definition 4.16.1. *Let $T \in \mathcal{K}(V)$. A function f is said to be locally left (resp. right) s-regular on $\overline{\sigma}_S(T)$ if there exists a T-admissible open set U such that f is left (resp. right) s-regular on U and at infinity. We will denote by $\mathcal{R}^L_{\overline{\sigma}_S(T)}$ (resp. $\mathcal{R}^R_{\overline{\sigma}_S(T)}$) the set of locally left (resp. right) s-regular functions on $\overline{\sigma}_S(T)$.*

Remark 4.16.2. As we have pointed out in Remark 4.8.9, the open set U related to $f \in \mathcal{R}^L_{\overline{\sigma}_S(T)}$ (resp. $\mathcal{R}^R_{\overline{\sigma}_S(T)}$) need not be connected. Moreover, as in the classical functional calculus, U in general depends on f and can be unbounded.

Definition 4.16.3. *Consider $k \in \mathbb{R}$ and the function $\Phi : \overline{\mathbb{H}} \to \overline{\mathbb{H}}$ defined by $p = \Phi(s) = (s-k)^{-1}$, $\Phi(\infty) = 0$, $\Phi(k) = \infty$.*

Definition 4.16.4. *Let $T \in \mathcal{K}(V)$ with $\rho_S(T) \cap \mathbb{R} \neq \emptyset$ and suppose that $f \in \mathcal{R}^L_{\overline{\sigma}_S(T)}$ (resp. $f \in \mathcal{R}^R_{\overline{\sigma}_S(T)}$). Let us consider*

$$\phi(p) := f(\Phi^{-1}(p))$$

and the bounded linear operator defined by

$$A := (T - k\mathcal{I})^{-1}, \quad \textit{for some} \quad k \in \rho_S(T) \cap \mathbb{R}.$$

We define, in both cases, the operator $f(T)$ as

$$f(T) = \phi(A). \tag{4.112}$$

Remark 4.16.5. Consider Φ, ϕ and k as above. Then:

(1) *The function $\Phi^{-1}(p) = p^{-1} + k$ has real coefficients so it is both left and right regular. So if f is left regular, then the function $\phi = f(\Phi^{-1}(p))$ is left regular, while if f is right regular, then the function $\phi = f(\Phi^{-1}(p))$ is right regular by Theorem 4.13.1.*

(2) *If $k \in \rho_S(T) \cap \mathbb{R}$ and $T \in \mathcal{K}^R(V)$, then*

$$(T - k\mathcal{I})^{-1}v = -\hat{S}_L^{-1}(k, T)v = -S_R^{-1}(k, T)v, \quad \text{for all} \quad v \in V.$$

(3) *If $k \in \rho_S(T) \cap \mathbb{R}$ and $T \in \mathcal{K}^L(V)$, then*

$$v(T - k\mathcal{I})^{-1} = -v\hat{S}_R^{-1}(k, T) = -vS_L^{-1}(k, T), \quad \text{for all} \quad v \in V.$$

Theorem 4.16.6. *Let $k \in \rho_S(T) \cap \mathbb{R} \neq \emptyset$ and Φ, ϕ are as above.*

(i) *Let $T \in \mathcal{K}^R(V)$, then $\Phi(\overline{\sigma}_S(T)) = \sigma_S(A)$ and $\phi(p) = f(\Phi^{-1}(p))$ determines a one-to-one correspondence between $f \in \mathcal{R}_{\overline{\sigma}_S(T)}$ and $\phi \in \mathcal{R}_{\overline{\sigma}_S(A)}$. Moreover we have*

$$\hat{S}_L^{-1}(s, T)v = p\mathcal{I}v - S_L^{-1}(p, A)p^2 v, \quad v \in V, \tag{4.113}$$

and

$$S_R^{-1}(s, T)v = p\mathcal{I}v - p^2 S_R^{-1}(p, A)v, \quad v \in V. \tag{4.114}$$

(ii) *Let $T \in \mathcal{K}^L(V)$, then $\Phi(\overline{\sigma}_S(T)) = \sigma_S(A)$ and $\phi(p) = f(\Phi^{-1}(p))$ determines a one-to-one correspondence between $f \in \mathcal{R}_{\overline{\sigma}_S(T)}$ and $\phi \in \mathcal{R}_{\overline{\sigma}_S(A)}$. Moreover we have*

$$vS_L^{-1}(s, T) = vp\mathcal{I} - vS_L^{-1}(p, A)p^2, \quad v \in V, \tag{4.115}$$

and

$$v\hat{S}_R^{-1}(s, T) = vp\mathcal{I} - vp^2 S_R^{-1}(p, A), \quad v \in V. \tag{4.116}$$

Proof. Let s, $p \in \mathbb{H}$ and $k \in \mathbb{R}$ such that $p = (s-k)^{-1}$. Then the identities

$$s_0|p|^2 = k|p|^2 + p_0, \tag{4.117}$$

$$|p|^2|s|^2 = k^2|p|^2 + 2p_0k + 1, \tag{4.118}$$

$$(2k\overline{p} - 2s_0\overline{p} + 1)\frac{1}{|p|^2} = -p^{-2}, \tag{4.119}$$

$$\frac{k^2\overline{p} - |s|^2\overline{p} + k}{|p|^2} = -\overline{s}p^{-2} \tag{4.120}$$

can be verified by direct calculations. Let us prove (4.113) in Point (i). We prove that $\Phi(\overline{\sigma}_S(T)) = \sigma_S(A)$. We recall that

$$\hat{S}_L^{-1}(s,T) = Q_s(T)\overline{s} - TQ_s(T) : V \to \mathcal{D}(T) \quad \text{for all } s \in \rho_S(T).$$

From the definition of A we also have, for $k \in \rho_S(T) \cap \mathbb{R} \neq 0$,

$$A := (T - k\mathcal{I})^{-1} : V \to \mathcal{D}(T), \qquad \text{and} \quad A^{-1} = T - k\mathcal{I} : \mathcal{D}(T) \to V,$$

$$A^2 := (T^2 - 2kT + k^2\mathcal{I})^{-2} : V \to \mathcal{D}(T^2),$$

and

$$A^{-2} = T^2 - 2kT + k^2\mathcal{I} : \mathcal{D}(T^2) \to V.$$

Observe that, for $p \in \rho_S(A)$,

$$Q_p(A) := (A^2 - 2p_0A + |p|^2\mathcal{I})^{-1} \in \mathcal{B}(V)$$

and

$$S_L^{-1}(p,A) = Q_p(A)\overline{p} - AQ_p(A).$$

Let us consider the relation

$$\begin{aligned}
Q_p(A) &= \Big[(T-k\mathcal{I})^{-2} - 2p_0(T-k\mathcal{I})^{-1} + |p|^2\mathcal{I}\Big]^{-1} \\
&= \Big[[\mathcal{I} - 2p_0(T-k\mathcal{I}) + |p|^2(T-k\mathcal{I})^2](T-k\mathcal{I})^{-2}\Big]^{-1} \\
&= (T-k\mathcal{I})^2[\mathcal{I} - 2p_0(T-k\mathcal{I}) + |p|^2(T-k\mathcal{I})^2]^{-1} \\
&= |p|^{-2}(T-k\mathcal{I})^2[T^2 - 2(k + p_0|p|^{-2})T + (k^2|p|^2 + 2p_0k + 1)|p|^{-2}\mathcal{I}]^{-1};
\end{aligned}$$

for (4.117) and (4.118) we get

$$Q_p(A) = |p|^{-2}(T-k\mathcal{I})^2[T^2 - 2s_0T + |s|^2\mathcal{I}]^{-1} : \ V \to V,$$

that is

$$Q_p(A) = |p|^{-2}(T-k\mathcal{I})^2Q_s(T). \tag{4.121}$$

Since A is a bounded operator, then

$$S^{-1}(p,A) = Q_p(A)\overline{p} - AQ_s(A) : V \to V,$$

so we have

$$\begin{aligned} S^{-1}(p,A) &= |p|^{-2}(T-k\mathcal{I})^2 Q_s(T)\overline{p} - |p|^{-2}(T-k\mathcal{I})Q_s(T) \\ &= |p|^{-2}\Big[(T^2 - 2kT + k^2\mathcal{I})Q_s(T)\overline{p} - (T-k)Q_s(T)\Big] \\ &= |p|^{-2}\Big[(T^2 - 2s_0T + |s|^2\mathcal{I})Q_s(T)\overline{p} \\ &\quad + (-2kT + k^2 + 2s_0T - |s|^2)Q_s(T)\overline{p} - (T-k\mathcal{I})Q_s(T)\Big] \\ &= |p|^{-2}\Big[\mathcal{I}\overline{p} + Q_s(T)[k^2\overline{p} - |s|^2\overline{p} + k] - TQ_s(T)[2k\overline{p} - 2s_0\overline{p} + 1]\Big] \\ &= \Big[\mathcal{I}p^{-1} + Q_s(T)\frac{k^2\overline{p} - |s|^2\overline{p} + k}{|p|^2} - TQ_s(T)\frac{2k\overline{p} - 2s_0\overline{p} + 1}{|p|^2}\Big]. \end{aligned}$$

Now we use the identities (4.119) and (4.120) to get

$$S_L^{-1}(p,A) = \mathcal{I}p^{-1} - Q_s(T)\overline{s}p^{-2} + TQ_s(T)p^{-2}$$

and finally

$$S_L^{-1}(p,A) = \mathcal{I}p^{-1} - \hat{S}_L^{-1}(s,T)p^{-2}. \tag{4.122}$$

So $p \in \rho_S(A)$, $p \neq 0$, then $s \in \rho_S(T)$.

Now take $s \in \rho_S(T)$. We verify that

$$\hat{S}_L^{-1}(s,T) = -AS_L^{-1}(p,A)p$$

holds. Indeed, by (4.121) we get the equalities

$$\begin{aligned} &-AS_L^{-1}(p,A)p = -A[Q_p(A)\overline{p} - AQ_p(A)]p \\ &= -(T-k\mathcal{I})^{-1}\Big[[|p|^{-2}(T-k\mathcal{I})^2Q_s(T)]\overline{p} - (T-k\mathcal{I})^{-1}[|p|^{-2}(T-k\mathcal{I})^2Q_s(T)]\Big]p \\ &= -TQ_s(T) + Q_s(T)(\frac{p}{|p|^2} + k) = \hat{S}_L^{-1}(s,T). \end{aligned}$$

So if $s \in \rho_S(T)$, then $p \in \rho_S(A)$, $p \neq 0$.

The point $p = 0$ belongs to $\sigma_S(A)$ since $S_L^{-1}(0,A) = A^{-1} = T - k\mathcal{I}$ is unbounded.

The fact that $\phi(p) = f(\Phi^{-1}(p))$ determines a one-to-one correspondence between $f \in \mathcal{R}_{\overline{\sigma}_S(T)}$ and $\phi \in \mathcal{R}_{\overline{\sigma}_S(A)}$, as is evident from the definition of Φ.

It remains to prove relation (4.114). We recall that

$$S_R^{-1}(s,T) = -(T - \overline{s}\mathcal{I})Q_s(T) : V \to \mathcal{D}(T) \quad \text{for all } s \in \rho_S(T).$$

From the definition of A, which is a bounded operator, when $k \in \rho_S(T) \cap \mathbb{R} \neq 0$ we have

$$A := (T - k\mathcal{I})^{-1} : V \to \mathcal{D}(T), \qquad A^{-1} = T - k\mathcal{I} : \mathcal{D}(T) \to V,$$

$$A^2 := (T^2 - 2kT + k^2\mathcal{I})^{-1} : \quad V \to \mathcal{D}(T^2),$$

and

$$A^{-2} = T^2 - 2kT + k^2\mathcal{I} : \quad \mathcal{D}(T^2) \to V.$$

Observe that, for $p \in \rho_S(A)$,

$$Q_p(A) := (A^2 - 2p_0A + |p|^2\mathcal{I})^{-1} \in \mathcal{B}^R(V).$$

By the relation between $Q_p(A)$ and $Q_s(T)$ in (4.121) and since A is a bounded operator, we get $S_R^{-1}(p, A) = -(A - \overline{p}\mathcal{I})Q_s(A) : V \to V$; therefore, using (4.121), we have

$$\begin{aligned}
&S_R^{-1}(p, A) = -[(T - k\mathcal{I})^{-1} - \overline{p}\mathcal{I}]\, |p|^{-2}\, (T - k\mathcal{I})^2 Q_s(T) \\
&= -|p|^{-2}\Big[T - k\mathcal{I} - \overline{p}(T - k\mathcal{I})^2\Big] Q_s(T) \\
&= -|p|^{-2}\overline{p}\Big[\overline{p}^{\,-1}T - \overline{p}^{\,-1}k\mathcal{I} - T^2 + 2kT - k^2\mathcal{I}\Big] Q_s(T) \\
&= p^{-1}\Big[(T^2 - 2s_0T + |s|^2\mathcal{I}) + (2s_0 - 2k - \overline{p}^{\,-1})T + (\overline{p}^{\,-1}k + k^2 - |s|^2)\mathcal{I}\Big] Q_s(T) \\
&= p^{-1}\mathcal{I} + \Big[p^{-1}(2s_0 - 2k - \overline{p}^{\,-1})T - p^{-1}(|s|^2 - k^2 - k\overline{p}^{\,-1})\mathcal{I}\Big] Q_s(T).
\end{aligned}$$

By the identities

$$2s_0 - 2k - \overline{p}^{\,-1} = p^{-1}, \qquad |s|^2 - k^2 - k\overline{p}^{\,-1} = p^{-1}\overline{s}$$

we finally get

$$S_R^{-1}(p, A) = p^{-1}\mathcal{I} - p^{-2}S_R^{-1}(s, T), \tag{4.123}$$

from which we obtain (4.114). So, if $p \in \rho_S(A)$ and $p \neq 0$, then $s \in \rho_S(T)$.

Now take $s \in \rho_S(T)$. We verify that

$$S_R^{-1}(s, T) = -pS_R^{-1}(p, A)A$$

holds. In fact, by (4.121) we get the equalities

$$\begin{aligned}
pS_R^{-1}(p, A)A &= -p(A - \overline{p}\mathcal{I})Q_p(A)A = -p(A - \overline{p}\mathcal{I})AQ_p(A) \\
&= (-pA^2 + |p|^2A)|p|^{-2}(T - k\mathcal{I})^2 Q_s(T) \\
&= (-p(T - k\mathcal{I})^{-2} + |p|^2(T - k\mathcal{I})^{-1})|p|^{-2}(T - k\mathcal{I})^2 Q_s(T) \\
&= \Big[T - \Big(k + p|p|^{-2}\Big)\mathcal{I}\Big] Q_s(T) = -S_R^{-1}(s, T).
\end{aligned}$$

So if $s \in \rho_S(T)$, then $p \in \rho_S(A)$, $p \neq 0$. Thus $p = 0$ belongs to $\sigma_S(A)$ since $S_R^{-1}(0, A) = A^{-1} = T - k\mathcal{I}$ is unbounded.

Point (ii): The fact that $\phi(p) = f(\Phi^{-1}(p))$ determines a one-to-one correspondence between $f \in \mathcal{R}_{\overline{\sigma}_S(T)}$ and $\phi \in \mathcal{R}_{\overline{\sigma}_S(A)}$ follows from the definition of Φ. We can prove the equalities (4.115) and (4.116) with techniques similar to those used to prove (4.113) and (4.114), with obvious different interpretation of the symbols since $T \in \mathcal{K}^L(V)$. □

Theorem 4.16.7. *Let V be a two-sided quaternionic Banach space and let W be a T-admissible open set.*

(i) *Let $T \in \mathcal{K}^R(V)$ with $\rho_S(T) \cap \mathbb{R} \neq \emptyset$. Then the operator $f(T)$ defined in (4.112) is independent of $k \in \rho_S(T) \cap \mathbb{R}$, and, for $f \in \mathcal{R}^L_{\overline{\sigma}_S(T)}$ and $v \in V$, we have*

$$f(T)v = f(\infty)\mathcal{I}v + \frac{1}{2\pi} \int_{\partial(W\cap\mathbb{C}_I)} \hat{S}_L^{-1}(s,T)\ ds_I\ f(s)v, \tag{4.124}$$

and for $f \in \mathcal{R}^R_{\overline{\sigma}_S(T)}$ and $v \in V$, we have

$$f(T)v = f(\infty)\mathcal{I}v + \frac{1}{2\pi} \int_{\partial(W\cap\mathbb{C}_I)} f(s)\ ds_I\ S_R^{-1}(s,T)v. \tag{4.125}$$

(ii) *Let $T \in \mathcal{K}^L(V)$ with $\rho_S(T) \cap \mathbb{R} \neq \emptyset$. Then the operator $f(T)$ defined in (4.112) is independent of $k \in \rho_S(T) \cap \mathbb{R}$, and, for $f \in \mathcal{R}^L_{\overline{\sigma}_S(T)}$ and $v \in V$, we have*

$$vf(T) = vf(\infty)\mathcal{I} + \frac{1}{2\pi} \int_{\partial(W\cap\mathbb{C}_I)} v\, S_L^{-1}(s,T)\ ds_I\ f(s), \tag{4.126}$$

and for $f \in \mathcal{R}^R_{\overline{\sigma}_S(T)}$ and $v \in V$, we have

$$vf(T) = vf(\infty)\mathcal{I} + \frac{1}{2\pi} \int_{\partial(W\cap\mathbb{C}_I)} v\, f(s)\ ds_I\ \hat{S}_R^{-1}(s,T). \tag{4.127}$$

Proof. The fact that the operator $f(T)$ defined in (4.112) is independent of $k \in \rho_S(T) \cap \mathbb{R}$ follows from the validity of formulas (4.124)-(4.127) since the integrals are independent of k.

Consider $k \in \rho_S(T) \cap \mathbb{R}$, and assume that the set W is such that $k \notin \overline{(W \cap \mathbb{C}_I)}$, $\forall I \in \mathbb{S}$. Otherwise, by the Cauchy theorem, we can replace W by W', on which f is regular, such that $k \notin \overline{(W' \cap \mathbb{C}_I)}$, without altering the value of the integral (4.125). Moreover, the integral (4.125) is independent of the choice of $I \in \mathbb{S}$.

We have that $\mathcal{V} \cap \mathbb{C}_I := \Phi^{-1}(W \cap \mathbb{C}_I)$ is an open set that contains $\sigma_S(T)$ and its boundary $\partial(\mathcal{V} \cap \mathbb{C}_I) = \Phi^{-1}(\partial(W \cap \mathbb{C}_I))$ is positively oriented and consists of a finite number of continuously differentiable Jordan curves.

Let us prove formula (4.124) in Point (i). Using the relation (4.113) we have

$$\begin{aligned}
&\frac{1}{2\pi}\int_{\partial(W\cap\mathbb{C}_I)} \hat{S}_L^{-1}(s,T)ds_I f(s)\\
&= -\frac{1}{2\pi}\int_{\partial(\mathcal{V}\cap\mathbb{C}_I)} \left(p\mathcal{I} - S_L^{-1}(p,A)p^2\right)p^{-2}dp_I\phi(p)\\
&= -\frac{1}{2\pi}\int_{\partial(\mathcal{V}\cap\mathbb{C}_I)} p^{-1}dp_I\phi(p) + \frac{1}{2\pi}\int_{\partial(\mathcal{V}\cap\mathbb{C}_I)} S_L^{-1}(p,A)dp_I\phi(p)\\
&= -\mathcal{I}\phi(0) + \phi(A).
\end{aligned}$$

Now by definition $\phi(A) = f(T)$ and $\phi(0) = f(\infty)$ we obtain

$$\frac{1}{2\pi}\int_{\partial(W\cap\mathbb{C}_I)} S^{-1}(s,T)ds_I f(s) = -\mathcal{I}f(\infty) + f(T),$$

so we get (4.124). Now, using the relation (4.114) we have

$$\begin{aligned}
&\frac{1}{2\pi}\int_{\partial(W\cap\mathbb{C}_I)} f(s)\ ds_I\ S_R^{-1}(s,T)v\\
&= -\frac{1}{2\pi}\int_{\partial(\mathcal{V}\cap\mathbb{C}_I)} \phi(p)\ dp_I\ p^{-2}\left(p\mathcal{I} - p^2 S_R^{-1}(p,A)\right)v\\
&= -\frac{1}{2\pi}\int_{\partial(\mathcal{V}\cap\mathbb{C}_I)} \phi(p)\ dp_I\ p^{-1}v + \frac{1}{2\pi}\int_{\partial(\mathcal{V}\cap\mathbb{C}_I)} \phi(p)\ dp_I\ S_R^{-1}(p,A)v\\
&= -\mathcal{I}\phi(0)v + \phi(A)v.
\end{aligned}$$

Now by definition $\phi(A) = f(T)$ and $\phi(0) = f(\infty)$ we obtain

$$\frac{1}{2\pi}\int_{\partial(W\cap\mathbb{C}_I)} f(s)\ ds_I\ S_R^{-1}(s,T)v = -\mathcal{I}f(\infty)v + f(T)v.$$

Formulas (4.126) and (4.127) in point (ii) can be proved following the same arguments with suitable modifications and interpretations of the symbols. □

In the following theorem we show some algebraic properties that can be deduced easily. We state the results for functions $f \in \mathcal{R}^L_{\sigma_S(T)}$, but analogous results hold for $f \in \mathcal{R}^R_{\sigma_S(T)}$.

Theorem 4.16.8. *Let V be a two-sided quaternionic Banach space and let $T \in \mathcal{K}(V)$ with $\rho_S(T)\cap\mathbb{R} \neq \emptyset$. If f and $g \in \mathcal{R}^L_{\sigma_S(T)}$, then*

$$(f+g)(T) = f(T) + g(T).$$

If $g \in \mathcal{R}^L_{\sigma_S(T)}$ and $f \in \mathcal{N}_{\overline{\sigma}_S(T)}$, then

$$(fg)(T) = f(T)g(T).$$

Proof. Observe that $fg \in \mathcal{R}^R_{\sigma_S(T)}$ thanks to Lemma 4.11.5. Let $\phi(\mu) = f(\Phi^{-1}(\mu))$ and $\psi(\mu) = g(\Phi^{-1}(\mu))$. Lemma 4.11.5 and Lemma 4.13.1 give that the product $\phi\psi$ is s-regular. By definition we have

$$f(T) = \phi(A), \qquad g(T) = \psi(A),$$

thus by Theorem 4.11.6 we get

$$(\phi + \psi)(A) = \phi(A) + \psi(A), \qquad (\phi\psi)(A) = \phi(A)\psi(A).$$

The statement follows. □

Theorem 4.16.9. *Let V be a two-sided quaternionic Banach space and let $T \in \mathcal{K}(V)$ with $\rho_S(T) \cap \mathbb{R} \neq \emptyset$. If $f \in \mathcal{N}_{\sigma_S(T)}$, then*

$$\sigma_S(f(T)) = f(\overline{\sigma}_S(T)).$$

Proof. Let $\phi(\mu) = f(\Phi^{-1}(\mu))$. For the S-spectral mapping theorem we have $\phi(\sigma_S(A)) = \sigma_S(\phi(A))$ and for Theorem 4.16.6 we also have $\Phi(\overline{\sigma}_S(T)) = \sigma_S(A)$. So we obtain

$$\phi(\Phi(\overline{\sigma}_S(T)) = \phi(\sigma_S(A)) = \sigma_S(\phi(A)) = \sigma_S(f(T)).$$

On the other hand,

$$\phi(\Phi(\overline{\sigma}_S(T)) = f(\Phi^{-1}(\Phi(\overline{\sigma}_S(T))) = f(\overline{\sigma}_S(T)).$$ □

4.17 An application: uniformly continuous quaternionic semigroups

We generalize to the quaternionic setting the classical result that a semigroup has a bounded infinitesimal generator if and only if it is uniformly continuous. To start with, we recall the definition of uniformly continuous and of strongly continuous semigroups and some preliminary results useful in the sequel. Note that to develop our theory we will make use of the functional calculus based on left regular functions.

Definition 4.17.1. *Let V be a two-sided quaternionic Banach space and $t \in \mathbb{R}$. A family $\{\mathcal{U}(t)\}_{t\geq 0}$ of linear bounded quaternionic operators in V will be called a* strongly continuous quaternionic semigroup *if*

(1) $\mathcal{U}(t+\tau) = \mathcal{U}(t)\mathcal{U}(\tau), \quad t, \tau \geq 0,$

(2) $\mathcal{U}(0) = \mathcal{I},$

(3) *for every $v \in V$, $\mathcal{U}(t)v$ is continuous in $t \in [0, \infty]$.*

If, in addition,

(4) *the map* $t \to \mathcal{U}(t)$ *is continuous in the uniform operator topology,*

then the family $\{\mathcal{U}(t)\}_{t\geq 0}$ *is called a* uniformly continuous quaternionic semigroup *in* $\mathcal{B}(V)$.

From the functional calculus in Definition 4.10.4, it is clear that for any operator $T \in \mathcal{B}(V)$, e^{tT} is a uniformly continuous quaternionic semigroup in $\mathcal{B}(V)$. The following theorem shows that also the converse is true, i.e., every uniformly continuous quaternionic semigroup is of this form.

Theorem 4.17.2. *Let* $\{\mathcal{U}(t)\}_{t\geq 0}$ *be a uniformly continuous quaternionic semigroup in* $\mathcal{B}(V)$. *Then:*

(1) *there exists a bounded linear quaternionic operator* T *such that* $\mathcal{U}(t) = e^{tT}$;

(2) *the quaternionic operator* T *is given by the formula*

$$T = \lim_{h\to 0} \frac{\mathcal{U}(h) - \mathcal{U}(0)}{h};$$

(3) *we have the relation*

$$\frac{d}{dt} e^{tT} = T\, e^{tT} = e^{tT}\, T.$$

Proof. The proof follows the lines of the proof of the analogous result in the classical case. However, since we are working in a noncommutative setting, it is necessary to check that all the computations can be performed over the quaternions. We start by proving (1). Let us consider the logarithmic function $\ln q$ defined on $\mathbb{H} \setminus \{q \in \mathbb{R} \,:\, q \leq 0\}$ by extending the principal branch of the function $\ln q$. Since $\mathcal{U}(0) = I$ whose S-spectrum is reduced to the real point 1, it follows that we can apply the perturbation theorem of the S-resolvent operator, see Theorem 4.14.14, to the operators $\mathcal{U}(0) = I, \mathcal{U}(\delta)$ for a suitable $\delta > 0$, using the function $\ln q$. Thus, there exists $\varepsilon > 0$ such that $P(t) = \ln \mathcal{U}(t)$ is defined and continuous for $t \in [0, \varepsilon]$. If $nt \leq \varepsilon$, then, by the semigroup properties, we have

$$P(nt) = \ln \mathcal{U}(nt) = \ln (\mathcal{U}(t))^n = n\, P(t)$$

thus

$$P(t) = n\, P(t/n) \ \text{ for every } t \in [0, \varepsilon].$$

As a consequence, for each rational number m/n such that $m/n \in [0,1]$ and for each $t \in [0, \varepsilon]$, we have

$$\frac{m}{n} \mathcal{U}(t) = m\mathcal{U}(t/n) = \mathcal{U}(mt/n),$$

and so

$$\frac{m}{n} \mathcal{U}(\varepsilon) = \mathcal{U}(m\varepsilon/n).$$

By continuity, we get

$$tP(\varepsilon) = P(t\varepsilon) \quad \text{for every } t \in [0,1],$$

and

$$P(t) = \frac{t}{\varepsilon} P(\varepsilon) \quad \text{for every } t \in [0,\varepsilon].$$

If we set

$$T := \frac{1}{\varepsilon} P(\varepsilon),$$

we obtain

$$\mathcal{U}(t) = e^{t\,T} \quad \text{for every } t \in [0,\varepsilon].$$

If $t > 0$ is arbitrary, then $t/n < \varepsilon$ for sufficiently large n, and so we obtain

$$e^{t\,T} = (e^{(t/n)\,T})^n = [\mathcal{U}(t/n)]^n = \mathcal{U}(t).$$

This proves the representation of the semigroup. To prove point (2) let $h > 0$ and observe that the limit $\lim_{h\to 0^+}(e^{h\,q} - 1)/h = q$ and the sequence $(e^{h\,q} - \mathcal{I})/h$ converges uniformly in any bounded set of $\mathbb{H}$. So, by Theorem 4.10.6, the limit $\lim_{h\to 0^+}(e^{h\,T} - \mathcal{I})/h$ converges to T. Point (3) can be deduced by the functional calculus. In fact, taking $h \in \mathbb{R}$, we get

$$\frac{e^{(t+h)T} - e^{tT}}{h} = \frac{1}{2\pi} \int_{\partial(U\cap\mathbb{C}_I)} S_L^{-1}(s,T)\,ds_I \frac{(e^{(t+h)s} - e^{ts})s}{h} s^{-1}.$$

Now consider the fact that for any t, $h \in \mathbb{R}$ and for any quaternion $s \in \mathbb{H}$ we have that e^{ts} and e^{hs} commute between themselves and with s, moreover $e^{(t+h)s} = e^{ts}e^{hs}$ holds. Thus we have

$$\frac{e^{(t+h)T} - e^{tT}}{h} = \frac{1}{2\pi} \int_{\partial(U\cap\mathbb{C}_I)} S_L^{-1}(s,T)\,ds_I\; s\,e^{ts}\, \frac{(e^{hs} - 1)}{h} s^{-1}.$$

Taking the limit for $h \to 0$ we get

$$\lim_{h\to 0} \frac{e^{(t+h)T} - e^{tT}}{h} = Te^{tT}. \qquad \square$$

We now want to generalize the important result that the Laplace transform of a semigroup e^{tB} of a bounded linear complex operator B is the usual resolvent operator $(\lambda I - B)^{-1}$. The generalization we obtain is somewhat surprising. Both the left and the right S-resolvent operators $S_L^{-1}(s,T)$ and $S_R^{-1}(s,T)$ are the Laplace transform of the semigroup according to two different possible definitions of the Laplace transform according to the two possible integrands $e^{tT}e^{-ts}$ and $e^{-ts}e^{tT}$.

Theorem 4.17.3. *Let $T \in \mathcal{B}(V)$ and let $s_0 > \|T\|$. Then the left S-resolvent operator $S_L^{-1}(s,T)$ is given by*

$$S_L^{-1}(s,T) = \int_0^{+\infty} e^{tT}\, e^{-ts}\, dt. \tag{4.128}$$

Proof. We have to prove that

$$S_L(s,T) \int_0^{\infty} e^{tT}\, e^{-ts}\, dt = \mathcal{I},$$

where

$$S_L(s,T) = -(T - \overline{s}\mathcal{I})^{-1}(T^2 - 2s_0T + |s|^2\mathcal{I}).$$

Take $\theta > 0$ and consider

$$S_L(s,T) \int_0^{\theta} \mathcal{U}(t) e^{-ts}\, dt = -(T - \overline{s}\mathcal{I})^{-1}(T^2 - 2s_0T + |s|^2\mathcal{I}) \int_0^{\theta} e^{tT}\, e^{-ts}\, dt.$$

Since every bounded linear operator commutes with the integral, we get

$$S_L(s,T) \int_0^{\theta} e^{tT}\, e^{-ts}\, dt = -\int_0^{\theta} (T - \overline{s}\mathcal{I})^{-1}(T^2 - 2s_0T + |s|^2\mathcal{I}) e^{tT}\, e^{-ts}\, dt. \tag{4.129}$$

Thanks to Theorem 4.17.2 we obtain the identities

$$\begin{aligned}
&(T - \overline{s}\mathcal{I})^{-1}(T^2 - 2s_0T + |s|^2\mathcal{I}) e^{tT}\, e^{-ts} \\
&= (T - \overline{s}\mathcal{I})^{-1} e^{tT}(T^2 - T\overline{s} - Ts + \overline{s}s\mathcal{I})\, e^{-ts} \\
&= (T - \overline{s}\mathcal{I})^{-1}\Big\{ e^{tT} T (T - \overline{s}\mathcal{I})\, e^{-ts} - e^{tT}(T - \overline{s}\mathcal{I}) s\, e^{-ts} \Big\} \\
&= (T - \overline{s}\mathcal{I})^{-1}\Big\{ \frac{d}{dt} e^{tT} (T - \overline{s}\mathcal{I})\, e^{-ts} + e^{tT}(T - \overline{s}\mathcal{I}) \frac{d}{dt} e^{-ts} \Big\} \\
&= \frac{d}{dt}[(T - \overline{s}\mathcal{I})^{-1} e^{tT}(T - \overline{s}\mathcal{I})\, e^{-ts}].
\end{aligned} \tag{4.130}$$

So by the identity (4.130) we can write (4.129) as

$$\begin{aligned}
S_L(s,T) \int_0^{\theta} e^{tT}\, e^{-ts}\, dt &= -\int_0^{\theta} \frac{d}{dt}[(T - \overline{s}\mathcal{I})^{-1} e^{tT}(T - \overline{s}\mathcal{I})\, e^{-ts}] dt \\
&= \mathcal{I} - (T - \overline{s}\mathcal{I})^{-1} e^{\theta T}(T - \overline{s}\mathcal{I})\, e^{-s\theta}.
\end{aligned}$$

Observe that

$$\begin{aligned}
&\|(T - \overline{s}\mathcal{I})^{-1} e^{\theta T}(T - \overline{s}\mathcal{I})\, e^{-s\theta}\| \\
&\leq \|(T - \overline{s}\mathcal{I})^{-1}\| \|e^{\theta T}\| \|(T - \overline{s}\mathcal{I})\|\, \|e^{-s\theta}\| \\
&\leq \|(T - \overline{s}\mathcal{I})^{-1}\| \|(T - \overline{s}\mathcal{I})\|\, e^{\theta\|T\|} e^{-s_0\theta} \to 0
\end{aligned}$$

for $\theta \to +\infty$ because we have assumed $s_0 > \|T\|$. So we get the statement. □

The case of $S_R^{-1}(s,T)$ is similar: it will be treated in Theorem 4.18.8 in the Notes. In the sequel we will use the following result.

Proposition 4.17.4. *Let V be a quaternionic Banach space. Let $\{\mathcal{U}(t)\}$ be a family of bounded linear quaternionic operators defined on a finite closed interval $[a,b]$ such that $\mathcal{U}(t)v$ is continuous in t for each $v \in V$; then $\|\mathcal{U}(\cdot)\|$ is measurable and bounded on $[a,b]$. Conversely, if $\{\mathcal{U}(t)\}_{t\geq 0}$ is a semigroup of bounded linear quaternionic operators in V and if $\mathcal{U}(\cdot)v$ is measurable on $(0,\infty)$ for each $v \in V$, then $\mathcal{U}(\cdot)v$ is continuous at every point in $(0,\infty)$.*

Proof. The statement follows by adapting the arguments in the proof of Lemma 3 p. 616 in [35]. In fact, under the hypotheses in the first part of the statement, the boundedness of $\|\mathcal{U}(\cdot)\|$ follows from the Uniform Boundedness Principle and the fact that $\|\mathcal{U}(\cdot)\|$ is measurable follows from Theorem III.6.10 in [35]. To show the second part, we can assume at the beginning that $\|\mathcal{U}(\cdot)\|$ is bounded over each interval of the form $[\delta, 1/\delta]$, $\delta > 0$. Under this assumption, if one repeats the computations in the proof of Lemma 3, p. 616 one gets that $\|\mathcal{U}(\cdot)\|$ is continuous at each point $t_0 > 0$ for any $v \in V$. Finally, it is sufficient to show that, if $\|\mathcal{U}(\cdot)\|v$ is measurable on $(0,\infty)$ for all $v \in V$, then it is bounded on $[\delta, 1/\delta]$, $\delta > 0$. □

Proposition 4.17.5. *Let $\{\mathcal{U}(t)\}_{t\geq 0}$ be a family of bounded linear quaternionic operators on the quaternionic Banach space V. If*

$$p(t) := \ln \|\mathcal{U}(t)\|$$

is bounded from the above on the interval $(0,a)$ for every positive $a \in \mathbb{R}$, then

$$\lim_{t\to+\infty} t^{-1} \ln \|\mathcal{U}(t)\| = \inf_{t>0} t^{-1} \ln \|\mathcal{U}(t)\|.$$

Proof. The proof is similar to the one of the analogous result in the complex case. In fact, it is immediate to check that all the computations can be repeated over the quaternions. For the sake of completeness we recall the main points. Observe that

$$p(t+\tau) = \ln \|\mathcal{U}(t+\tau)\| \leq \ln \|\mathcal{U}(t)\|\|\mathcal{U}(\tau)\| \leq p(t) + p(\tau), \qquad t, \tau \geq 0.$$

Set $\alpha := \inf_{t>0} t^{-1} \ln \|\mathcal{U}(t)\|$ finite or $-\infty$. Suppose α finite. We choose for any $\varepsilon > 0$, a positive number $a > 0$ in such a way that $p(a) \leq (\alpha+\varepsilon)a$. Let $t > a$ and $n = n(t)$ be an integer such that $na \leq t < (n+1)a$. Then we have the chain of inequalities

$$\begin{aligned}\alpha \leq \frac{p(t)}{t} &\leq \frac{p(na)}{t} + \frac{p(t-na)}{t} \leq \frac{na}{t}\frac{p(a)}{a} + \frac{p(t-na)}{t}\\ &\leq \frac{na}{t}(\alpha+\varepsilon) + \frac{p(t-na)}{t}.\end{aligned}$$

By hypothesis $p(t-na)$ is bounded from above as $t \to +\infty$. Thus letting $t \to +\infty$ in the above inequality we obtain $\lim_{t\to+\infty} t^{-1}p(t) = \alpha$. In a similar way we treat the case $\alpha = -\infty$. □

A direct consequence of Proposition 4.17.5 is the following important result.

Proposition 4.17.6. *Let* $\{\mathcal{U}(t)\}_{t\geq 0}$ *be a family of bounded linear quaternionic operators on a quaternionic Banach space* V*. Then:*

(1) *the limit* $\omega_0 := \lim_{t\to+\infty} t^{-1} \ln \|\mathcal{U}(t)\|$ *exists;*

(2) *for each* $\delta > \omega_0$ *there exists a positive constant* M_δ *such that* $\|\mathcal{U}(t)\| \leq M_\delta e^{\delta t}, \quad \forall t \geq 0$.

Proof. Point (1) is immediate from Proposition 4.17.5. To show point (2) define, for $t \geq 0$, the function $p(t) := \ln \|\mathcal{U}(t)\|$ and observe that p is subadditive since $p(t_1 + t_2) \leq p(t_1) + p(t_2)$. So the result follows from Propositions 4.17.4 and 4.17.5 . □

Definition 4.17.7 (Quaternionic infinitesimal generator). *Let* $\{\mathcal{U}(t)\}_{t\geq 0}$ *be a family of bounded linear quaternionic operators on a quaternionic Banach space* V*.*

(1) *For each* $h > 0$ *define the linear quaternionic operator*

$$T_h v = \frac{\mathcal{U}(h)v - v}{h}, \qquad v \in V.$$

(2) *Set* $\mathcal{D}(T) := \{v \in V \ : \ \lim_{h\to 0^+} T_h v$ *exists in* $V\}$ *and define the quaternionic operator* T *with domain* $\mathcal{D}(T)$ *by the formula*

$$Tv = \lim_{h\to 0^+} T_h v, \quad v \in \mathcal{D}(T).$$

The operator T*, with domain* $\mathcal{D}(T)$*, is called the infinitesimal quaternionic generator of the quaternionic semigroup* $\mathcal{U}(t)$*.*

Proposition 4.17.8. *Let* T *be the infinitesimal quaternionic generator of the quaternionic semigroup* $\mathcal{U}(t)$ *and let* $\mathcal{D}(T)$ *be its domain. Then:*

(1) *the set* $\mathcal{D}(T)$ *is a linear subspace of* V *and* T *is linear on* $\mathcal{D}(T)$*;*

(2) *if* $v \in V$*, then* $\mathcal{U}(t)v \in \mathcal{D}(T)$ *for* $t \geq 0$*. Moreover,*

$$\frac{d}{dt}\mathcal{U}(t)v = T\mathcal{U}(t)v = \mathcal{U}(t)Tv, \quad v \in \mathcal{D}(T);$$

(3) *if* $v \in \mathcal{D}(T)$*, then*

$$\mathcal{U}(t)v - \mathcal{U}(\tau)v = \int_\tau^t \mathcal{U}(\theta)\, T\, v\, d\theta, \quad 0 \leq \tau < t < \infty;$$

(4) *let* $g : [0,\infty] \to \mathbb{H}$ *be a Lebesgue integrable function, continuous at* $t \in [0,\infty]$*, then*

$$\lim_{h\to 0^+} \frac{1}{h} \int_t^{t+h} \mathcal{U}(\theta)\, g(\theta)\, v\, d\theta = \mathcal{U}(t)\, g(t)\, v.$$

Proof. Point (1) follows from the definition. Let us show point (2). Set $h > 0$, $t \geq 0$ and $v \in \mathcal{D}(T)$. Then we can write

$$\mathcal{U}(t)T_h v = T_h \mathcal{U}(t)v.$$

Passing to the limit

$$\lim_{h\to 0^+} \mathcal{U}(t)T_h v = \lim_{h\to 0^+} T_h \mathcal{U}(t)v$$

we have that $\mathcal{U}(t)v \in \mathcal{D}(T)$, so by definition

$$T\mathcal{U}(t)v = \lim_{h\to 0^+} T_h \mathcal{U}(t)v,$$

and thus

$$\mathcal{U}(t)Tv = T\mathcal{U}(t)v, \quad v \in \mathcal{D}(T).$$

This proves that $\mathcal{U}(t)v \in \mathcal{D}(T)$, for all $t \geq 0$. If $t > 0$ and $h > 0$, then, considering the limit

$$L = \lim_{h\to 0^+} \Big(\frac{\mathcal{U}(t)v - \mathcal{U}(t-h)v}{h} - \mathcal{U}(t)Tv\Big),$$

by the semigroup properties and the definition of T_h we have

$$\begin{aligned}
\frac{\mathcal{U}(t)v - \mathcal{U}(t-h)v}{h} - \mathcal{U}(t)Tv &= \mathcal{U}(t-h)\frac{\mathcal{U}(h)v - v}{h} - \mathcal{U}(t)Tv \\
&= \mathcal{U}(t-h)\frac{\mathcal{U}(h)v - v}{h} - \mathcal{U}(t-h)Tv + \mathcal{U}(t-h)Tv - \mathcal{U}(t)Tv \\
&= \mathcal{U}(t-h)(T_h v - Tv) + [\mathcal{U}(t-h) - \mathcal{U}(t)]Tv.
\end{aligned}$$

Taking the limit for $h \to 0^+$ we get $L = 0$ since the semigroup is uniformly continuous in $\mathcal{B}(V)$ and thanks to Proposition 4.17.4. On the other hand we have that

$$\frac{\mathcal{U}(t+h)v - \mathcal{U}(t)v}{h} = \mathcal{U}(t)T_h v.$$

Taking the limit for $h \to 0^+$ we get

$$\frac{d}{dt}\mathcal{U}(t)v = T\mathcal{U}(t)v = \mathcal{U}(t)Tv, \quad \text{for all} \ \ v \in \mathcal{D}(T).$$

We have thus proved that the derivation formula holds for all $v \in \mathcal{D}(T)$. Point (3): observe that, for all linear and continuous functionals $\varphi \in V'$, from point (2) we have

$$\langle \varphi, \frac{d}{d\tau}\mathcal{U}(\tau)v\rangle = \langle \varphi, \mathcal{U}(\tau)Tv\rangle.$$

Now we integrate

$$\int_s^t \langle \varphi, \frac{d}{d\tau}\mathcal{U}(\tau)\, v\rangle \, d\tau = \int_s^t \langle \varphi, \mathcal{U}(\tau)Tv\rangle \, d\tau$$

so we get

$$\langle \varphi, \mathcal{U}(t)v - \mathcal{U}(s)v \rangle = \langle \varphi, \int_s^t \mathcal{U}(\tau)\, Tv\, d\tau \rangle \quad \text{for all} \quad \varphi \in V'$$

from which we deduce (3). Finally point (4) follows from Theorem III.12.8 in [35] which holds also in this setting, with obvious modifications. □

Lemma 4.17.9. *The linear subspace*

$$\mathcal{D}(T) := \{v \in V \ : \ \lim_{h \to 0^+} T_h v \ \ \text{exists in } V\}$$

is dense in V and T is closed on $\mathcal{D}(T)$.

Proof. Let T_h be as in Definition 4.17.7, take $v \in V$ and, for $h > 0$ and $t > 0$, consider

$$\begin{aligned}
& T_h \int_0^t \mathcal{U}(\tau) v d\tau = \frac{1}{h} \int_0^t [\mathcal{U}(h+\tau)v - \mathcal{U}(\tau)v] d\tau \\
& = \frac{1}{h} \int_h^{h+t} \mathcal{U}(\tau)v\, d\tau - \frac{1}{h} \int_0^t \mathcal{U}(\tau)v\, d\tau \\
& = \frac{1}{h} \int_t^h \mathcal{U}(\tau)v\, d\tau + \frac{1}{h} \int_h^{h+t} \mathcal{U}(\tau)v\, d\tau - \frac{1}{h} \int_0^t \mathcal{U}(\tau)v\, d\tau - \frac{1}{h} \int_t^h \mathcal{U}(\tau)v\, d\tau \\
& = \frac{1}{h} \int_t^{h+t} \mathcal{U}(\tau)v\, d\tau - \frac{1}{h} \int_0^h \mathcal{U}(\tau)v\, d\tau.
\end{aligned}$$

By Proposition 4.17.8 point (4) we get

$$\lim_{h \to 0^+} T_h \int_0^t \mathcal{U}(\tau) v d\tau = \mathcal{U}(t)v - v,$$

so $\int_0^t \mathcal{U}(\tau) v d\tau \in \mathcal{D}(T)$ and since

$$v = \lim_{t \to 0^+} \frac{1}{t} \int_0^t \mathcal{U}(\tau) v d\tau$$

we conclude that $\mathcal{D}(T)$ is dense in V. We now prove that T is closed. Let us take a sequence $\{v_n\}_{n\in\mathbb{N}} \subset \mathcal{D}(T)$ such that $\lim_{n\to\infty} v_n = v_0$ and $\lim_{n\to\infty} Tv_n = y_0$. Thanks to Proposition 4.17.8 point (3) we have

$$\mathcal{U}(t)v_0 - v_0 = \lim_{n\to\infty} [\mathcal{U}(t)v_n - v_n] = \lim_{n\to\infty} \int_0^t \mathcal{U}(\tau) T v_n\, d\tau = \int_0^t \mathcal{U}(\tau) y_0\, d\tau$$

where we have used the fact that

$$\lim_{n\to\infty} \mathcal{U}(\tau) T v_n = \mathcal{U}(\tau) y_0, \quad \text{uniformly in} \quad [0,t].$$

So we get, thanks to Proposition 4.17.8 point (4)

$$\lim_{t\to 0^+} T_t v_0 = \lim_{t\to 0^+} \frac{1}{t}\int_0^t \mathcal{U}(\tau)y_0\, d\tau = y_0.$$

This implies that $v_0 \in \mathcal{D}(T)$ and $Tv_0 = y_0$ this means that T is closed. □

We can now prove the following characterization result.

Theorem 4.17.10. *Let $\mathcal{U}(t)$ be a quaternionic semigroup on a quaternionic Banach space V. Then $\mathcal{U}(t)$ has a bounded infinitesimal quaternionic generator if and only if it is uniformly continuous.*

Proof. If $\mathcal{U}(t)$ is a uniformly continuous semigroup, then by Theorem 4.17.2 it has a bounded infinitesimal quaternionic generator. To prove the other implication, we suppose that $\mathcal{U}(t)$ has a bounded infinitesimal quaternionic generator T. It follows from Lemma 4.17.9 that T is defined everywhere. Applying Proposition 4.17.4 we have that for every $\tau \geq 0$ there exists a positive constant $C(\tau)$ such that

$$\|\mathcal{U}(t)\| \leq C(\tau), \quad \text{for} \quad \tau \geq 0, \quad |t-\tau| \leq 1$$

by the semigroup properties

$$\mathcal{U}(t) - \mathcal{U}(\tau) = \mathcal{U}(\tau)[\mathcal{U}(t-\tau) - \mathcal{I}] = (t-\tau)\mathcal{U}(\tau)\, T_{t-\tau}, \quad \text{for} \quad t > \tau \tag{4.131}$$

and

$$\mathcal{U}(t) - \mathcal{U}(\tau) = -\mathcal{U}(t)[\mathcal{U}(\tau - t) - \mathcal{I}] = -(\tau - t)\mathcal{U}(t)\, T_{\tau-t}, \quad \text{for} \quad \tau > t, \tag{4.132}$$

where $T_{\tau-t}$ and $T_{t-\tau}$ are as in Definition 4.17.7. Using Proposition 4.17.4 and the Principle of Uniform Boundedness we have

$$\sup_{\tau > t \ | \ \tau - t \leq 1} \|T_{\tau-t}\| = K < +\infty,$$

so, taking the norm of (4.131) and (4.132), we get

$$\|\mathcal{U}(t) - \mathcal{U}(\tau)\| \leq C(\tau)K|t-\tau|, \quad \text{for} \quad t \geq 0, \quad |t-\tau| \leq 1,$$

which proves that $\mathcal{U}(t)$ is a uniformly continuous quaternionic semigroup. □

4.18 Notes

Note 4.18.1. Historical notes and further readings. The most successful theory of quaternionic functions, which are the analog of the holomorphic functions in one complex variable, is the one due to Fueter [44] and the one due to Moisil and

Theodorescu [79]. The theory we will consider to our purposes is the one due to Fueter, who introduced the differential operator

$$\frac{\partial}{\partial \overline{q}} = \frac{1}{4}\left(\frac{\partial}{\partial x_0} + i\frac{\partial}{\partial x_1} + j\frac{\partial}{\partial x_2} + k\frac{\partial}{\partial x_3}\right) \tag{4.133}$$

and defined the space of regular functions as the space of its nullsolutions.

These functions are nowadays known as Cauchy–Fueter (or Fueter) regular functions. It is interesting to note that one of the motivations for which Fueter introduced this class of functions was the study of functions in two complex variables, in an attempt to find an integral representation for them. This function theory is beautifully illustrated in the papers [33] and [98] and is successful in replicating the most important properties of holomorphic functions (and not only in one variable, see [23]). One of the reasons of the richness of results of this theory is that the Cauchy–Fueter operator factorizes the 4-dimensional Laplacian (up to a constant), in fact the nice behavior of the quaternionic conjugation gives

$$\frac{\partial}{\partial \overline{q}}\frac{\partial}{\partial q} = \frac{\partial}{\partial q}\frac{\partial}{\partial \overline{q}} = \frac{1}{16}\Delta_4.$$

In this sense, the Fueter operator is the "nearest" generalization of the Cauchy–Riemann operator and the theory of regular functions is the "nearest" generalization of the theory of holomorphic functions in one complex variable. Exactly as in the case of monogenic functions, Fueter regular functions can be expanded into power series in terms of suitable monomials, see [43], [98]. This series expansion serves perfectly to show the quaternionic analogs of the results for power series in the complex variable. However, this series expansion is not suitable if one wishes to formally substitute a quaternion with a linear operator T. As we already observed, the powers $f(q) = q^n$, $n \in \mathbb{N}$, in particular the identity function $f(q) = q$, and therefore polynomials and series, fail to be regular in this sense so, given a quaternionic linear operator T, it is not possible, for example, to obtain T^n using a function calculus (see Note 4.18.9).

For quaternion-valued functions, one can also study the functions regular in the sense of Moisil-Theodorescu, see [79], the theory of monogenic functions, i.e., functions in the kernel of the Dirac operator $i\partial_{x_1} + j\partial_{x_2}$ or the Weyl operator $\partial_{x_0} + i\partial_{x_1} + j\partial_{x_2}$, see [7], [34], and [74]. This last class consists of all the solutions of a generalized Cauchy–Riemann system of equations, it contains the natural polynomials, and supports the series expansion of its elements as well.

There are other paths to define quaternionic-valued "holomorphic" functions which has led to many attempts that, in a sense, have failed. For example, a natural attempt to define a notion of quaternionic holomorphicity would be based on the basis of the existence in $\mathbb{H}$ of the limit

$$\lim_{q\to q_0} (q-q_0)^{-1}(f(q)-f(q_0)) \qquad \text{(resp.} \quad \lim_{q\to q_0} (f(q)-f(q_0))(q-q_0)^{-1}).$$

It turns out that if the limit exists, then necessarily $f(q) = qa + b$ (resp. $f(q) = aq + b$) for some $a, b \in \mathbb{H}$ and therefore this definition is not viable. In order to obtain a meaningful theory, it is necessary to restrict to 3-dimensional increments $q - q_0$, as Mitelman and Shapiro did in [78] in order to develop their theory.

A second natural attempt could be to consider the class of functions which admit (local) series expansions of the form

$$\sum a_0 \cdot q \cdot a_1 \ldots a_{s-1} \cdot q \cdot a_s,$$

where they converge. However, writing $q = x_0 + ix_1 + jx_2 + kx_3$, it is very easy to verify that

$$\begin{aligned}
x_0 &= \frac{1}{4}(q - iqi - jqj - kqk), \\
x_1 &= \frac{1}{4i}(q - iqi + jqj + kqk), \\
x_2 &= \frac{1}{4j}(q + iqi - jqj + kqk), \\
x_3 &= \frac{1}{4k}(q + iqi + jqj - kqk),
\end{aligned}$$

so that the class of maps considered coincides with the class of real analytic maps of $\mathbb{R}^4$ in $\mathbb{R}^4$.

Another definition was given by Cullen in [32] on the basis of the notion of intrinsic functions as developed in [89]. This definition has the advantage that polynomials and even power series of the form $\sum_{n\geq 0} q^n a_n$, with real coefficients a_n, are regular in this sense. This theory was already envisioned by Fueter who in [43] used a subclass of these functions to construct nullsolutions of the Cauchy–Fueter operator by applying the Laplacian to them.

The theory of s-regular functions arises with the works of Gentili and Struppa, see [48], [49], and it is fully embedded in this field of research. The theory is inspired by the work of Cullen, but it is slightly different since the definition of s-regular function requires that a function be "holomorphic" on each complex plane and, as a consequence, it includes polynomials and even power series with quaternionic coefficients. The theory developed allows us to recover several classical properties of quaternionic polynomials. For example the fundamental theorem of algebra, [37], [81] has been proved also in [52]. Moreover, some well-known properties of zeros, see for example [6], [8], [56], [71], [84], [85], [86], [102] can be proved in this framework, see [51], and can be generalized to power series [45]. The description which we have given in this chapter is the most up-to-date version and includes the results in [9], [12], [49], but we recall also [30], [46]. Recently, Ghiloni and Perotti [53], [55] proposed an approach which, in the spirit of Cullen, allows a general treatment of s-regular (and s-monogenic functions) as we explained in Notes of Chapter 2.

We note that none of the various definitions of regularity on $\mathbb{H}$ leads to a natural functional calculus for quaternionic linear operators (see for example, our Note 4.18.9 that deals with the case of Fueter regularity). The case of s-regularity, on the other hand, allows a very natural construction, because of the discovery of a suitable Cauchy formula proved by Colombo and Sabadini, first developed in connection with the theory of s-monogenicity [15] and [18], and then adapted to the quaternionic case in [9].

Note 4.18.2. The construction of the quaternionic functional calculus offered in this chapter is mainly based on some works by Colombo and Sabadini, see [16], [17], and [21], which put in the necessary generality some ideas first introduced in [10], [13], and [14].

Note 4.18.3. Niven's Algorithm. Niven's algorithm, see [81], gives a method to determine the zeros of a quaternionic polynomial with coefficients on one side in terms of the coefficients of the polynomial. The method is explicit in the case of quadratic polynomials while for higher degrees it relies on the solution of a system of two equations which may not be available in closed form. Even though in this book we have treated the case of polynomials with coefficients on the right (and thus are s-regular) we illustrate the algorithm in the case of polynomials with left coefficients. In fact, the corresponding algorithm in the case of right coefficients gives an expression for the Cauchy kernel which is not suitable for the functional calculus of operators with noncommuting components.

Let us consider a monic quaternionic polynomial $A_n(q)$ with coefficients on the left:

$$A_n(q) = q^n - \sum_{s=0}^{n-1} a_s q^s, \quad a_s \in \mathbb{H}.$$

It is always possible to divide $A_n(q)$ by a second-degree polynomial $C_2(q) = q^2 - c_1 q - c_0$, with real coefficients and to obtain a quotient $B_{n-2}(q)$ and a degree-one remainder $D_1(q)$ given by

$$B_{n-2} := q^{n-2} - \sum_{s=0}^{n-3} b_s q^s, \qquad D_1(q) = d_1 q + d_0, \qquad b_s,\ d_1,\ d_0 \in \mathbb{H}$$

such that

$$A_n(q) = B_{n-2}(q) C_2(q) - D_1(q).$$

Now note that if p is a solution to the polynomial equation $A_n(p) = 0$, if we choose the coefficients of the polynomial C_2 to be

$$c_0 = -|p|^2, \quad c_1 = 2Re[p],$$

then p is also a root of C_2, i.e.,

$$C_2(p) = 0.$$

This fact implies that p is also a root of D_1 and so

$$D_1(p) = 0, \quad i.e., \quad d_1 p + d_0 = 0 \quad \Rightarrow \quad p = -d_1^{-1} d_0.$$

The strategy behind Niven's algorithm consists of two steps. First one determines d_0 and d_1 in terms of c_0, c_1 and $a_0,\dots,a_{n-1}$; then one obtains two coupled real equations which allow us to calculate c_0 and c_1. To clarify how this works, we consider the case in which the polynomial A_n is of second degree (this being the case of interest for our applications):

$$A_2(q) = q^2 - a_1 q - a_0, \quad a_s \in \mathbb{H}.$$

In this case, by reason of degree, we have

$$B_0(q) = 1$$

so we have

$$B_0(q)C_2(q) = C_2(q) = q^2 - c_1 q - c_0$$

and

$$A_2(q) + D_1(q) = q^2 - a_1 q - a_0 + d_1 p + d_0.$$

Since

$$A_2(q) + D_1(q) = B(q)C_2(q)$$

we get

$$q^2 - a_1 q - a_0 + d_1 p + d_0 = q^2 - c_1 q - c_0$$

from which we have

$$d_1 = a_1 - c_1, \qquad d_0 = a_0 - c_0.$$

The system to compute $c_0 \in \mathbb{R}$ and $c_1 \in \mathbb{R}$, is given by

$$\begin{cases} c_0|a_1 - c_1|^2 + |a_0 - c_0|^2 = 0, \\ c_1|a_1 - c_1|^2 + 2Re[(\overline{a}_1 - c_1)(a_0 - c_0)] = 0. \end{cases}$$

If we are able to solve the system we get $\hat{c}_0$ and $\hat{c}_1$ so that we determine

$$\hat{d}_1 = a_1 - \hat{c}_1, \qquad \hat{d}_0 = a_0 - \hat{c}_0$$

and we finally get the solution $p = -\hat{d}_1^{-1}\hat{d}_0$.

The solution to the equation $S^2 + Sq - sS = 0$. We now consider the specific case of the equation $S^2 + Sq - sS = 0$. First of all, we write it with coefficients on the left by setting $S := W - q$. We obtain the equation

$$W^2 - (s + q)W + sq = 0.$$

With the positions

$$a_1 = s + q \qquad a_0 = -sq$$

we get the system

$$\begin{cases} c_0|s+q-c_1|^2+|-sq-c_0|^2=0, \\ c_1|s+q-c_1|^2+2Re[(\overline{s+q}-c_1)(-sq-c_0)]=0. \end{cases} \tag{4.134}$$

If we can find real solutions $\hat{c}_0$ and $\hat{c}_1$ we have

$$\hat{d}_1=s+q-\hat{c}_1, \qquad \hat{d}_0=-sq-\hat{c}_0,$$

$$\hat{d}_1W+\hat{d}_0=0.$$

We obtain

$$W=-\hat{d}_1^{-1}\hat{d}_0=(s+q-\hat{c}_1)^{-1}(sq+\hat{c}_0)$$

from which we have

$$S:=W-q=(s+q-\hat{c}_1)^{-1}(sq+\hat{c}_0)-q.$$

We overcome the calculation for the solution of system (4.134) reasoning as follows: we know that when $sq=qs$, then $R=s-q$ satisfies $R^2+Rq-sR=0$. Thus in the case $q\in\mathbb{R}$ and $s\in\mathbb{H}$, we must have

$$(s+q-\hat{c}_1)^{-1}(sq+\hat{c}_0)=s$$

that gives

$$sq+\hat{c}_0=(s+q-\hat{c}_1)s$$

and

$$sq+\hat{c}_0=s^2+qs-\hat{c}_1s;$$

finally we have

$$s^2-\hat{c}_1s-\hat{c}_0=0.$$

Using the identity

$$s^2-2\ sRe[s]+|s|^2=0$$

we get

$$\hat{c}_1=2\ Re[s], \qquad \hat{c}_0=-|s|^2,$$

so we obtain the solution

$$S(s,q)=(s+q-2\ Re[s])^{-1}(sq-|s|^2)-q=(q-\bar{s})^{-1}s(q-\bar{s})-q.$$

Note 4.18.4. A simple proof of Theorem 4.10.3. We now provide a proof of Theorem 4.10.3 which is of limited validity, but follows by a direct computation. It applies only in the case the functions we consider admit power series expansions on U. We recall that s-regular functions admit Taylor series expansions only on balls centered at real points and they admit Laurent series expansions only on spherical shells centered at real points.

Let us consider the case in which the domain U is contained in a ball $B(\alpha,r)\subset\mathbb{H}$ centered in a real point α and of radius $r>0$ in which the s-regular function f admits a power series expansion.

Proposition 4.18.5. *Let V be a two-sided quaternionic Banach space, $T \in \mathcal{B}(V)$. Suppose that f is an s-regular function such that*

$$f(s) = \sum_{m\geq 0} (s-\alpha)^m a_m, \quad \forall s \in B(\alpha, r), \quad \alpha \in \mathbb{R}, \quad a_m \in \mathbb{H}, \quad r > 0 \tag{4.135}$$

and assume that $\sigma_S(T) \subset U \subset B(\alpha, r)$ where U is a T-admissible open set. Then

$$\frac{1}{2\pi} \int_{\partial(U\cap\mathbb{C}_I)} S^{-1}(s,T)\ ds_I\ \ f(s) \tag{4.136}$$

does not depend on the choice of the imaginary unit $I \in \mathbb{S}$ and on U.

Proof. In $B(\alpha, r)$ the Taylor expansion of f has the form (4.135) where the elements a_m are fixed quaternions and do not depend on the particular plane $\mathbb{C}_I$. Now observe that

$$f(s) = \sum_{m\geq 0} (s-\alpha)^m a_m = \sum_{m\geq 0} \sum_{j=0}^{m} \binom{m}{j} s^j (-\alpha)^{m-j} a_m.$$

Consider the integral (4.136) and replace the power series expansion for f. By the absolute and uniform convergence we get

$$\begin{aligned} &\frac{1}{2\pi} \int_{\partial(U\cap\mathbb{C}_I)} S^{-1}(s,T)\ ds_I\ f(s) \\ &= \frac{1}{2\pi} \sum_{m\geq 0} \sum_{j=0}^{m} \binom{m}{j} \Big(\int_{\partial(U\cap\mathbb{C}_I)} S^{-1}(s,T)\ ds_I\ s^j \Big) (-\alpha)^{m-j} a_m. \end{aligned} \tag{4.137}$$

Now consider the integral

$$\int_{\partial(U\cap\mathbb{C}_I)} S^{-1}(s,T)\ ds_I\ s^j$$

and observe that s^j is s-regular everywhere so we can deform the integration path in such a way that $S^{-1}(s,T)$ admits the power series expansion (3.3) in a suitable ball $B(0,r)$. We have:

$$\frac{1}{2\pi} \sum_{n\geq 0} T^n \int_{\partial(B(0,r)\cap\mathbb{C}_I)} s^{-1-n+j}\ ds_I = T^j, \tag{4.138}$$

since

$$\begin{aligned} &\int_{\partial(B(0,r)\cap\mathbb{C}_I)} ds_I s^{-n-1+j} = 0 \ \ if \ \ n \neq j, \\ &\int_{\partial(B(0,r)\cap\mathbb{C}_I)} ds_I s^{-n-1+j} = 2\pi \ \ if \ \ n = j. \end{aligned} \tag{4.139}$$

The standard Cauchy theorem on the complex plane $\mathbb{C}_I$ shows that the above integral (4.138) is not affected if we replace $\partial(B(0,r)\cap\mathbb{C}_I)$ by $\partial(U\cap\mathbb{C}_I)$, so

$$\frac{1}{2\pi}\int_{\partial(U\cap\mathbb{C}_I)} S^{-1}(s,T)\; ds_I\; s^j \; = T^j.$$

We conclude that the integral (4.137) does not depend on U and on $I\in\mathbb{S}$ because the coefficients $(-\alpha)^{j-m}a_m$ are independent of $I\in\mathbb{S}$. □

More in general, we can consider the open sets $U\subset\mathbb{H}$ that contain the S-spectrum of T, and such that

(a) $\partial(U\cap\mathbb{C}_I)$ is the union of a finite number of continuously differentiable Jordan curves for every $I\in\mathbb{S}$,

(b) $\sigma_S(T)$ is contained in a finite union of open balls $B_i\subset U$ with center in real points and of spherical shells $A_j=\{q\in\mathbb{H}\mid r_j<|q-\alpha_j|<R_j,\ r_j,R_j\in\mathbb{R}^+\}\subset U$ with center in real points α_j, and whose boundaries do not intersect $\sigma_S(T)$.

Since an analog of Proposition 4.18.5 holds also for Laurent power series expansions, we can prove that, for open sets $U\supset\sigma_S(T)$ satisfying (a) and (b), the integral (4.136) does not depend on the choice of the imaginary unit $I\in\mathbb{S}$ and on U.

Note 4.18.6. Some comments on the evolution operator. Since the exponential function is both left and right regular, the evolution operator can be introduced and studied also using the right version of the quaternionic functional calculus. For example, point (3) in Theorem 4.17.2 can also be proved as in the next result:

Theorem 4.18.7. *Let $\{\mathcal{U}(t)\}_{t\geq 0}$ be a uniformly continuous quaternionic semigroup in $\mathcal{B}(V)$. Then,*

$$\frac{d}{dt}e^{tT}=T\,e^{tT}=e^{tT}\,T.$$

Proof.

$$e^{tT}=\frac{1}{2\pi}\int_{\partial(U\cap\mathbb{C}_I)} e^{ts}\; ds_I\; S_R^{-1}(s,T)$$

where U is a T-admissible open set containing the S-spectrum of the bounded operator T, which is a closed and bounded set in $\mathbb{H}$ thanks to Theorem 5.4 in [13], so

$$\frac{e^{(t+h)T}-e^{tT}}{h}=\frac{1}{2\pi}\int_{\partial(U\cap\mathbb{C}_I)}\frac{(e^{(t+h)s}-e^{ts})\,s}{h}\,s^{-1}\,ds_I\,S_R^{-1}(s,T)$$

and taking the limit we get

$$\frac{d}{dt}e^{tT}=\lim_{h\to 0}\frac{e^{(t+h)T}-e^{tT}}{h}=Te^{tT}.$$

From the formula

$$\frac{(e^{(t+h)s}-e^{ts})s}{h}s^{-1}=s^{-1}\frac{s(e^{(t+h)s}-e^{ts})}{h}$$

we also have that

$$T\,e^{tT}=e^{tT}\,T.$$

□

Let us introduce the Laplace transform that gives the right S-resolvent operator.

Theorem 4.18.8. *Let $T\in\mathcal{B}(V)$ and let $s_0>\|T\|$. Then the right S-resolvent operator $S_R^{-1}(s,T)$ is given by*

$$S_R^{-1}(s,T)=\int_0^{+\infty} e^{-t\,s}\,e^{t\,T}\,dt. \tag{4.140}$$

Proof. Consider, for $\overline{s}\notin\sigma_L(T)$,

$$e^{-t\,s}\,e^{t\,T}S_R(s,T)=-e^{-t\,s}\,e^{t\,T}(T^2-2s_0T+|s|^2\mathcal{I})(T-\overline{s}\mathcal{I})^{-1};$$

since T and $e^{t\,T}$ commute we have

$$\begin{aligned}
e^{-t\,s}\,e^{t\,T}S_R(s,T)&=-e^{-t\,s}\,(T^2-2s_0T+|s|^2\mathcal{I})e^{t\,T}(T-\overline{s}\mathcal{I})^{-1}\\
&=-\frac{d}{dt}[e^{-t\,s}(T-\overline{s}\mathcal{I})e^{t\,T}](T-\overline{s}\mathcal{I})^{-1}\\
&=-\frac{d}{dt}[e^{-t\,s}(T-\overline{s}\mathcal{I})e^{t\,T}(T-\overline{s}\mathcal{I})^{-1}].
\end{aligned}$$

For $\theta>0$ we have

$$\begin{aligned}
\int_0^{\theta} e^{-t\,s}\,e^{t\,T}S_R(s,T)dt&=-\int_0^{\theta}\frac{d}{dt}[e^{-t\,s}(T-\overline{s}\mathcal{I})e^{t\,T}(T-\overline{s}\mathcal{I})^{-1}]dt\\
&=\mathcal{I}-e^{-\theta s}(T-\overline{s}\mathcal{I})e^{\theta\,T}(T-\overline{s}\mathcal{I})^{-1}.
\end{aligned}$$

Since we have assumed $s_0>\|T\|$, for $\theta\to+\infty$, we get

$$\begin{aligned}
&\|e^{-\theta\,s}(T-\overline{s}\mathcal{I})e^{\theta\,T}(T-\overline{s}\mathcal{I})^{-1}\|\\
&\leq e^{-\theta\,s_0}e^{\theta\|T\|}\|(T-\overline{s}\mathcal{I})\|\,\|(T-\overline{s}\mathcal{I})^{-1}\|\to 0
\end{aligned}$$

so we obtain the statement. □

Note 4.18.9. The Fueter regularity and its functional calculus. From now on, we will consider only linear bounded operators and we will follow the ideas in [65]. We introduce a regular function which is related to the resolvent operator and is regular where defined. The idea is to generalize what happens in the complex setting: classically, one considers the Cauchy–Riemann kernel $g(z)=(z-\xi)^{-1}$

defined for $z \neq \xi$ and introduces $R(z,T) = (zI - T)^{-1}$ which is defined for z not in the spectrum of T. Let $\mathcal{G}(q)$ be the standard Cauchy–Fueter kernel

$$\mathcal{G}(q) = \frac{\bar{q}}{|q|^4} = \frac{q^{-1}}{|q|^2} = q^{-2}\bar{q}^{-1}$$

which is both left and right regular on $\mathbb{H}\setminus\{0\}$.

We have the following proposition:

Proposition 4.18.10. *The expansions*

$$\mathcal{G}(q,p) := \mathcal{G}(q-p) = \sum_{n\geq 0}\sum_{\nu\in\sigma_n} P_\nu(p)\mathcal{G}_\nu(q) = \sum_{n\geq 0}\sum_{\nu\in\sigma_n} \mathcal{G}_\nu(q)P_\nu(p)$$

hold for $|p| < |q|$.

Theorem 4.18.11. *Let* $f : U \subseteq \mathbb{H} \to \mathbb{H}$, f *Fueter regular on* U. *Let* $q_0 \in U$ *and* $\delta <$ dist $(q_0, \partial U)$. *Then there exists an open ball* $B = \{q \in \mathbb{H} : |q - q_0| < \delta\}$ *such that* $f(q)$ *can be represented by the uniformly convergent series*

$$f(q) = \sum_{n\geq 0}\sum_{\nu\in\sigma_n} P_\nu(q - q_0)a_\nu,$$

where

$$a_\nu = (-1)^n\partial_\nu f(q_0) = \frac{1}{2\pi^2}\int_{|q-q_0|=\delta} \mathcal{G}_\nu(q-q_0)Dqf(q),$$

and

$$\mathcal{G}_\nu(q) := \frac{\partial^n}{\partial x_1^{n_1}\partial x_2^{n_2}\partial x_3^{n_3}}\mathcal{G}(q).$$

Moreover we have

$$\int_S \mathcal{G}_\mu(q)\, Dq\, P_\nu(q) = 2\pi^2\,\delta_{\mu\nu}$$

where S *is any sphere containing the origin,* $\nu = (n_1, n_2, n_3)$, $n_1 + n_2 + n_3 = n$ *and* $\delta_{\mu\nu}$ *denotes the Kronecker delta.*

Let T be a bounded linear quaternionic operator with commuting components on a two-sided quaternionic Banach space V. The set of such operators will be denoted by $\mathcal{BC}(V)$. In this case, we consider the function $\mathcal{G}(q,p)$ written in series expansion as (replacing p by T):

$$\mathcal{G}(q,T) = \sum_{n\geq 0}\sum_{\nu\in\sigma_n} P_\nu(T)\mathcal{G}_\nu(q) = \sum_{n\geq 0}\sum_{\nu\in\sigma_n} \mathcal{G}_\nu(q)P_\nu(T). \tag{4.141}$$

The expansions hold for $\|T\| < |q|$ (cfr. Proposition 4.18.10) and define a bounded operator. It is natural to give the following definition:

Definition 4.18.12. *The maximal open set $\rho(T)$ in $\mathbb{H}$ on which the series (4.141) converges in the operator norm topology to a bounded operator is called the resolvent set of T. The spectral set $\sigma(T)$ of T is defined as the complement set in $\mathbb{H}$ of the resolvent set.*

Definition 4.18.13. *A function $f : \mathbb{H} \to \mathbb{H}$ is said to be locally right Fueter regular on the spectral set $\sigma(T)$ of an operator $T \in \mathcal{BC}(V)$ if there is an open set $U \subset \mathbb{H}$ containing $\sigma(T)$ whose boundary ∂U is a rectifiable 3-cell and such that f is regular in every connected component of U. We will denote by $\mathcal{R}_{r,\sigma(T)}$ the set of locally right Fueter regular functions on $\sigma(T)$.*

Definition 4.18.14. *Let $f \in \mathcal{R}_{r,\sigma(T)}$ and $T \in \mathcal{BC}(V)$ and set*

$$f(T) := \frac{1}{2\pi^2} \int_{\partial U} f(q) Dq \mathcal{G}(q, T),$$

where U is an open set in $\mathbb{H}$ containing $\sigma(T)$.

The definition is well posed since the integral does not depend on the open set U. The following proposition holds.

Proposition 4.18.15. *The map $F : \mathcal{R}_{r,\sigma(T)} \to \mathcal{BC}(V)$ defined by $F(f) = f(T)$ is a left vector space homomorphism.*

Theorem 4.18.16. *Let $T \in \mathcal{BC}(V)$ and consider*

$$f(q) = \sum_{n=0}^{N} \sum_{\nu \in \sigma_n} a_\nu P_\nu(q)$$

to be a right Fueter regular polynomial. Let U be a ball with center in the origin and radius $r > \|T\|$. Then

$$f(T) = \sum_{n=0}^{N} \sum_{\nu \in \sigma_n} a_\nu P_\nu(T).$$

Proof. Let U be an open set in $\mathbb{H}$ containing $\sigma(T)$. We have

$$\begin{aligned} f(T) &= \frac{1}{2\pi^2} \int_{\partial U} \sum_{n=0}^{N} \sum_{\nu \in \sigma_n} a_\nu P_\nu(q) Dq \mathcal{G}(q, T) \\ &= \frac{1}{2\pi^2} \sum_{n=0}^{N} \sum_{\nu \in \sigma_n} \int_{\partial U} a_\nu P_\nu(q) Dq \mathcal{G}(q, T) \end{aligned}$$

We have, by Proposition 4.141,

$$\begin{aligned} \int_{\partial U} a_\nu P_\nu(q) Dq \mathcal{G}(q, T) &= \int_{\partial U} a_\nu P_\nu(q) Dq \sum_{n \geq 0} \sum_{\mu \in \sigma_n} \mathcal{G}_\mu(q) P_\mu(T) \\ &= \sum_{n \geq 0} \sum_{\mu \in \sigma_n} a_\nu \int_{\partial U} P_\nu(q) Dq \mathcal{G}_\mu(q) P_\mu(T) = 2\pi^2 a_\nu P_\nu(T) \end{aligned}$$

which gives $f(T) = \sum_{n=0}^{N} \sum_{\nu \in \sigma_n} a_\nu P_\nu(T)$. □

Proposition 4.18.17. *Let $T \in \mathcal{BC}(V)$. For any open set U with piecewise smooth boundary which does not contain $\sigma(T)$ and for any $f \in \mathcal{R}(U)$ we have*

$$\int_{\partial U} f(q) Dq \mathcal{G}(q,T) = 0.$$

Thanks to this proposition, we can replace a ball with center in the origin and suitable radius by any open set containing $\sigma(T)$ and, by the density of polynomials $P_\nu(Q)$ in the set of regular functions, we obtain:

Theorem 4.18.18. *Let $T \in \mathcal{BC}(V)$. If the right Fueter regular function*

$$f(q) = \sum_{n \geq 0} \sum_{\nu \in \sigma_n} \alpha_\nu P_\nu(q),$$

converges in a neighborhood U_0 of $\sigma(T)$, then

$$f(T) = \sum_{n \geq 0} \sum_{\nu \in \sigma_n} \alpha_\nu P_\nu(T),$$

converges in the operator norm topology.

Proof. As U_0 is an open set, it contains a circle

$$U_\delta = \{q : |q| \leq \rho(T) + \delta \ \}, \quad \delta > 0$$

in its interior. Since the series $f(q) = \sum_{n \geq 0} \sum_{\nu \in \sigma_n} \alpha_\nu P_\nu(q)$, converges uniformly in the circle U_δ for some $\delta > 0$, by the Cauchy integral formula we have

$$\begin{aligned} f(T) &= \frac{1}{2\pi^2} \int_{\partial U_\delta} f(q)\, Dq\, \mathcal{G}(q,T) \\ &= \frac{1}{2\pi^2} \int_{\partial U_\delta} \sum_{n \geq 0} \sum_{\nu \in \sigma_n} \alpha_\nu P_\nu(q)\, Dq\, \mathcal{G}(q,T) \\ &= \frac{1}{2\pi^2} \sum_{n \geq 0} \sum_{\mu \in \sigma_n} \int_{\partial U_\delta} \alpha_\nu P_\nu(q)\, Dq\, \mathcal{G}(q,T) \\ &= \sum_{n \geq 0} \sum_{\nu \in \sigma_n} \alpha_\nu P_\nu(T). \end{aligned}$$

□

Note 4.18.19. Some further comments and open problems.

(1) The properties which can be proved for the functional calculus defined in [65] and [66] can be demonstrated also in this case. One may also think to generalize the functional calculus as in [62]. However, this functional calculus possesses a strong limitation: even when considering the simplest case of a

regular function, i.e., a regular (symmetric) polynomial, we have that this function is formed by using the components of a given operator T, not the operator T itself. For example, $P(q) = x_0 i - x_1$ is a regular polynomial and $P(T) = iT_0 - T_1$ for any bounded operator $T = T_0 + iT_1 + jT_2 + kT_3$. Note also that this feature of the functional calculus does not seem to have physical interest when considering a linear quaternionic operator T.

(2) As we have shown, the sum of the series $\sum_{n\geq 0} q^n s^{-1-n}$ equals $-(q^2 - 2qRe[s] + |s|^2)^{-1}(q - \overline{s})$ for $|q| < |s|$ and it does not depend on the commutativity of the components of q so that, when one replaces q by an operator T with noncommuting components, the sum remains the same. In this case: what is the sum $G(q, T)$ of

$$\mathcal{G}(q,p) = \sum_{n\geq 0} \sum_{\nu\in\sigma_n} P_\nu(p)\mathcal{G}_\nu(q) \tag{4.142}$$

when one replaces p by operator T with noncommuting components?

(3) In the case in which the components of T commute, the sum $G(q, T)$ is

$$\mathcal{G}(q,T) = (q\mathcal{I} - T)^{-2}(\overline{q\mathcal{I} - T})^{-1}.$$

The knowledge of the sum $\mathcal{G}(q, T)$ in the general case would naturally lead to a notion of spectrum of the operator T in the case of Fueter regularity.

(4) When we consider unbounded operators, the series

$$\sum_{n\geq 0} \sum_{\nu\in\sigma_n} P_\nu(T)\mathcal{G}_\nu(q)$$

does not converge. So it is crucial to manage the sum of such a series in order to extend the functional calculus to the case of unbounded operators with noncommuting components.

Chapter 5

Appendix: The Riesz–Dunford functional calculus

In this Appendix we collect some basic material on the Riesz–Dunford functional calculus useful for the readers who are not familiar with this subject. This background, with all the details, can be found in [35] and [91].

5.1 Vector-valued functions of a complex variable

We start by recalling some basic results in the theory of complex functions with values in a Banach space.

Definition 5.1.1. *Let X and Y be two complex Banach spaces.*

(1) *We will call a map $T : X \to Y$ such that*

$$T(\lambda x + \mu y) = \lambda T x + \mu T y, \quad \text{for all } \ x, y \in X, \quad \lambda, \mu \in \mathbb{C},$$

a linear operator.

(2) *A linear operator $T : X \to Y$ is said to be bounded if there exists $k \geq 0$ such that*

$$\|Tx\| \leq k\|x\|, \quad \forall x \in X.$$

(3) *The set of all bounded linear operators $T : X \to Y$ with the norm*

$$\|T\| := \sup_{x \neq 0} \frac{\|Tx\|}{\|x\|}$$

is denoted by $\mathcal{B}(X, Y)$. We set $\mathcal{B}(X) = \mathcal{B}(X, X)$.

Definition 5.1.2. *Let X be a Banach space, $f : \mathbb{C} \to X$ be a function and let $z_0 \in \mathbb{C}$. We say that f is holomorphic in z_0 if there exists an open disc $D(z_0, r)$, $r > 0$ such that f admits the power series expansion*

$$f(z) = \sum_{n\geq 0} T_n(z - z_0)^n, \quad T_n \in \mathcal{B}(X), \quad n \in \mathbb{N},$$

converging in the norm of X in $D(z_0, r)$.

The classical Cauchy theorems can be generalized to functions with values in a normed vector space X. To this purpose, we will make use of the following result which is a corollary of the Hahn-Banach theorem:

Corollary 5.1.3 (Hahn-Banach Theorem). *Let X be a normed vector space and let $x \in X$. If for any continuous linear functional x' acting on X it is $\langle x, x'\rangle = 0$, then $x = 0$.*

We can now state and prove, for the sake of completeness, the vectorial version of the Cauchy theorem and of the Cauchy integral formula:

Theorem 5.1.4. *Let U be an open bounded set in $\mathbb{C}$ such that ∂U is a finite union of continuously differentiable Jordan curves . Let $f : U \cup \partial U \to X$ be a holomorphic function. Then*

$$\int_{\partial U} f(z)dz = 0.$$

Proof. First observe that for every bounded linear functional x' on X the crochet $\langle f, x'\rangle$ is holomorphic on $U \cup \partial U$ so by the Cauchy theorem

$$\langle \int_{\partial U} f d\eta, x'\rangle = \int_{\partial U} \langle f, x'\rangle d\eta = 0.$$

By Corollary 5.1.3 we get

$$\int_{\partial U} f d\eta = 0.$$ □

Theorem 5.1.5. *Let U be an open bounded set in $\mathbb{C}$. Let $f : U \to X$ be holomorphic. Suppose that $V \subset U$ such that $\partial V \cup V \subseteq U$ such that ∂V is a finite union of continuously differentiable Jordan curves. Then for each $z_0 \in V$ we have*

$$f(z_0) = \frac{1}{2\pi i} \int_{\partial V} f(z)(z - z_0)^{-1} dz.$$

Proof. For every bounded linear functional x' on X the function $\langle f, x'\rangle$ is holomorphic in U. So

$$\begin{aligned}\langle f(\eta_0), x'\rangle &= \frac{1}{2\pi i} \int_{\partial UV} \langle f(\eta), x'\rangle(\eta - \eta_0)^{-1} d\eta \\ &= \langle \frac{1}{2\pi i} \int_{\partial V} f(\eta)(\eta - \eta_0)^{-1} d\eta, x'\rangle.\end{aligned}$$

By Corollary 5.1.3 we get the statement. □

5.2 The functional calculus for linear bounded operators

Definition 5.2.1. *Let X be a complex Banach space and $T \in \mathcal{B}(X)$. We give the following definitions.*

(1) *The resolvent set $\rho(T)$ of T is the set of complex numbers λ for which $(\lambda\mathcal{I} - T)^{-1}$ exists as a linear bounded operator with domain X.*

(2) *The spectrum $\sigma(T)$ of T is the complement of $\rho(T)$.*

(3) *The function $R(\lambda, T) = (\lambda\mathcal{I} - T)^{-1}$, defined on $\rho(T)$, is called the resolvent of T.*

(4) *The number*

$$r(T) = \sup\{|\lambda| \ : \ \lambda \in \sigma(T)\}$$

is called the spectral radius of T.

We have the following properties:

Proposition 5.2.2. *Let $T \in \mathcal{B}(X)$, where $X \neq \{0\}$.*

(1) *The resolvent set $\rho(T)$ is open.*

(2) *The function $R(\lambda, T)$ is analytic on $\rho(T)$.*

(3) *The closed set $\sigma(T)$ is compact and nonempty.*

Proof. Let $\lambda \in \rho(T)$ and let μ be any complex number with $|\mu|\|R(\lambda, T)\| < 1$. The inverse of

$$(\lambda + \mu)\mathcal{I} - T = \mu\mathcal{I} + (\lambda\mathcal{I} - T)$$

is given by the series

$$\Gamma(\mu) = \sum_{n\geq 0}(-\mu)^n[R(\lambda, T)]^{n+1}$$

which converges since, by assumption, $|\mu|\|R(\lambda, T)\| < 1$. Observe that $\Gamma(\mu)$ commutes with T and

$$[(\lambda + \mu)\mathcal{I} - T]\Gamma(\mu) = \mathcal{I}.$$

So we have that

$$\lambda + \mu \in \rho(T)$$

and this proves point (1).

To prove (2) it is enough to note that $R(\lambda+\mu, T) = \Gamma(\mu)$ is analytic at $\mu = 0$.

Finally, we prove point (3). We consider the series

$$F(\lambda) = \sum_{n\geq 0} T^n\lambda^{-1-n}, \quad F : D \to \mathcal{B}(X)$$

whose domain of convergence is $D = \{\lambda \in \mathbb{C} \;:\; |\lambda| > \|T\|\}$. It is easy to verify that

$$(\lambda\mathcal{I} - T)F(\lambda) = F(\lambda)(\lambda\mathcal{I} - T) = \mathcal{I}, \quad \lambda \in D,$$

so $D \subseteq \rho(T)$. This proves that $\sigma(T)$ is bounded. It is closed since $\rho(T)$ is open, so $\sigma(T)$ is compact. We prove that it is nonempty. Let us suppose the contrary. If $\sigma(T) = \emptyset$, then for every linear and continuous functional $x' \in X'$ the function $\theta : \mathbb{C} \to \mathbb{C}$,

$$\theta(\lambda) := \langle R(\lambda, T)x, x'\rangle,$$

is entire $\forall x \in X$. Since x' is linear and continuous, we have

$$\langle R(\lambda, T)x, x'\rangle = \sum_{n\geq 0} \lambda^{-1-n}\langle T^n x, x'\rangle.$$

If $\lambda \to \infty$, then $\langle R(\lambda, T)x, x'\rangle \to 0$ so $\theta(\lambda)$ is also bounded thus, by the Liouville theorem, it is constant and equal to zero. By Corollary 5.1.3 $R(\lambda, T)x = 0$, for every $x \in X$ and for every $\lambda \in \mathbb{C}$ so $R(\lambda, T)$ is the zero operator. This contradicts the fact that

$$R(\lambda, T)(\lambda\mathcal{I} - T) = \mathcal{I}, \quad X \neq \{0\}$$

so the spectrum is nonempty. □

Proposition 5.2.3. *For every pair λ, $\mu \in \rho(T)$ we have:*

(1) $R(\lambda, T)R(\mu, T) = R(\mu, T)R(\lambda, T)$.

(2) *(Resolvent equation)* $R(\lambda, T) - R(\mu, T) = (\mu - \lambda)R(\lambda, T)R(\mu, T)$.

Proof. Point (1) follows from the identity $TR(\lambda, T) = R(\lambda, T)T$ and simple algebraic computation.

Point (2). It is immediate to verify that

$$\lambda R(\lambda, T) - TR(\lambda, T) = \mathcal{I}, \qquad \mu R(\mu, T) - TR(\mu, T) = \mathcal{I}$$

and also

$$R(\lambda, T)R(\mu, T) = R(\mu, T)R(\lambda, T).$$

Now multiply the first equality by $R(\mu, T)$ and the second one by $R(\lambda, T)$ to get

$$\lambda R(\lambda, T)R(\mu, T) - TR(\lambda, T)R(\mu, T) = R(\mu, T)$$

and

$$\mu R(\mu, T)R(\lambda, T) - TR(\mu, T)R(\lambda, T) = R(\lambda, T).$$

By taking the difference of the two equations and, thanks to point (1), we obtain the resolvent equation. □

Definition 5.2.4. *Let $T \in \mathcal{B}(X)$. By $\mathcal{F}(T)$ we denote the family of functions f which are analytic on some neighborhood of $\sigma(T)$.*

Definition 5.2.5. *Let $f \in \mathcal{F}(T)$, and let U be an open set whose boundary ∂U is a finite union of continuously differentiable Jordan curves, oriented in the positive sense. Suppose that $\sigma(T) \subseteq U$ and that $U \cup \partial U$ is contained in the domain of analyticity of f. Then the operator $f(T)$ is defined by*

$$f(T) = \frac{1}{2\pi i}\int_{\partial U} R(\lambda, T)\, f(\lambda)\, d\lambda. \tag{5.1}$$

Remark 5.2.6. The integral (5.1) depends only on f and does not depend on the open set U.

Theorem 5.2.7. *Let f, $g \in \mathcal{F}(T)$, α_1, $\alpha_2 \in \mathbb{C}$. Then*

(1) $\alpha_1 f + \alpha_2 g \in \mathcal{F}(T)$ *and* $(\alpha_1 f + \alpha_2 g)(T) = \alpha_1 f(T) + \alpha_2 g(T)$.

(2) $f \cdot g \in \mathcal{F}(T)$ *and* $f(T)g(T) = (f \cdot g)(T)$.

(3) *If* $f(\lambda) = \sum_{n\geq 0} \alpha_n \lambda^n$ *converges in a neighborhood of* $\sigma(T)$, *then* $f(T) = \sum_{n\geq 0} \alpha_n T^n$.

Proof. Point (1) follows from the definition.

Let us prove point (2). Since $f, g \in \mathcal{F}(T)$ it is obvious that$f \cdot g \in \mathcal{F}(T)$. Let U_1 and U_2 be two neighborhoods of $\sigma(T)$ whose boundaries ∂U_1 and ∂U_2 are finite unions of continuously differentiable Jordan curves . Let us assume that

(a) $U_1 \cup \partial U_1 \subseteq U_2$.

(b) $U_2 \cup \partial U_2$ is contained in a common region of analyticity of f and g.

We have

$$\begin{aligned}
f(T)g(T) &= -\frac{1}{4\pi^2}\int_{\partial U_1} f(\lambda)R(\lambda,T)d\lambda \int_{\partial U_2} g(\mu)R(\mu,T)d\mu \\
&= -\frac{1}{4\pi^2}\int_{\partial U_1}\int_{\partial U_2} f(\lambda)g(\mu)R(\lambda,T)R(\mu,T)d\mu d\lambda \\
&= -\frac{1}{4\pi^2}\int_{\partial U_1}\int_{\partial U_2} f(\lambda)g(\mu)\frac{R(\lambda,T)-R(\mu,T)}{\mu-\lambda}d\mu d\lambda \\
&= -\frac{1}{4\pi^2}\int_{\partial U_1} f(\lambda)R(\lambda,T)\Big(\int_{\partial U_2}\frac{g(\mu)}{\mu-\lambda}d\mu\Big)d\lambda \\
&\quad + \frac{1}{4\pi^2}\int_{\partial U_2} g(\mu)R(\mu,T)\Big(\int_{\partial U_1}\frac{f(\lambda)}{\mu-\lambda}d\lambda\Big)d\mu \\
&= \frac{1}{2\pi i}\int_{\partial U_1} f(\lambda)g(\lambda)R(\lambda,T)d\lambda \\
&= (f\cdot g)(T).
\end{aligned}$$

Point (3) follows from the fact that the series $\sum_{n\geq 0} \alpha_k \lambda^k$ converges uniformly on the set $C_\varepsilon = \{\lambda \in \mathbb{C} \,:\, |\lambda| \leq r(T) + \varepsilon\}$ for suitable $\varepsilon > 0$. We have

$$\begin{aligned} f(T) &= \frac{1}{2\pi i} \int_{C_\varepsilon} \sum_{n\geq 0} \alpha_n \lambda^n R(\lambda, T) d\lambda \\ &= \frac{1}{2\pi i} \sum_{n\geq 0} \alpha_n \int_{C_\varepsilon} \lambda^n R(\lambda, T) d\lambda \\ &= \frac{1}{2\pi i} \sum_{n\geq 0} \alpha_n \int_{C_\varepsilon} \Big(\sum_{n\geq 0} \lambda^{-1-n} T^n \Big) \lambda^n d\lambda \\ &= \sum_{n\geq 0} \alpha_n T^n. \end{aligned}$$

□

We now show a particular case of an important result, the Spectral Mapping Theorem, which is crucial to compute the spectral radius.

Proposition 5.2.8. *Let $T \in \mathcal{B}(X)$, where $X \neq \{0\}$.*

(1) $\sigma(T^n) = [\sigma(T)]^n := \{\lambda^n \,:\, \lambda \in \sigma(T)\}$.

(2) $r(T) = \lim_{n\to\infty} \sqrt[n]{T^n}$.

Proof. Point (1). Since we are in a commutative setting we can write

$$\begin{aligned} \lambda^n \mathcal{I} - T^n &= (\lambda \mathcal{I} - T)(\lambda^{n-1}\mathcal{I} + \lambda^{n-2} T + \ldots + T^{n-1}) \\ &= (\lambda^{n-1}\mathcal{I} + \lambda^{n-2} T + \ldots + T^{n-1})(\lambda \mathcal{I} - T). \end{aligned}$$

So if $\lambda\mathcal{I} - T$ is not injective also $\lambda^n \mathcal{I} - T^n$ is not injective. This proves $\sigma(T^n) \supseteq [\sigma(T)]^n$.

If $\nu \notin [\sigma(T)]^n$, then, by the Fundamental Theorem of Algebra, we get

$$(\nu - \lambda^n) = (-1)^{n+1}(\lambda_1 - \lambda)\ldots(\lambda_n - \lambda)$$

where λ_1, λ_2,…,λ_n are the roots of ν and $\lambda_i \neq \lambda_j$ if and only if $i \neq j$. Replacing the operator T, thanks to Theorem 5.2.7 we get

$$(\nu\mathcal{I} - T^n) = (-1)^{n+1}(\lambda_1 \mathcal{I} - T)\ldots(\lambda_n \mathcal{I} - T);$$

since $\nu\mathcal{I} - T^n$ is not invertible there exists an i such that $\lambda_i \mathcal{I} - T$ is not invertible. So $\lambda_i \in \sigma(T)$. This means $\sigma(T^n) \subseteq [\sigma(T)]^n$, i.e., our first assertion.
Point (2). For every λ such that $|\lambda| > r(T)$ the series $\sum_{n\geq 0} T^n \lambda^{-1-n}$ converges in the norm of $B(X)$ to $R(\lambda, T)$ so the sequence $T^n \lambda^{-1-n}$ is bounded. We have

$$\limsup_{n\to\infty} \|T^n\|^{1/n} \leq r(T).$$

Using point (1) we get

$$r(T) \leq \liminf_{n\to\infty} \|T^n\|^{1/n}.$$

As a consequence, we have

$$r(T) \leq \liminf_{n\to\infty} \|T^n\|^{1/n} \leq \limsup_{n\to\infty} \|T^n\|^{1/n} \leq r(T). \qquad \square$$

We now prove the general version of the Spectral Mapping Theorem:

Theorem 5.2.9 (The Spectral Mapping Theorem). *If $f \in \mathcal{F}(T)$, then $f(\sigma(T)) = \sigma[f(T)]$.*

Proof. Let $\lambda \in \sigma(T)$. Define the function

$$\gamma(\xi) = \frac{f(\lambda) - f(\xi)}{\lambda - \xi},$$

whose domain is the domain of definition of f. By point 2) of Theorem 5.2.7 we get

$$f(\lambda)\mathcal{I} - f(T) = (\lambda\mathcal{I} - T)\gamma(T).$$

If $f(\lambda)\mathcal{I} - f(T)$ had a bounded everywhere defined inverse $[f(\lambda)\mathcal{I} - f(T)]^{-1}$, then $\gamma(T)[f(\lambda)\mathcal{I} - f(T)]^{-1}$ would be a bounded everywhere inverse of $\lambda\mathcal{I} - T$. This means that $f(\lambda) \in \sigma[f(T)]$.

Conversely let $\mu \in \sigma[f(T)]$, and suppose that $\mu \notin \sigma[f(T)]$. Then the function

$$\eta(\xi) = \frac{1}{f(\xi) - \mu} \in \mathcal{F}(T).$$

Applying again Theorem 5.2.7 we have

$$\eta(T)(f(T) - \mu\mathcal{I}) = \mathcal{I}$$

which contradicts the assumption $\mu \in \sigma[f(T)]$. $\square$

Thanks to the Spectral Mapping Theorem we can prove the following theorem.

Theorem 5.2.10. *Let $f \in \mathcal{F}(T)$, $g \in \mathcal{F}(f(T))$, and $F(\lambda) = g(f(\lambda))$. Then we have*

1. *$F \in \mathcal{F}(T)$,*
2. *$F(T) = g(f(T))$.*

Proof. Point (1) follows from the Spectral Mapping Theorem.

To prove point (2) let us consider a set U which is a neighborhood of $\sigma[f(T)]$. Assume that the boundary ∂U of U is a finite union of continuously differentiable Jordan curves and that the domain of analyticity of g contains $U \cup \partial U$. Next, consider a neighborhood V of $\sigma(T)$ such that the boundary ∂V of V is a finite union of continuously differentiable Jordan curves. Suppose that the domain of analyticity of f contains $V \cup \partial V$ and that $f(V \cup \partial V) \subseteq U$. Thanks to Theorem 5.2.7 the operator

$$E(\lambda) = \frac{1}{2\pi i} \int_{\partial V} \frac{R(\xi, T)}{\lambda - f(\xi)} \, d\xi$$

satisfies the equation

$$[\lambda\mathcal{I} - f(T)]E(\lambda) = E(\lambda)[\lambda\mathcal{I} - f(T)] = \mathcal{I},$$

which implies that

$$E(\lambda) = R(\lambda, f(T)).$$

So we obtain

$$\begin{aligned} g(f(T)) &= \frac{1}{2\pi i}\int_{\partial U} g(\lambda)R(\lambda, f(T))\,d\lambda \\ &= -\frac{1}{4\pi^2}\int_{\partial U}\int_{\partial V} g(\lambda)\frac{R(\xi,T)}{\lambda - f(\xi)}\,d\lambda\,d\xi \\ &= \frac{1}{2\pi i}\int_{\partial V} R(\xi,T)g(f(\xi))\,d\xi = F(T) \end{aligned}$$

which proves the assertion. □

We conclude by giving an idea of the proof of the perturbation of the functional calculus.

Theorem 5.2.11. *Let T be a linear bounded operator, $f \in \mathcal{F}(T)$ and $\varepsilon > 0$. Then there is a $\delta > 0$ such that if T_1 is a bounded operator and $\|T_1 - T\| < \delta$, then $f \in \mathcal{F}(T_1)$ and $\|f(T_1) - f(T)\| < \varepsilon$.*

Proof. Let us introduce the notation. We denote by $\mathcal{N}(\sigma(T), \varepsilon)$, for $\varepsilon > 0$, the ε-neighborhood of $\sigma(T)$, i.e., the set

$$\mathcal{N}(\sigma(T), \varepsilon) := \{\lambda \in \mathbb{C} : \inf_{\mu\in\sigma_S(T)} |\mu - \lambda| < \varepsilon\}.$$

The proof is based on the following fact. Let $T \in B(X)$, let $\varepsilon > 0$. Then there exists a $\delta > 0$ such that if $T_1 \in B(X)$ and $\|T_1 - T\| < \delta$, then $\sigma(T_1) \subseteq \mathcal{N}(\sigma(T), \varepsilon)$ and

$$\|R(\lambda, T_1) - R(\lambda, T)\| < \varepsilon, \quad \lambda \notin \mathcal{N}(\sigma(T), \varepsilon).$$

We leave the details to the reader. □

5.3 The functional calculus for unbounded operators

In the case of unbounded operators the spectrum may be a bounded set, an unbounded set, the empty set, or even the whole plane. To our purposes, we suppose $\rho(T) \neq \emptyset$.

Definition 5.3.1. *By $\mathcal{F}_\infty(T)$ we denote the family of functions f which are analytic on some neighborhood of $\sigma(T)$ and at ∞.*

The neighborhood need not be connected and can depend on f. Let $\alpha \in \rho(T)$ and define

$$A = (T - \alpha\mathcal{I})^{-1} = -R(\alpha, T).$$

The operator A defines a one-to-one mapping on X onto the domain $D(T)$ of T and

$$TAx = \alpha Ax + x, \quad x \in X$$

and

$$ATx = \alpha Ax + x, \quad x \in D(T).$$

We can now define a functional calculus for the unbounded operator T in terms of the bounded operator A.

Denote by $\mathcal{K}$ the complex sphere, with its usual topology, and define the homeomorphism

$$\mu = \Phi(\lambda) = (\lambda - \alpha)^{-1}, \qquad \Phi(\infty) = 0, \qquad \Phi(\alpha) = \infty.$$

Theorem 5.3.2. *Let $\alpha \in \rho(T)$. Then $\Phi(\sigma(T) \cup \{\infty\}) = \sigma(A)$ and the relation*

$$\phi(\mu) = f(\Phi^{-1}(\mu))$$

determines a one-to-one correspondence between $f \in \mathcal{F}_\infty(T)$ and $\phi \in \mathcal{F}(A)$.

Proof. Take $\lambda \in \rho(T)$, so $0 \neq \mu = \Phi(\lambda) = (\lambda - \alpha)^{-1}$ and

$$(T - \alpha\mathcal{I})(T - \lambda\mathcal{I})^{-1} = \mathcal{I} + \mu^{-1}(T - \lambda\mathcal{I})^{-1}.$$

We can also write

$$(T - \alpha\mathcal{I})(T - \lambda\mathcal{I})^{-1} = A^{-1}\Big((T - \alpha\mathcal{I}) - \mu^{-1}\mathcal{I}\Big) = \mu(\mu\mathcal{I} - A)^{-1}.$$

So we obtain

$$(T - \lambda\mathcal{I})^{-1} = \mu^2(\mu\mathcal{I} - A)^{-1} - \mu\mathcal{I}$$

which is the relation between the resolvent operators

$$R(\lambda, T) = \mu^2 R(\mu, A) + \mu\mathcal{I}. \tag{5.2}$$

This implies that $\mu \in \rho(A)$. Now if $\mu \in \rho(A)$ and $\mu \neq 0$, then

$$(\mu\mathcal{I} - A)^{-1}A = \frac{1}{\mu}(T - \lambda\mathcal{I})^{-1}.$$

This shows that $\lambda \in \rho(T)$. The point $m = 0$ is in $\sigma(A)$ since $A^{-1} = T - \alpha\mathcal{I}$ is unbounded. The last part of the theorem is trivial and follows from the definition of Φ. □

We can now define the functional calculus for an unbounded operator T:

Definition 5.3.3. *Let $f \in \mathcal{F}_\infty(T)$. We define*

$$f(T) = \phi(A),$$

where $\phi \in \mathcal{F}(A)$ and $\phi(\mu) = f(\Phi^{-1}(\mu))$.

The next two results describe the main properties of the functional calculus.

Theorem 5.3.4. *Let $f \in \mathcal{F}_\infty(T)$. Then $f(T)$ is independent of the choice of $\alpha \in \rho(T)$. Let $V \supset \sigma(T)$ be an open set whose boundary ∂V is a finite union of continuously differentiable Jordan curves. Let f be analytic on $V \cup \partial V$ and suppose that ∂V has positive orientation with respect to the set V. Then*

$$f(T) = f(\infty)\mathcal{I} + \frac{1}{2\pi i}\int_{\partial V} R(\lambda, T)\, f(\lambda)\, d\lambda. \tag{5.3}$$

Proof. We just have to prove formula (5.3) since the integral is independent of α. Take $\alpha \in \rho(T)$. Thanks to the analyticity of $R(\lambda, T)$ and the Cauchy theorem we can assume $\alpha \notin V \cup \partial V$. The set $U = \Phi^{-1}(V)$ is open and contains $\sigma(T)$. The boundary $\partial U = \Phi^{-1}(\partial V)$ is positively oriented and it is a finite union of continuously differentiable Jordan curves . The function $\phi(\mu) = f(\Phi^{-1}(\mu))$ is analytic on $U \cup \partial U$. Since $\phi(0) = f(\infty)$ and $0 \in \sigma(A)$ from relation (5.2) and Definition 5.3.3 we get

$$\begin{aligned}\frac{1}{2\pi i}\int_{\partial V} f(\lambda)R(\lambda, T)\, d\lambda = \frac{1}{2\pi i}\int_{\partial U} \phi(\mu)[R(\mu, A) - \mu^{-1}\mathcal{I}]d\mu& \\ = \phi(A) - \phi(0)\mathcal{I}& \\ = f(T) - f(\infty)\mathcal{I}.& \quad \square\end{aligned}$$

For unbounded operators the algebraic rules become as follows.

Theorem 5.3.5. *Let $f,\ g \in \mathcal{F}_\infty(T)$, $\alpha_1,\ \alpha_2 \in \mathbb{C}$. Then:*

(1) *$\alpha_1 f + \alpha_2 g \in \mathcal{F}(T)$ and $(\alpha_1 f + \alpha_2 g)(T) = \alpha_1 f(T) + \alpha_2 g(T)$.*

(2) *$f \cdot g \in \mathcal{F}(T)$ and $f(T)g(T) = (f \cdot g)(T)$.*

(3) *$\sigma(f(T)) = f(\sigma(T) \cup \{\infty\})$.*

(4) *Let $f \in \mathcal{F}_\infty(T)$, $g \in \mathcal{F}_\infty(f(T))$, and $F(\lambda) = g(f(\lambda))$. Then $F \in \mathcal{F}_\infty(T)$ and $F(T) = g(f(T))$.*

Bibliography

[1] S. Adler, Quaternionic Quantum Field Theory, Oxford University Press, 1995.

[2] L. Alhfors, Complex Analysis, McGraw-Hill, New York, 1966.

[3] F.W. Anderson, K.R. Fuller, Rings and Categories of Modules, 2nd edition, Springer Verlag, New York, 1992.

[4] R.F.V. Anderson, The Weyl functional calculus, J. Funct. Anal., **4** (1969), 240–267.

[5] M. F. Atiyah, R. Bott, A. Shapiro, Clifford modules, Topology, **3** (1964), 3–38.

[6] B. Beck, Sur les équations polynomiales dans les quaternions, Enseign. Math., **25** (1979), 193–201.

[7] F. Brackx, R. Delanghe, F. Sommen, Clifford Analysis, Pitman Res. Notes in Math. 76, 1982.

[8] U. Bray, G. Whaples, Polynomials with coefficients from a division ring, Canad. J. Math., **35** (1983), 509–515.

[9] F. Colombo, G. Gentili, I. Sabadini, A Cauchy kernel for slice regular functions, Ann. Global Anal. Geom., **37** (2010), 361–378.

[10] F. Colombo, G. Gentili, I. Sabadini, D.C. Struppa, A functional calculus in a noncommutative setting, Electron. Res. Announc. Math. Sci., **14** (2007), 60–68.

[11] F. Colombo, G. Gentili, I. Sabadini, D.C. Struppa, An overview of functional calculus in different settings, Hypercomplex Analysis, Trends in Mathematics, Birkhäuser, 69–99, 2009.

[12] F. Colombo, G. Gentili, I. Sabadini, D.C. Struppa, Extension results for slice regular functions of a quaternionic variable, Adv. Math., **222** (2009), 1793–1808.

[13] F. Colombo, G. Gentili, I. Sabadini, D.C. Struppa, Noncommutative functional calculus: bounded operators, Complex Anal. Oper. Theory, **4** (2010), 821–843.

[14] F. Colombo, G. Gentili, I. Sabadini, D.C. Struppa, Noncommutative functional calculus: unbounded operators, J. Geom. Phys., **60** (2010), 251–259.

[15] F. Colombo, I. Sabadini, A structure formula for slice monogenic functions and some of its consequences, in Hypercomplex Analysis, Trends in Mathematics, Birkhäuser, 101–114, 2009.

[16] F. Colombo, I. Sabadini, On some properties of the quaternionic functional calculus, J. Geom. Anal., **19** (2009), 601–627.

[17] F. Colombo, I. Sabadini, On the formulations of the quaternionic functional calculus, J. Geom. Phys., **60** (2010), 1490–1508.

[18] F. Colombo, I. Sabadini, The Cauchy formula with s-monogenic kernel and a functional calculus for noncommuting operators, J. Math. Anal. Appl., **373** (2011), 655–679.

[19] F. Colombo, I. Sabadini, Some remarks on the $\mathcal{S}$-spectrum for a noncommutative functional calculus, to appear in Complex Var. Elliptic Equ. (2011).

[20] F. Colombo, I. Sabadini, Bounded perturbations of the resolvent operators associated to the $\mathcal{F}$-spectrum, Hypercomplex Analysis and applications, Trends in Mathematics, Birkhäuser, (2011), 13–28.

[21] F. Colombo, I. Sabadini, The quaternionic evolution operator, submitted.

[22] F. Colombo, I. Sabadini, The $\mathcal{F}$-spectrum and the $\mathcal{SC}$-functional calculus, submitted.

[23] F. Colombo, I. Sabadini, F. Sommen, D.C. Struppa, Analysis of Dirac Systems and Computational Algebra, Progress in Mathematical Physics, Vol. 39, Birkhäuser, Boston, 2004.

[24] F. Colombo, I. Sabadini, F. Sommen, The Fueter mapping theorem in integral form and the $\mathcal{F}$-functional calculus, Math. Methods Appl. Sci., **33** (2010), 2050–2066.

[25] F. Colombo, I. Sabadini, D.C. Struppa, A new functional calculus for noncommuting operators, J. Funct. Anal., **254** (2008), 2255–2274.

[26] F. Colombo, I. Sabadini, D.C. Struppa, Slice monogenic functions, Israel J. Math., **171** (2009), 385–403.

[27] F. Colombo, I. Sabadini, D.C. Struppa, An extension theorem for slice monogenic functions and some of its consequences, Israel J. Math., **177** (2010), 369–389.

[28] F. Colombo, I. Sabadini, D. C. Struppa, Duality theorems for slice hyperholomorphic functions, J. Reine Angew. Math., **645** (2010), 85–104.

[29] F. Colombo, I. Sabadini, D. C. Struppa, The Pompeiu formula for slice hyperholomorphic functions, to appear in Michigan Math. J. (2011).

[30] F. Colombo, I. Sabadini, D. C. Struppa, The Runge theorem for slice hyperholomorphic functions, to appear in Proc. Amer. Math. Soc. (2011).

[31] J. Cnops, An introduction to Dirac operators on manifolds, Progress in Mathematical Physics, Vol. 24, Birkhäuser, Boston, 2002.

[32] C.G. Cullen, An integral theorem for analytic intrinsic functions on quaternions, Duke Math. J., **32** (1965), 139–148.

[33] C.A. Deavours, The quaternion calculus, Amer. Math. Monthly, **80** (1973), 995–1008.

[34] R. Delanghe, F. Sommen, V. Souček, Clifford Algebra and Spinor-valued Functions, Mathematics and Its Applications 53, Kluwer Academic Publishers, 1992.

[35] N. Dunford, J. Schwartz, Linear Operators, part I: General Theory, J. Wiley and Sons, 1988.

[36] S. Eilenberg, I. Niven, The "fundamental theorem of algebra" for quaternions, Bull. Amer. Math. Soc., **50** (1944), 246–248.

[37] S. Eilenberg, N. Steenrod, Foundations of algebraic topology, Princeton University Press, Princeton, New Jersey, 1952.

[38] S. L. Eriksson-Bique, H. Leutwiler, Hypermonogenic functions. In Clifford Algebras and their Applications in Mathematical Physics, Vol. 2, Birkhäuser, Boston, (2000), 287–302.

[39] S. L. Eriksson, H. Leutwiler, Hypermonogenic functions and their Cauchy-type theorems. In Trends in Mathematics: Advances in Analysis and Geometry, Birkhäuser, Basel, (2004), 97–112.

[40] S. L. Eriksson, H. Leutwiler, Contributions to the theory of hypermonogenic functions, Complex Var. Elliptic Equ., **51** (2006), 547–561.

[41] A. Fabiano, G. Gentili, D.C. Struppa, Sheaves of quaternionic hyperfunctions and microfunctions, Compl. Variab. Theory and Appl., **24** (1994), 161–184.

[42] T. Friedrich, Dirac operators in Riemannian geometry. Translated from the 1997 German original by Andreas Nestke. Graduate Studies in Mathematics, 25. American Mathematical Society, Providence, 2000.

[43] R. Fueter, Die Funktionentheorie der Differentialgleichungen $\triangle u = 0$ und $\triangle\triangle u = 0$ mit vier reellen Variablen. Comm. Math. Helv., **7** (1934), 307–330.

[44] R. Fueter, Über eine Hartogs'schen Satz., Comment. Math. Helv., **12** (1939/40), 75–80.

[45] G. Gentili, C. Stoppato, Zeros of regular functions and polynomials of a quaternionic variable, Michigan Math. J., **56** (2008), 655–667.

[46] G. Gentili, C. Stoppato, The open mapping theorem for regular functions, Ann. Sc. Norm. Super. Pisa Cl. Sci., **8** (2009), 805–815.

[47] G. Gentili, C. Stoppato, D. C. Struppa, F. Vlacci, Recent developments for regular function of a hypercomplex variable, in Hypercomplex Analysis, Trends in Mathematics, Birkhäuser, 165–185, 2009.

[48] G. Gentili, D.C. Struppa, A new approach to Cullen-regular functions of a quaternionic variable, C.R. Acad. Sci. Paris, **342** (2006), 741–744.

[49] G. Gentili, D.C. Struppa, A new theory of regular functions of a quaternionic variable, Adv. Math., **216** (2007), 279–301.

[50] G. Gentili, D. C. Struppa, Regular functions on a Clifford algebra, Complex Var. Elliptic Equ., **53** (2008), 475–483.

[51] G. Gentili, D.C. Struppa, The multiplicity of the zeros of polynomials with quaternionic coefficients, Milan J. Math., **216** (2008), 15–25.

[52] G. Gentili, D. C. Struppa, F. Vlacci, The fundamental theorem of algebra for Hamilton and Cayley numbers, Math. Z., **259** (2008), 895–902.

[53] R. Ghiloni, A. Perotti, Slice regular functions on real alternative algebras, Adv. Math., **226** (2011), 1662–1691.

[54] R. Ghiloni, A. Perotti, Zeros of regular functions of quaternionic and octonionic variable: a division lemma and the camshaft effect, to appear in Ann. Mat. Pura e Appl., 2011.

[55] R. Ghiloni, A. Perotti, A new approach to slice regularity on real algebras, in Hypercomplex Analysis and its Applications, Trends in Mathematics, Birkhäuser, (2011), 109–124.

[56] B. Gordon, T. S. Motzkin, On the zeros of polynomials over division rings, Trans. Amer. Math. Soc., **116** (1965), 218–226.

[57] A. Grothendieck, Sur certain espaces de fonctions holomorphes, I, II, J. Reine Angew. Math., **192** (1953), 35–64, 77–95.

[58] K. Gürlebeck, K. Habetha, W. Sprößig, Holomorphic Functions in the Plane and n-Dimensional Space, Birkhäuser, Basel, 2008.

[59] L. Hörmander, The Analysis of Linear Partial Differential Operators I, Springer Verlag, Berlin Heidelberg, 1983.

[60] B. Jefferies, A. McIntosh, The Weyl calculus and Clifford analysis, Bull. Austral. Math. Soc., **57** (1998), 329–341.

[61] B. Jefferies, A. McIntosh, J. Picton-Warlow, The monogenic functional calculus, Studia Math., **136** (1999), 99–119.

[62] B. Jefferies, Spectral Properties of Noncommuting Operators, Lecture Notes in Mathematics, 1843, Springer-Verlag, Berlin, 2004.

[63] A. Kaneko, Introduction to Hyperfunctions, Kluwer, 1994.

[64] G. Kato, D.C. Struppa, Fundamentals of Algebraic Microlocal Analysis, Marcel Dekker, New York, 1999.

[65] V.V. Kisil, E. Ramirez de Arellano, The Riesz–Clifford functional calculus for non-commuting operators and quantum field theory, Math. Methods Appl. Sci., **19** (1996), 593–605.

[66] V.V. Kisil, E. Ramirez de Arellano, A functional model for quantum mechanics: unbounded operators, Math. Methods Appl. Sci., **20** (1997), 745–757.

[67] G. Köthe, Dualität in der Funktionentheorie, J. Reine Angew. Math., **191** (1953), 30–49.

[68] K. I. Kou, T. Qian, F. Sommen, Generalizations of Fueter's theorem, Meth. Appl. Anal., **9** (2002), 273–290.

[69] V.V. Kravchenko, M.V. Shapiro, Integral Representations for Spatial Models of Mathematical Physics, Pitman Research Notes in Mathematics Series, 351, Longman, Harlow, 1996.

[70] A. G. Kurosh, Lectures in General Algebra, International Series of Monographs in Pure and Applied Mathematics, Vol. 70 Pergamon Press, Oxford-Edinburgh-New York, 1965.

[71] T.Y. Lam, A First Course in Noncommutative Rings, Graduate Texts in Mathematics, 123, Springer-Verlag, New York, (1991).

[72] G. Laville, I. Ramadanoff, Holomorphic Cliffordian functions, Adv. Appl. Clifford Algebras, **8** (1998), 323–340.

[73] G. Laville, I. Ramadanoff, Elliptic Cliffordian functions, Compl. Variab. Theory and Appl., **45** (2001), 297–318.

[74] H. Leutwiler, Modified quaternionic analysis in $\mathbb{R}^3$, Compl. Variab. Theory and Appl., **20** (1992), 19–51.

[75] P. Lounesto, Clifford Algebras and Spinors, London, Math. Soc. Lecture Notes Series, Vol. 286, Cambridge Univ. Press, Cambridge, 2001.

[76] A. Martineau, Indicatrices des fonctionelles analitiques et transformée de Laplace, in Oeuvres de André Martineau, Éditions du Centre National de la Recherche Scientifique (CNRS), Paris, (1977), 595–608.

[77] A. McIntosh, A. Pryde, A functional calculus for several commuting operators, Indiana U. Math. J., **36** (1987), 421–439.

[78] I.M. Mitelman, M. Shapiro, Differentiation of the Martinelli-Bochner integrals and the notion of hyperderivability, Math. Nachr., **172** (1995), 211–238.

[79] G. Moisil, N. Theodorescu, Functions holomorphes dans l'espace, Mathematica (Cluj), **5** (1931), 142–159.

[80] M.A.M. Murray, The Cauchy integral, Calderon commutators and conjugations of singular integrals in $\mathbb{R}^m$, Trans. Amer. Math. Soc., **289** (1985), 497–518.

[81] I. Niven, Equations in quaternions. Amer. Math. Monthly, **48** (1941), 654–661.

[82] I. Niven, The roots of a quaternion, Amer. Math. Monthly, **49** (1942), 386–388.

[83] D. Peña-Peña, T. Qian, F. Sommen, An alternative proof of Fueter's theorem, Complex Var. Elliptic Equ., **51** (2006), 913–922.

[84] R. J. Pereira, Quaternionic Polynomials and Behavioral Systems, PhD thesis, Universidade de Aveiro, 2006.

[85] A. Pogorui, M. V. Shapiro, On the structure of the set of zeros of quaternionic polynomials, Complex Var. Theory Appl., **49** (2004), 379–389.

[86] S. Pumplün, S. Walcher, On the zeros of polynomials over quaternions, Comm. Algebra, **30** (2002), 4007–4018.

[87] T. Qian, Generalization of Fueter's result to $\mathbb{R}^{n+1}$, Rend. Mat. Acc. Lincei, **8** (1997), 111–117.

[88] T. Qian, Fourier Analysis on Starlike Lipschitz Surfaces, J. Funct. Anal., **183** (2001), 370–412.

[89] R.F. Rinehart, Elements of a theory of intrinsic functions on algebras, Duke Math. J., **27** (1960), 1–19.

[90] R. Rocha-Chavez, M. Shapiro, F. Sommen, Integral theorems for functions and differential forms in C^m, Chapman Hall/CRC Research Notes in Mathematics, 428. Chapman Hall/CRC, Boca Raton, 2002.

[91] W. Rudin, Functional Analysis, McGraw-Hill Series in Higher Mathematics, McGraw-Hill Book Co., New York-Düsseldorf-Johannesburg, 1973.

[92] J. Ryan, Complexified Clifford analysis, Compl. Variab. Theory and Appl., **1** (1982), 119–149.

[93] I. Sabadini, D.C. Struppa Topologies on quaternionic hyperfunctions and duality theorems, Compl. Variab. Theory and Appl., **30** (1996), 19–34.

[94] M. Sce, Osservazioni sulle serie di potenze nei moduli quadratici, Atti Acc. Lincei Rend. Fisica, **23** (1957), 220–225.

[95] F. Sommen, Microfunctions with values in a Clifford algebra II, Sci. Papers College Arts Sci. Univ. Tokyo, **36** (1986), 15–37.

[96] F. Sommen, On a generalization of Fueter's theorem, Zeit. Anal. Anwen., **19** (2000), 899–902.

[97] G.A. Suchomlinov, On the extension of linear functionals in linear normed spaces and linear quaternionic spaces (Russian), Mat. Sbornik, Ser. 2, (1938), 353–358.

[98] A. Sudbery, Quaternionic analysis, Math. Proc. Camb. Phil. Soc., **85** (1979), 199–225.

[99] J.L. Taylor, The analytic-functional calculus for several commuting operators, Acta Math., **125** (1970), 1–38.

[100] J.L. Taylor, A general framework for a multi-operator functional calculus, Adv. Math., **9** (1972), 183–252.

[101] M.E. Taylor, Functions of several self-adjoint operators, Proc. Amer. Math. Soc., **19** (1968), 91–98.

[102] J.H.M. Wedderburn, On division algebras, Trans. Amer. Math. Soc., **22** (1921), 129–135.

[103] H. Weyl, The Theory of Groups and Quantum Mechanics, Dover Inc. Publications, 1950, (English translation of Gruppentheorie und Quantenmechanik, 1931).

[104] Y. Yang and T. Qian. On sets of zeroes of Clifford algebra-valued polynomials, to appear in Acta Math. Sin.

Index

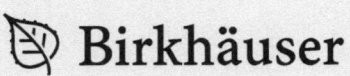